BARRON'S

AP®

BIOLOGY

3RD EDITION

Deborah T. Goldberg, M.S.
Former AP Biology Teacher
Lawrence High School
Cedarhurst, New York

BARRON'S

About the Author

Deborah Goldberg earned her B.S. and M.S. degrees at Long Island University. For fourteen years, she studied cells using electron microscopy at NYU Medical Center and New York Medical College. For the following twenty-two years, she taught Chemistry, Biology, and Forensic Science at Lawrence High School on Long Island.

Dedication

I dedicate this book to my children, Michael and Sara Boilen, to my husband, Howard Blue, and to all my AP Biology students at Lawrence High School, who are every teacher's dream.

Acknowledgments

I wish to thank—

—My husband, Howard Blue, for his constant love and support
—My sister, Rachel, for her mastery of the English language and her eagerness to share it
—Pat Hunter, my editor, for her expert guidance
—My students at Lawrence High School on Long Island, New York, who make teaching the best job in the world
—Michele Sandifer, for her expert copyediting
—Reviewers, who made some great recommendations

© Copyright 2010, 2007 by Barron's Educational Series, Inc.
Previous edition © copyright 2004 under the title *How to Prepare for AP Biology* by Barron's Educational Series, Inc.

All inquiries should be addressed to:
Barron's Educational Series, Inc.
250 Wireless Boulevard
Hauppauge, New York 11788
www.barronseduc.com

ISBN-13: 978-0-7641-4051-8 (book)
ISBN-10: 0-7641-4051-5 (book)

ISBN-13: 978-0-7641-9524-2 (book/CD-ROM package)
ISBN-10: 0-7641-9524-7 (book/CD-ROM package)

Library of Congress Control Number: 2009940367

10%
POST-CONSUMER
WASTE
Paper contains a minimum of 10% post-consumer waste (PCW). Paper used in this book was derived from certified, sustainable forestlands.

PRINTED IN THE UNITED STATES OF AMERICA

9 8 7

Contents

7 Cell Division   143

8 Heredity   159

9 The Molecular Basis of Inheritance   183

10 Classification   215

11 Evolution   233

Why Should I Buy This Book?

This book includes:

- More than 400 pages of subject review
- More than 350 multiple-choice questions and answers with explanations in the content area
- One diagnostic test to help you identify your strengths and weaknesses
- Two practice tests with questions that mimic the actual AP Biology Exams
- Review of the 12 required AP labs
- Presentation of 5 themes to help you write a great essay on the AP exam
- Explanation of how to grade an essay the way the College Board does

INTRODUCTION

Introduction to the Exam

CHAPTER 1

GENERAL INFORMATION

The AP Biology course is designed to be the equivalent of a two-semester, college introductory biology course and is meant to be taken after a first-year high school biology course. This course is rigorous. The textbook you are using probably has more than twelve hundred pages. You have nine and a half months to complete it and to prepare for the exam, which is in mid-May. You will be reading volumes of text, devising and carrying out sophisticated experiments, and writing lots of essays.

This book has fifteen chapters of subject area review, which most likely follow the same order as the textbook you are using. All key words are in **bold**, and vocabulary terms are defined in the glossary. After each review section, the book provides sample multiple-choice questions with answers and explanations. There are many sample essays throughout the book, which will give you plenty of practice answering free-response questions.

The College Board requires AP Biology students to complete twelve college-level lab exercises before the examination. This book includes a complete review of all twelve labs and gives you detailed guidelines on how to devise an experiment.

During the school year, study from your textbook and your notes from class and then review from this book. This book is tailored to help you prepare for the AP Exam as well as for exams during the school year.

Good luck.

> **SCORING CHANGE**
>
> The College Board has announced a change to how the AP Biology exam is scored. Beginning with the May 2011 exam, there will be no penalty for wrong answers on the multiple-choice section. So it pays to guess—but guess wisely.

THE AP EXAM

The AP Exam is three hours long and is composed of two parts. Part I consists of an 80-minute, 100-item multiple-choice section, which tests all content areas and counts for 60 percent of the exam grade. Part II begins with a 10-minute reading interval in which you have the opportunity to read the four required free-response questions, gather your thoughts and jot down key words, and prepare to write your essays. After the reading interval, you have 90 minutes in which to write the essays in the exam booklets. These essays count for 40 percent of the exam and consist of four mandatory questions that encompass broader topics than those in Part I. You get a short break between Part I and Part II.

Here is an example of the different types of questions you might find in each section. Whereas a Part I question might ask for a simple recall of a fact about muscle cells, the free-response question asks you to explain particular details and to *make connections* between separate broad themes.

- **Sample Part I Question**
  Which of the following is not involved in the regulation of blood sugar?
  (A) adrenaline
  (B) insulin
  (C) glucagon
  (D) cortisol
  (E) estrogen

- **Sample Part II Question**
  Regulation is a major theme in biology. Discuss one example of regulation at each of the following levels: molecular, cellular, organismal, and population.

Here is a breakdown of the topics and percentages covered in this course. The AP Exam seeks to be representative of these same percentages.

TABLE 1.1

AP Biology Exam

Topics	Percent of Course
I. Molecules and Cells	25%
A. Chemistry of Life	
Water	
Organic molecules in organisms	
Free energy changes	
Enzymes	
B. Cells	
Prokaryotes and eukaryotes	
Membranes	
Subcellular organization	
Cell cycle and its regulation	
C. Cellular Energetics	
Coupled reactions	
Fermentation and cellular respiration	
Photosynthesis	
II. Heredity and Evolution	25%
A. Heredity	
Meiosis and gametogenesis	
Eukaryotic chromosomes	
Inheritance patterns	
B. Molecular Genetics	
RNA and DNA structure and function	
Gene regulation	
Mutation	
Viral structure and replication	
Nucleic acid technology and applications	
C. Evolutionary biology	
Early evolution of life	
Evidence of evolution	
Mechanisms of evolution	
III. Organisms and Populations	50%
A. Diversity of Organisms	
Evolutionary patterns	
Survey of the diversity of life	
Phylogenetic classification	
Evolutionary relationships	
B. Structure and Function of Plants and Animals	
Reproduction, growth, and development	
Structural, physiological, and behavioral adaptations	
Response to the environment	
C. Ecology	
Population dynamics	
Communities and ecosystems	
Global issues	

GRADES ON THE EXAM

Advanced placement and/or college credit is awarded by the college or university you will attend. Different institutions observe different guidelines about awarding AP credit. Success on the AP Exam may allow you to take a more advanced course and bypass an introductory course, or it might qualify you for 8 credits of advanced standing and tuition credit. The best source of specific up-to-date information about an institution's policy is its catalog or web site.

Exams are graded on a scale from 1 to 5, with 5 being the best. The total raw score on the exam is translated to the AP's 5-point scale.

AP Grade	Qualification
5	Extremely Well Qualified
4	Well Qualified
3	Qualified
2	Possibly Qualified
1	No Recommendation

Here are the grade distributions for all the students who recently sat for the exam. These numbers tend to be consistent from year to year.

Exam Grade	Student Scoring that Grade
5	19.5%
4	15.5%
3	15.8%
2	15.1%
1	34%

Part I questions have always been designed so that the mean score (50 percent) is based on getting about half the questions correct. That means that if you got half the questions in Part I correct, you got a 3 on that section.

Part II questions are also designed to achieve a mean score of 50 percent, but scores vary significantly among the four questions. On the 1999 exam, mean scores ranged from 2.51 to 3.82 (out of a possible 10 points) for each of the four questions. This is consistent with results from other years.

Both Part I and Part II questions are designed to be difficult. In spite of their difficulty, though, about 65 percent of the candidates earn an AP grade of 3 or higher.

HINTS FOR TAKING THE MULTIPLE-CHOICE SECTION

BE NEAT

Improperly erased pencil marks can cause the machine to misgrade your paper. On the other hand, you may write or draw anywhere in the question booklet.

PACE YOURSELF

The first 60 questions are easy to read and straightforward. The remaining questions are more time consuming. They require interpreting and analyzing data. So work quickly at the beginning and leave time for the more involved questions.

ANSWER EVERY QUESTION

Do not leave any blanks! You should guess whenever you are not sure about any question. As of May 2011, the College Board will no longer penalize you for incorrect answers. You gain points for every correct answer.

READ CAREFULLY AND WATCH OUT FOR TRICKY WORDS

Questions with EXCEPT or NOT often trip students up. Also, pay attention to "Which of the following is FALSE?" or "Which of the following is TRUE?"

HINTS FOR TAKING THE FREE-RESPONSE SECTION

This section includes four essays. **You must answer all of them.** The four essays are drawn from three topics: molecules and cells, heredity and evolution, and organisms and populations. At least one of the essays will be lab based and will test your ability to interpret scientific data and/or design a controlled experiment.

The free-response questions are probably different from any standardized test questions that you have ever taken. You must approach them in a special way. Of greatest importance is what the readers look for and how they grade an essay; see chapter 20, "How to Grade an Essay."

CHAPTER 19 IS A GREAT REFERENCE FOR YOU

Just as an Olympic athlete must anticipate what the judges want to see, you must be prepared to give the exam readers what they want to read. If you can do that on the AP Exam, you will get a high score.

Here are things the readers **do not** particularly care about:

Spelling
Penmanship (unless they cannot read the paper)
Grammar
Wrong information—**You do not get points off for incorrect statements**.

Here are the things the graders **do** care about:

The answer must be in essay form, not an outline.
Lots of correct information—so write, write, and write!

YOU DO NOT LOSE POINTS FOR GIVING INCORRECT INFORMATION

You start out with zero points and you gain points as you state correct things that answer the question. The reader does **not take off points for any reason**, not even incorrect information. He or she only adds points. The reader is like the person who stands at the entrance to a concert and uses a clicker to count the number of people entering. He or she reads your essay. Every time you state a correct piece of information that answers the question, you get a click; that is, you get credit. It is straightforward.

ANSWER ALL FOUR ESSAYS IN ANY ORDER

Each essay is worth 10 points. If you answer only three essays, even if they are masterpieces and you get full credit on each, you still will get only 30 out of 40 points.

ANSWER EVERY PART OF THE ESSAY

Each essay is worth 10 points. If the essay is divided into two parts, each section is worth 5 points. If you write a ten-page masterpiece on the first part of the essay but you leave out the second part of the essay, you still get only 5 points. If there are three parts to the essay, each part will be worth 3 points with an extra point given to the section where you have demonstrated extra depth of understanding, for a total of 10 points.

BRING A WATCH AND BUDGET YOUR TIME

You have 90 minutes total; a 10-minute reading period and 20 minutes for each essay. The exam proctors will **not** announce when it is time to move from one essay to another. You must monitor the time. One essay may take you 30 minutes, but you may find that you have answered another essay satisfactorily in 10 minutes.

WATCH OUT FOR TRICKY WORDS

Read the question and determine what you must do: "Describe," "Explain," "Compare," or "Contrast." Pay particular attention to the word *or*. Some years ago, the students were asked to discuss "*either* the nitrogen cycle *or* the carbon cycle." Some AP students, trying to give the readers more than was asked, mentioned the nitrogen cycle before diving into their intended topic, which was the carbon cycle. Since the students had begun writing about the nitrogen cycle first, the readers were required to grade the essay based on the little that the students had written on the nitrogen cycle. The students had not read carefully and did not do well as a result.

TAKE TIME TO ORGANIZE YOUR THOUGHTS

Before you are allowed to begin to write your essays, you have a 10-minute reading period to think, analyze, and generally prepare to write the essay. Brainstorm and write down all the **key words** you can think of that relate to each topic. Then look over the key words, eliminate the ones that are not related, and prioritize the ones you will be writing about. Present your ideas, in order, *from the general to the particular*. After the reading period is over, you may begin to write your essay.

DO NOT LEAVE OUT BASIC MATERIAL

Many students think that a college-level essay should contain only the most complex ideas. This is incorrect. Include everything you can think of that is related to the topic and answers the question. Remember, you are trying to accumulate points by presenting relevant, correct statements.

DO NOT CONTRADICT YOURSELF

No points will be given if you give contradictory information. For example, you will receive no credit if you state, "The Calvin cycle occurs in the stroma of chloroplasts," and you also write, "the Calvin cycle occurs in the grana of chloroplasts."

LABEL YOUR ANSWERS

Number each essay—1, 2, 3, and 4—and label all parts, such as 1a, 1b, and 1c. For readability, leave at least one line between essays. If the reader cannot find your answer, you cannot get any credit.

YOU MAY INCLUDE DRAWINGS

Drawings must be *titled* and *labeled*. They must be near the text they relate to. You may *not* use drawings instead of writing an essay.

DO NOT WRITE FORMAL ESSAYS

You do not need to include an introduction, a body, and a conclusion. Doing so is not expected and may take up too much time. Jump right in and answer the question.

DO NOT WORRY

You do not have to include every piece of information about the topic to get full credit. Usually, each essay question is very broad, and there are plenty of ways to get full credit. Remember the reader with the clicker, so just write, write, and write!

DIAGNOSTIC TEST

Answer Sheet
DIAGNOSTIC TEST

1 Ⓐ Ⓑ Ⓒ Ⓓ Ⓔ	26 Ⓐ Ⓑ Ⓒ Ⓓ Ⓔ	51 Ⓐ Ⓑ Ⓒ Ⓓ Ⓔ	76 Ⓐ Ⓑ Ⓒ Ⓓ Ⓔ
2 Ⓐ Ⓑ Ⓒ Ⓓ Ⓔ	27 Ⓐ Ⓑ Ⓒ Ⓓ Ⓔ	52 Ⓐ Ⓑ Ⓒ Ⓓ Ⓔ	77 Ⓐ Ⓑ Ⓒ Ⓓ Ⓔ
3 Ⓐ Ⓑ Ⓒ Ⓓ Ⓔ	28 Ⓐ Ⓑ Ⓒ Ⓓ Ⓔ	53 Ⓐ Ⓑ Ⓒ Ⓓ Ⓔ	78 Ⓐ Ⓑ Ⓒ Ⓓ Ⓔ
4 Ⓐ Ⓑ Ⓒ Ⓓ Ⓔ	29 Ⓐ Ⓑ Ⓒ Ⓓ Ⓔ	54 Ⓐ Ⓑ Ⓒ Ⓓ Ⓔ	79 Ⓐ Ⓑ Ⓒ Ⓓ Ⓔ
5 Ⓐ Ⓑ Ⓒ Ⓓ Ⓔ	30 Ⓐ Ⓑ Ⓒ Ⓓ Ⓔ	55 Ⓐ Ⓑ Ⓒ Ⓓ Ⓔ	80 Ⓐ Ⓑ Ⓒ Ⓓ Ⓔ
6 Ⓐ Ⓑ Ⓒ Ⓓ Ⓔ	31 Ⓐ Ⓑ Ⓒ Ⓓ Ⓔ	56 Ⓐ Ⓑ Ⓒ Ⓓ Ⓔ	81 Ⓐ Ⓑ Ⓒ Ⓓ Ⓔ
7 Ⓐ Ⓑ Ⓒ Ⓓ Ⓔ	32 Ⓐ Ⓑ Ⓒ Ⓓ Ⓔ	57 Ⓐ Ⓑ Ⓒ Ⓓ Ⓔ	82 Ⓐ Ⓑ Ⓒ Ⓓ Ⓔ
8 Ⓐ Ⓑ Ⓒ Ⓓ Ⓔ	33 Ⓐ Ⓑ Ⓒ Ⓓ Ⓔ	58 Ⓐ Ⓑ Ⓒ Ⓓ Ⓔ	83 Ⓐ Ⓑ Ⓒ Ⓓ Ⓔ
9 Ⓐ Ⓑ Ⓒ Ⓓ Ⓔ	34 Ⓐ Ⓑ Ⓒ Ⓓ Ⓔ	59 Ⓐ Ⓑ Ⓒ Ⓓ Ⓔ	84 Ⓐ Ⓑ Ⓒ Ⓓ Ⓔ
10 Ⓐ Ⓑ Ⓒ Ⓓ Ⓔ	35 Ⓐ Ⓑ Ⓒ Ⓓ Ⓔ	60 Ⓐ Ⓑ Ⓒ Ⓓ Ⓔ	85 Ⓐ Ⓑ Ⓒ Ⓓ Ⓔ
11 Ⓐ Ⓑ Ⓒ Ⓓ Ⓔ	36 Ⓐ Ⓑ Ⓒ Ⓓ Ⓔ	61 Ⓐ Ⓑ Ⓒ Ⓓ Ⓔ	86 Ⓐ Ⓑ Ⓒ Ⓓ Ⓔ
12 Ⓐ Ⓑ Ⓒ Ⓓ Ⓔ	37 Ⓐ Ⓑ Ⓒ Ⓓ Ⓔ	62 Ⓐ Ⓑ Ⓒ Ⓓ Ⓔ	87 Ⓐ Ⓑ Ⓒ Ⓓ Ⓔ
13 Ⓐ Ⓑ Ⓒ Ⓓ Ⓔ	38 Ⓐ Ⓑ Ⓒ Ⓓ Ⓔ	63 Ⓐ Ⓑ Ⓒ Ⓓ Ⓔ	88 Ⓐ Ⓑ Ⓒ Ⓓ Ⓔ
14 Ⓐ Ⓑ Ⓒ Ⓓ Ⓔ	39 Ⓐ Ⓑ Ⓒ Ⓓ Ⓔ	64 Ⓐ Ⓑ Ⓒ Ⓓ Ⓔ	89 Ⓐ Ⓑ Ⓒ Ⓓ Ⓔ
15 Ⓐ Ⓑ Ⓒ Ⓓ Ⓔ	40 Ⓐ Ⓑ Ⓒ Ⓓ Ⓔ	65 Ⓐ Ⓑ Ⓒ Ⓓ Ⓔ	90 Ⓐ Ⓑ Ⓒ Ⓓ Ⓔ
16 Ⓐ Ⓑ Ⓒ Ⓓ Ⓔ	41 Ⓐ Ⓑ Ⓒ Ⓓ Ⓔ	66 Ⓐ Ⓑ Ⓒ Ⓓ Ⓔ	91 Ⓐ Ⓑ Ⓒ Ⓓ Ⓔ
17 Ⓐ Ⓑ Ⓒ Ⓓ Ⓔ	42 Ⓐ Ⓑ Ⓒ Ⓓ Ⓔ	67 Ⓐ Ⓑ Ⓒ Ⓓ Ⓔ	92 Ⓐ Ⓑ Ⓒ Ⓓ Ⓔ
18 Ⓐ Ⓑ Ⓒ Ⓓ Ⓔ	43 Ⓐ Ⓑ Ⓒ Ⓓ Ⓔ	68 Ⓐ Ⓑ Ⓒ Ⓓ Ⓔ	93 Ⓐ Ⓑ Ⓒ Ⓓ Ⓔ
19 Ⓐ Ⓑ Ⓒ Ⓓ Ⓔ	44 Ⓐ Ⓑ Ⓒ Ⓓ Ⓔ	69 Ⓐ Ⓑ Ⓒ Ⓓ Ⓔ	94 Ⓐ Ⓑ Ⓒ Ⓓ Ⓔ
20 Ⓐ Ⓑ Ⓒ Ⓓ Ⓔ	45 Ⓐ Ⓑ Ⓒ Ⓓ Ⓔ	70 Ⓐ Ⓑ Ⓒ Ⓓ Ⓔ	95 Ⓐ Ⓑ Ⓒ Ⓓ Ⓔ
21 Ⓐ Ⓑ Ⓒ Ⓓ Ⓔ	46 Ⓐ Ⓑ Ⓒ Ⓓ Ⓔ	71 Ⓐ Ⓑ Ⓒ Ⓓ Ⓔ	96 Ⓐ Ⓑ Ⓒ Ⓓ Ⓔ
22 Ⓐ Ⓑ Ⓒ Ⓓ Ⓔ	47 Ⓐ Ⓑ Ⓒ Ⓓ Ⓔ	72 Ⓐ Ⓑ Ⓒ Ⓓ Ⓔ	97 Ⓐ Ⓑ Ⓒ Ⓓ Ⓔ
23 Ⓐ Ⓑ Ⓒ Ⓓ Ⓔ	48 Ⓐ Ⓑ Ⓒ Ⓓ Ⓔ	73 Ⓐ Ⓑ Ⓒ Ⓓ Ⓔ	98 Ⓐ Ⓑ Ⓒ Ⓓ Ⓔ
24 Ⓐ Ⓑ Ⓒ Ⓓ Ⓔ	49 Ⓐ Ⓑ Ⓒ Ⓓ Ⓔ	74 Ⓐ Ⓑ Ⓒ Ⓓ Ⓔ	99 Ⓐ Ⓑ Ⓒ Ⓓ Ⓔ
25 Ⓐ Ⓑ Ⓒ Ⓓ Ⓔ	50 Ⓐ Ⓑ Ⓒ Ⓓ Ⓔ	75 Ⓐ Ⓑ Ⓒ Ⓓ Ⓔ	100 Ⓐ Ⓑ Ⓒ Ⓓ Ⓔ

Diagnostic Test

CHAPTER 2

MULTIPLE-CHOICE QUESTIONS

80 minutes
100 questions
60% of total grade

Directions: Select the best answer in each case.

1. Countercurrent exchange is a mechanism to

   (A) balance the pH of the blood
   (B) enhance the rate of diffusion
   (C) lower blood pressure
   (D) maximize the flow of blood within the heart
   (E) lower the acidity of the blood

2. Transpiration in plants requires all of the following EXCEPT

   (A) cohesion of water molecules
   (B) active transport of water molecules in the xylem
   (C) capillary action of water in the xylem
   (D) root pressure
   (E) evaporation of water

3. Concerning our understanding of scientific processes, which of the following does NOT occur by chance?

   (A) natural selection
   (B) crossing-over
   (C) variation
   (D) mutation
   (E) natural disasters

4. Which of the following is LEAST likely to result in a release of adrenaline from the adrenal glands?

   (A) walking down a dark, unfamiliar street alone
   (B) participating in a highly selective math competition
   (C) representing your school at the county track meet
   (D) hanging out with your friends after school
   (E) being called to your boss's office when you arrive late to work

5. Which is vascular tissue associated with phloem?

   (A) vessels
   (B) meristem
   (C) sieve
   (D) tracheids
   (E) sclerenchmya

6. Feather color in a species of bird is determined by two different genes. One gene controls the production of melanin (C), while the other gene controls the deposition of melanin (B). *C/_ B/_* is black; *C/_ b/b* is brown; *c/c B/_* is albino; and *c/c b/b* is albino. The inheritance of color in this organism is an example of

   (A) pleiotropy
   (B) expressivity
   (C) blending inheritance
   (D) multiple alleles
   (E) epistasis

7. Which nitrogenous waste requires the least water for its excretion?

   (A) ammonia
   (B) urea
   (C) nitrites
   (D) uric acid
   (E) all of the above require water for their safe removal from the body

8. The walls of arteries consist of

   (A) striated muscle and are under voluntary control
   (B) striated muscle and are not under voluntary control
   (C) smooth muscle and are not under voluntary control
   (D) smooth muscle and are under voluntary control
   (E) a mixture of smooth and striated muscle under voluntary control only

9. According to Hardy-Weinberg, the homozygous recessive organism is represented by

    (A) p
    (B) q
    (C) q^2
    (D) p^2
    (E) $2pq$

10. A salmon swims hundreds of miles to spawn where it was hatched. This is an example of

    (A) classical conditioning
    (B) habituation
    (C) imprinting
    (D) operant conditioning
    (E) fixed action pattern

11. Which of the following is the best example of K-strategists species?

    (A) fruit flies
    (B) fish
    (C) humans
    (D) beetles
    (E) desert annual flowers

12. Which of the following released by macrophages activates helper T cells?

    (A) interleukins
    (B) histamine
    (C) prostaglandin
    (D) pyrogens
    (E) complement

13. The only taxon that actually exits in nature as a natural unit is the

    (A) kingdom
    (B) phylum
    (C) class
    (D) genus
    (E) species

14. The role of oxygen in aerobic respiration is

    (A) the final hydrogen acceptor in the electron transport chain
    (B) in glycolysis
    (C) in the Krebs cycle
    (D) exactly the same as a cytochrome
    (E) to reduce NAD to NAD^+

15. To study the structure of internal membranes of mitochondria from many cells, the best tools to employ would be

    (A) scanning electron microscope and microdissection tools
    (B) scanning electron microscope and ultracentrifuge
    (C) transmission electron microscope and microdissection tools
    (D) transmission electron microscope and the ultracentrifuge
    (E) light microscope and the ultracentrifuge

16. Which of the following have cold, moist winters and short summers and are dominated by gymnosperms?

    (A) taiga
    (B) desert
    (C) tundra
    (D) deciduous forest
    (E) tropical rain forest

17. Cows belong to which trophic level?

    (A) primary consumers
    (B) producer
    (C) decomposers
    (D) carnivore
    (E) autotrophs

18. Which is the correct pathway of blood in humans?

    (A) right atrium-left atrium-pulmonary artery-lungs
    (B) pulmonary veins-vena cava-right atrium-right ventricle
    (C) right atrium-right ventricle-aorta-body cells
    (D) right atrium-right ventricle-pulmonary arteries-lungs
    (E) left atrium-left ventricle-pulmonary arteries-lungs

19. Which is NOT a normal function of the large intestine?

    (A) excretion
    (B) water retention and balance
    (C) vitamin production
    (D) removal of undigested waste
    (E) removal of fecal waste

20. Which of the following is NOT caused by a mutation in a gene?

    (A) cystic fibrosis
    (B) hemophilia
    (C) Huntington's disease
    (D) Tay-Sachs disease
    (E) Klinefelter's syndrome

21. The frequency of a crossing-over event occurring between any two linked genes is

    (A) proportional to the distance between them
    (B) greater if the genes are sex-linked
    (C) higher if they are dominant and lower if they are recessive
    (D) higher if the genes are in a male
    (E) higher if the genes are on separate chromosomes

22. All of the following are present in DNA EXCEPT

    (A) oxygen
    (B) nitrogen
    (C) sulfur
    (D) phosphorous
    (E) carbon

23. What kind of chemical bonds are found between the nitrogenous bases in a molecule of DNA?

    (A) covalent
    (B) van der Waals
    (C) ionic
    (D) hydrogen
    (E) disulfide

24. Acid rain is predominantly caused by which pollutants in the air?

    (A) nitrogen and phosphorous
    (B) phosphorous and nitrogen
    (C) nitrogen and sulfur
    (D) aluminum and carbon
    (E) carbon and sulfur

25. The major histocompatibility complex (MHC) is important because it

    (A) is an important part of the first line of defense
    (B) helps to lyse bacteria that have gotten past the first line of defense
    (C) produces memory cells that maintain immunological memory
    (D) consists of the IgG, the most abundant of the circulating antibodies
    (E) is a collection of markers on cell surfaces that identify cells as self

26. Which is CORRECT about prions?

    (A) They do not cause any diseases.
    (B) They are operons that have been found in mammals.
    (C) They are misfolded proteins.
    (D) They are jumping genes.
    (E) They are bacteria.

27. DNA from one strain of bacteria is assimilated by another strain. This is known as

    (A) generalized transduction
    (B) restricted transduction
    (C) translation
    (D) transformation
    (E) initiation

28. Contractile vacuoles would most likely be found in

    (A) freshwater bacteria
    (B) saltwater bacteria
    (C) freshwater protists
    (D) saltwater protists
    (E) freshwater plants

29. Who was the author and naturalist who developed a theory of natural selection independently of Darwin and who is often credited along with Darwin for the theory we accept today?

    (A) Linnaeus
    (B) Cuvier
    (C) Lyell
    (D) Wallace
    (E) Lamarck

30. In bacteria, genes for resistance to antibiotics are carried on

    (A) plasmids
    (B) prions
    (C) jumping genes
    (D) oncogenes
    (E) bacteriophage viruses

31. Which extraembryonic membrane in birds is analogous to the placenta in mammals?

    (A) chorion
    (B) amnion
    (C) eggshell
    (D) allantois
    (E) yolk sac

32. Genetic drift has the largest effect on allelic frequencies when

    (A) mating is random
    (B) the population is small
    (C) there is no migration, in or out
    (D) there is no natural selection
    (E) there is no mutation

33. A child has blood type O. Which could NOT be the blood type of the parents?

    (A) O; O
    (B) A; B
    (C) A; A
    (D) AB; O
    (E) B; O

34. When natural selection produces very similar phenotypes in two distantly related populations, this phenomenon is called

    (A) genetic drift
    (B) the founder effect
    (C) coevolution
    (D) stabilizing evolution
    (E) convergent evolution

35. Convert 3.5 mm to micrometers.

    (A) 0.035
    (B) 0.0035
    (C) 35.0
    (D) 350
    (E) 3,500

36. A species is defined in terms of

    (A) geographic isolation
    (B) reproductive isolation
    (C) common ancestry
    (D) location
    (E) analogous structures

37. All of the following are sources of variation in a population EXCEPT

    (A) mutation
    (B) crossing-over
    (C) sexual reproduction
    (D) convergent evolution
    (E) outbreeding

38. The wall of a fungus cell is composed of

    (A) cellulose
    (B) lignin
    (C) chitin
    (D) starch
    (E) fatty acid

39. All of the following are *true* of dicots EXCEPT

    (A) vascular bundles are scattered throughout the stem
    (B) the veins in the leaves are netlike
    (C) the floral parts are usually in 4s or 5s
    (D) roots are usually taproots
    (E) the cotyledon consists of two parts

40. Which of the following most accurately describes the way electrons flow during energy production?

    (A) electron transport chain → Krebs cycle → ATP
    (B) glycolysis → Krebs cycle → electron transport chain
    (C) Krebs cycle → pyruvate → electron transport chain
    (D) oxygen → Krebs cycle → electron transport chain
    (E) oxygen → pyruvate → glycolysis

41. A man and a woman who are both normally pigmented have a child who is albino, which is an autosomal recessive trait. The woman is pregnant again. What is the probability that this next child will have normal pigmentation?

    (A) $\frac{1}{16}$
    (B) $\frac{1}{4}$
    (C) $\frac{9}{16}$
    (D) $\frac{3}{4}$
    (E) 1

42. What is the relationship among DNA, a gene, and chromosomes?

    (A) A gene contains many chromosomes, which are composed of DNA.
    (B) A gene contains much DNA, which consists of chromosomes.
    (C) A chromosome consists of many genes, which are composed of DNA.
    (D) A chromosome consists of DNA, which consists of genes.
    (E) DNA consists of many genes, which each consist of one chromosome.

43. Which is true about these diagrams that depict the way in which DNA is replicated?

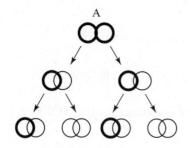 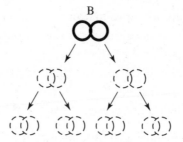 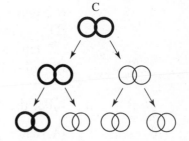

    (A) A depicts semiconservative replication which was proved by Meselson and Stahl.
    (B) B depicts semiconservative replication which was proved by Meselson and Stahl.
    (C) C depicts semiconservative replication which was proved by Meselson and Stahl.
    (D) A depicts semiconservative replication which was proved by F. Griffith in 1927.
    (E) C depicts semiconservative replication which was proved by F. Griffith in 1927.

44. Most of the DNA in eukaryotic chromosomes consists of

    (A) operons
    (B) sequences of DNA that never get transcribed
    (C) genes
    (D) exons
    (E) repetitive sequences that get transcribed over and over

45. Which kingdom consists of the most varied organisms?

    (A) Plantae
    (B) Fungi
    (C) Animalia
    (D) Protista
    (E) Bacteria

46. Which kingdom includes mosses?

    (A) Plantae
    (B) Fungi
    (C) Archaea
    (D) Animalia
    (E) Protista

47. The primary ecological role of prokaryotes is

    (A) causing infections and thereby limiting the population of predators
    (B) limiting the spread of extreme environments
    (C) decomposing organic matter
    (D) to limit the expansion of producers beyond the capacity of the environment
    (E) all of the above are reasons the prokaryotes are ecologically important

48. All of the following are evidence for the theory of endosymbiosis EXCEPT

    (A) chloroplasts have a double membrane
    (B) mitochondria have a double membrane
    (C) mitochondria contain DNA that is similar to bacterial DNA
    (D) mitochondria produce energy and chloroplasts do not
    (E) chloroplasts contain ribosomes that resemble those inside bacteria

49. Polymerase chain reaction would be carried out to accomplish which one of the following?

    (A) to prevent a virus from replicating
    (B) to amplify a region of DNA
    (C) to identify the location of a point mutation in a genome
    (D) to identify a chromosome mutation
    (E) to prepare a specimen of DNA for X-ray diffraction

50. Protozoans are generally classified according to

    (A) how they carry out nutrition
    (B) cell shape
    (C) locomotion
    (D) type of reproduction
    (E) the number of nuclei

51. All Protista are alike in that they are all

    (A) eukaryotes
    (B) autotrophic
    (C) heterotrophic
    (D) single-celled organisms
    (E) photosynthetic

52. All of the following structures are adaptations specifically for life on land EXCEPT

    (A) xylem
    (B) roots
    (C) waxy cuticle
    (D) seeds
    (E) cell walls

53. All of the following are characteristic of angiosperms EXCEPT

    (A) coevolution with pollinators
    (B) free-living gametophytes
    (C) internal fertilization
    (D) fruit
    (E) seeds

54. In the operon, the operator is the binding site for the

    (A) promoter
    (B) regulator gene
    (C) repressor
    (D) RNA polymerase
    (E) inducer

55. A plant cell is placed into pure water. Which of the following will occur?

    (A) Water will flow into the cell because the water potential inside the cell is higher than it is outside the cell.
    (B) Water will flow into the cell because the water potential inside the cell is lower than it is outside the cell.
    (C) Water will flow out of the cell because the water potential inside the cell is lower than it is outside the cell.
    (D) Water will flow out of the cell because the water potential inside the cell is higher than it is outside the cell.
    (E) Water will not flow into or out of the cell because the cell wall prevents water from diffusing in or out.

56. Cephalization is associated with all of the following EXCEPT

    (A) sessile existence
    (B) bilateral symmetry
    (C) a brain
    (D) sensory structures concentrated at the anterior end
    (E) a nerve cord and nervous system

57. All of the following are correct about enzyme-catalyzed reactions EXCEPT

    (A) glycolysis is inhibited by ATP
    (B) enzymes speed up reactions by supplying necessary energy
    (C) enzymes lower the energy of activation
    (D) enzymes often require the presence of minerals
    (E) all enzymes are not active at the same pH

58. Pinching back a plant, removing the new leaves at the top, results in the plant growing bushier. This happens because when the new leaves are removed, _____ is (are) removed as well.

    (A) gibberellins
    (B) cytokinins
    (C) auxins
    (D) ethylene
    (E) abscisic acid

59. All of the following are true about cellular respiration EXCEPT

    (A) NAD^+ is reduced to NADH during both glycolysis and the Krebs cycle
    (B) in the absence of NAD^+, glycolysis cannot occur
    (C) ATP is produced by oxidative phosphorylation
    (D) the most important product of glycolysis is ATP
    (E) lactic acid is converted back to pyruvic acid in the liver

Questions 60–63

Questions 60–63 refer to the structural formulas below of several important molecules.

(A)

(D)

(E)

(B)

(C)

60. Makes up channels within membranes

61. Gives plasma membranes stability and rigidity

62. Produced during cell respiration

63. Makes up the hydrophilic part of the plasma membrane

Questions 64–66

The following three questions refer to this experimental setup. Five dialysis bags (A–E), impermeable to sucrose, were filled with different concentrations of sucrose and then placed into separate beakers containing an initial concentration of 0.5 M sucrose solution. After 60 minutes, the bags were weighed. The data was collected and displayed on the graph below.

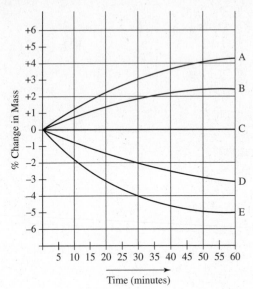

64. Which line represents the bag with the lowest initial concentration of sucrose?

    (A) A
    (B) B
    (C) C
    (D) D
    (E) E

65. Which line represents the bag containing a solution isotonic to the solution in the beaker?

    (A) A
    (B) B
    (C) C
    (D) D
    (E) E

66. All of the following statements about the chemistry of living things are correct EXCEPT

    (A) Water is a polar molecule and will dissolve a monosaccharide but not a lipid.
    (B) Elements containing the same number of protons but a different number of neutrons are called isomers.
    (C) In order for a molecule to form, electrons must be shared.
    (D) The secondary structure of a protein is due to intramolecular hydrogen bonding.
    (E) Chaperone proteins assist in the proper folding of proteins.

Questions 67–69

Questions 67–69 refer to the karyotype below.

67. Which is correct about the person from whom this karyotype was taken?

(A) The person has Klinefelter's syndrome.
(B) The person has a sex-linked condition.
(C) The person has Down's syndrome.
(D) The person has hemophilia.
(E) The person has Turner's syndrome.

68. These are the chromosomes of a person who is

(A) male
(B) female
(C) male with an extra X chromosome
(D) female with an extra chromosome
(E) female missing a chromosome

69. Which of the following is true about this condition?

(A) One chromosome has been deleted.
(B) This arose as a result of nondisjunction.
(C) There has been a translocation.
(D) These chromosomes exhibit polyploidy.
(E) This karyotype shows pleiotropy.

Questions 70–73

Questions 70–73 refer to the hormones below.

    (A) Auxins
    (B) Cytokinins
    (C) Gibberellins
    (D) Abscisic acid
    (E) Ethylene

70. Responsible for phototropism

71. Inhibits growth; counteracts breaking of dormancy

72. Ripens fruit

73. Is a main ingredient in rooting powder

Questions 74–77

Questions 74–77 refer to the four choices below.

    (A) Light-dependent reactions (light reactions)
    (B) Light-independent reactions (dark reactions)
    (C) Both the light and dark reactions
    (D) Neither the light nor dark reactions

74. Oxygen is released

75. Carbon fixation

76. Oxidative phosphorylation

77. Photolysis occurs

Questions 78–79

Questions 78–79 refer to the drawings below of chromosomes as they appear normally in stages of meiosis and mitosis.

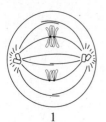 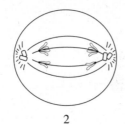 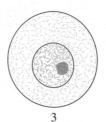  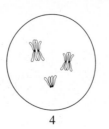

1 2 3 4 5

78. In which diagram are sister chromatids ready to separate?

    (A) 1
    (B) 2
    (C) 3
    (D) 4
    (E) 5

79. Which process is most responsible for increasing variation?

    (A) 1
    (B) 2
    (C) 3
    (D) 4
    (E) 5

Questions 80–83

Questions 80–83 refer to the pedigree which illustrates the inheritance pattern of sickle cell anemia.

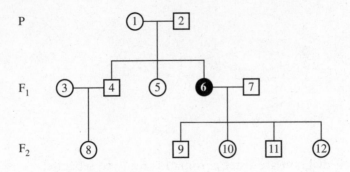

80. The genotype of individual 6 is

    (A) *X–X*
    (B) *X–X–*
    (C) *S/s*
    (D) *s/s*
    (E) cannot be determined

81. Which is correct about the offspring of parents 6 and 7?

    (A) Only the females are carriers of sickle cell.
    (B) Only the males are carriers of sickle cell.
    (C) The males have the condition, but the females are carriers.
    (D) All the children are carriers.
    (E) All the children inherited immunity against sickle cell disease from the parents.

82. What is the probability that parents 1 and 2 would have a child with sickle cell disease?

    (A) 1
    (B) ½
    (C) ¼
    (D) ¾
    (E) 0

83. What is the most likely reason that parents 6 and 7 did not have a child with sickle cell anemia?

    (A) The children did not inherit a sickle gene from their mother.
    (B) The children each inherited two healthy genes from their father.
    (C) Penetrance for sickle cell is not 100%. Even though they each inherited two sickle genes from their parents, they do not exhibit the trait.
    (D) The children each inherited one healthy gene from their father.
    (E) Sickle cell anemia is not inherited, and all the children received immunization against it immediately after birth.

Questions 84–86

Questions 84–86 refer to this graph which shows the progress of two chemical reactions. One is catalyzed by an enzyme, and one is not.

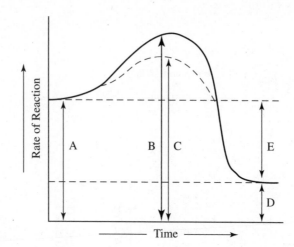

84. Which letter shows the energy needed to initiate the catalyzed reaction?

85. Which letter represents the energy level of the products of the uncatalyzed reaction?

86. Which is TRUE about the reaction shown here?

    (A) The forward reaction is exergonic.
    (B) The forward reaction absorbs more energy than the reverse reaction gives off.
    (C) The forward reaction gives off more energy than the reverse reaction absorbs.
    (D) The enzyme involved catalyzes only the forward reaction.
    (E) The potential energy at the end of this reaction is greater than the potential energy at the outset.

Questions 87–88

Questions 87–88 refer to this graph showing one species of the snail *Cepaea* from the fall of 1994 to the fall of 2000.

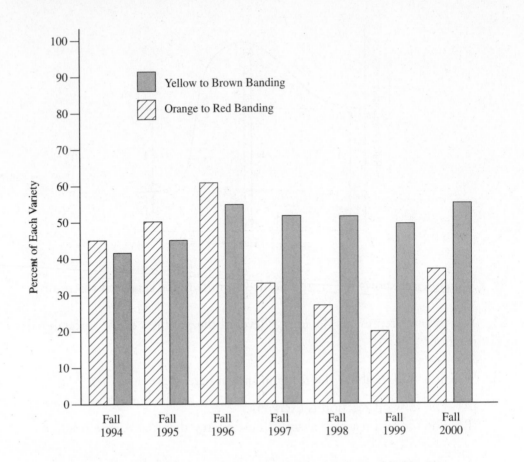

87. Which of the following statements about this population is correct?

    (A) There is no difference in the death rate.
    (B) The orange-banded snail is better adapted to the environment overall.
    (C) This is an example of a cline.
    (D) This is an example of balanced polymorphism.
    (E) There are not inadequate resources for both organisms in this ecosystem.

88. The most likely conclusion you can draw from this graph is

    (A) there has been increasingly more rain in 1999 and 2000
    (B) there are two different species of snail
    (C) one variety of snail is better adapted at one time and the other variety at another time
    (D) the gene for coloration in one variety is dominant over the gene for coloration in the other variety
    (E) a single factor in the environment is responsible for the superiority of one variety over the other

Questions 89–90

Questions 89–90 refer to this drawing of an embryo in its earliest stages.

89. Which of the following animals could NOT form from the embryo on the left?

   (A) flatworm
   (B) earthworm
   (C) shrimp
   (D) fish
   (E) fly

90. Which is TRUE of the embryo on the left?

   (A) The fate of these cells has not yet been determined.
   (B) The blastopore will become the mouth.
   (C) This development is characteristic of humans.
   (D) The four cells facing up make up the archenteron.
   (E) All these cells will give rise to the gametophyte.

Questions 91–92

Questions 91–92 refer to this sketch of part of a cell.

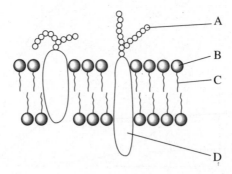

91. Which structure is most directly responsible for contact inhibition?

92. Which structure is hydrophobic?

Questions 93–96

Questions 93–96 refer to the graph below which shows the changes in a population size over time.

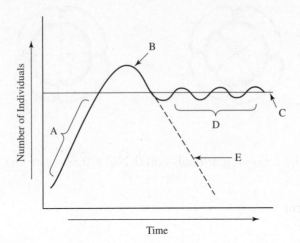

93. The exponential growth phase of a new population

94. The carrying capacity of the ecosystem

95. A population in unfavorable conditions

96. A well-established, stable population

Questions 97–98

Questions 97–98 refer to this graph showing survivorship curves

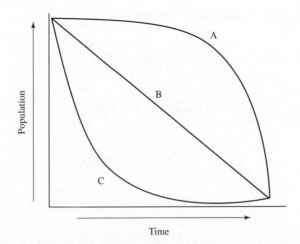

97. Which line describes human survivorship?

98. Which line best describes an organism that lays eggs in the water in which it lives?

Questions 99–100

Questions 99–100 refer to this graph showing stimulation of a muscle.

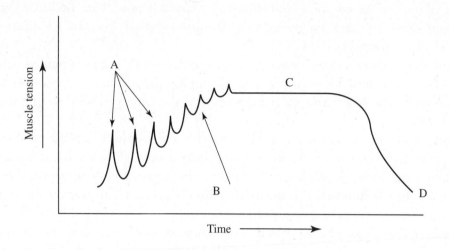

99. Shows a muscle contracting locally and briefly

100. Shows tetanus

STOP

If there is still time remaining, you may review your answers.

READING PERIOD

10 minutes

Read all four essays, organize your thoughts, make any notes you need to assist you on your question sheet. Following this 10-minute period, you will be directed to answer the four essays.

You must answer all four essays. Because each question will be weighted equally, you are advised to divide your time equally among them. Do not spend too much time on any one essay. You must keep track of the time yourself; no one will do that for you.

You are to write your answers in the answer booklet only. Use black or blue ink. Be sure to write clearly and legibly. If you make a mistake, you may save time by crossing it out with a single line rather than trying to erase it. Number each answer as the question is numbered in the examination. Begin each answer on a new page. *You may answer the essays in any order.*

Each answer should be organized, well-balanced, and as comprehensive as time permits. You must write in prose form; outline form is not acceptable. Do not spend time restating the questions, simply answer the questions. If a specific number of examples are asked for, no credit will be given for extra examples. Diagrams may be used to help explain your answer, but in no case will a diagram alone suffice.

WRITING PERIOD

1 hour and 30 minutes
Suggested writing time per question—22 minutes.

Question 1

Communication between and among cells is critical to the maintenance of healthy multicellular organisms. Choose 5 different specific examples demonstrating cells communicating with other cells and explain the mechanism of each. Examples can be in the animal and/or the plant kingdoms.

Question 2

Interdependence in nature is a major theme in biology. Discuss interdependence at each of the following levels.

   a. Molecular
   b. Cellular
   c. Organ/organism
   d. Population

Question 3

Explain the structure of the plasma membrane, and explain how a membrane would transport four of the following. Give examples to support your ideas.

   a. Glucose
   b. Sodium and potassium ions
   c. A small lipid hormone
   d. Large amounts of water
   e. Calcium ions
   f. Cholesterol

Question 4

Animals exhibit orientation behaviors. They move toward or away from stimuli. Choose an appropriate animal and devise an experiment to

   a. observe and demonstrate how the animal responds to two environmental stimuli
   b. explore any preference the animal has for different environments

If there is still time remaining, you may review your answers.

Answer Key
DIAGNOSTIC TEST

Multiple-Choice Questions

1. B	26. C	51. A	76. A
2. B	27. D	52. E	77. A
3. A	28. C	53. B	78. E
4. D	29. D	54. C	79. D
5. C	30. A	55. B	80. D
6. E	31. D	56. A	81. D
7. D	32. B	57. B	82. C
8. C	33. D	58. C	83. D
9. C	34. E	59. D	84. C
10. C	35. E	60. B	85. D
11. C	36. B	61. C	86. A
12. A	37. D	62. E	87. D
13. E	38. C	63. B	88. C
14. A	39. A	64. E	89. D
15. D	40. B	65. C	90. B
16. A	41. D	66. B	91. A
17. A	42. D	67. C	92. C
18. D	43. A	68. A	93. A
19. A	44. B	69. B	94. C
20. E	45. D	70. A	95. E
21. A	46. A	71. D	96. D
22. C	47. C	72. E	97. A
23. D	48. D	73. A	98. C
24. C	49. B	74. A	99. A
25. E	50. C	75. B	100. C

ANSWER EXPLANATIONS

Multiple-Choice Questions

1. **(B)** Countercurrent exchange occurs when two fluids flow next to each other in opposite directions. This maximizes the rate of diffusion of oxygen and carbon dioxide over fish gills. The blood in the gills is deoxygenated and flows toward the front of the fish, while the water flowing over it is oxygenated and flowing toward the back of the fish. Countercurrent exchange helps in heat retention because arteries carry warmer core blood out to the periphery of the body, while veins carry cooler blood from the periphery to the core of an animal. The two vessels travel next to each other, and heat moves from the arteries to the veins to warm the blood.

2. **(B)** The transport of water up a plant in the xylem does not require any energy. The major force moving water and minerals from the soil up the plant is described by transpirational pull-cohesion-tension theory. Root pressure is another force that helps to push water and minerals upward into the plant stem.

3. **(A)** Natural selection does not occur by chance. The environment directs the evolution of a population. A population in a stable environment does not change.

4. **(D)** Adrenaline is a hormone of fight or flight. It is released in times of stress and stimulates the heart and breathing rates.

5. **(C)** Companion and sieve make up phloem vessels. Tracheids and vessels make up the xylem.

6. **(E)** When inheritance involves epistasis, two separate genes control one trait, but one gene masks the expression of the other gene. The gene that masks the other gene is epistatic to the gene it masks. In the case in question, the animal with the genotype *c/c* cannot produce melanin. Therefore, the gene for deposition of melanin is irrelevant. If there is no melanin, no melanin can be deposited. The gene for production of melanin is epistatic to the gene for deposition of melanin.

   Pleiotropy is the inheritance pattern where one gene controls the expression of several traits. Frazzle trait in chickens and Marfan's syndrome are examples. Expressivity refers to how a gene varies in its expression. The dominant gene for polydactyly can express itself as a bent figure or as an extra finger. An example of multiple alleles is blood type, where there are three alleles: A, B, and O. Blending inheritance is incomplete dominance. An example can be seen in flowers where a red and a white parent produce a pink offspring.

7. **(D)** Uric acid is not soluble in water and is excreted as a dry crystal by the animal's digestive system. Ammonia is the most toxic nitrogenous waste and must be excreted in large volumes of water.

8. **(C)** Arteries are lined with smooth muscle cells and are under the control of the autonomic nervous system; they are not under voluntary control. Striated muscle is also called skeletal muscle and is under the voluntary control of the somatic nervous system.

9. **(C)** Lowercase q represents the recessive allele, and q^2 represents the homozygous recessive organism. Lowercase p represents the dominant allele, and p^2 represents the homozygous dominant organism. The heterozygous organism is represented by $2pq$.

10. **(C)** Imprinting is learning that occurs during a critical period in the early life of an animal and is generally irreversible. Somehow the place where the salmon hatches is imprinted in the salmon's memories, and it returns there to spawn.

11. **(C)** K-strategists, like mammals, have few, large young but reproduce more than once. They invest a great deal of energy in parenting. In contrast, r-strategists are opportunistic, reproducing rapidly when resources are plentiful. Fertilization is external and the females release hundreds of thousands of eggs. This applies to fish, beetles, and fruit flies from the question. Annual desert flowers are also opportunistic, reproducing prodigiously during the very short spring season.

12. **(A)** Interleukin-1 is released by macrophages and activates helper T cells. Helper T cells, in turn, release interleukin-2 which activates T_c and B cells. Histamine is released by basophils and mast cells and induces the inflammatory response, increasing blood supply to an area and causing sneezing and runny nose and eyes. Prostaglandins enhance the inflammatory response and also increase blood flow to a region. Pyrogens, released by certain leukocytes, increase body temperature and thus speed up the immune system to help fight invading microbes. Complement is a group of about twenty proteins that assist in lysing cells.

13. **(E)** All the taxa except species consist of arbitrarily defined groups.

14. **(A)** The role of oxygen in cell respiration is limited to the electron transport chain where it acts as the final hydrogen acceptor. Oxygen is not required in any other stage of cell respiration.

15. **(D)** A transmission electron microscope (TEM) has higher magnification and better resolution than a light microscope. Unlike the scanning electron microscope, which is used to study cell surfaces, the TEM is used to study internal membranes. To study many mitochondria, one would homogenize tissue in a blender and spin it in a centrifuge, which would separate the different layers of organelles by density. When tissue is differentially spun, the densest structures, the nuclei, sink to the bottom, the next layer contains mitochondria, and the next contains ribosomes and layers of membranes. One would extract the layer containing mitochondria and prepare them for study under the transmission electron microscope.

16. **(A)** The taiga is the largest terrestrial biome and covers much of the northern regions of the world. It is dominated by conifers, and the landscape is dotted with lakes, ponds, and bogs. The taiga is characterized by heavy snowfall with large mammals like moose and elk and lots of insects and birds in the summer.

17. **(A)** Cows eat grass or corn, which makes them primary consumers and herbivores.

18. **(D)** The normal pathway of blood is Right atrium → Right ventricle → Pulmonary arteries → Lungs → Pulmonary veins → Left atrium → Left ventricle → Aorta → Cells of the body

19. **(A)** Excretion is the removal of metabolic wastes, including the nitrogenous wastes ammonia, urea, and uric acid. It is carried out by the kidneys. The large intestine removes undigested waste, a process known as *egestion.*

20. **(E)** Klinefelter's syndrome is caused by a mutation in the chromosomes, not in the genes. It results from nondisjunction. A person with Klinefelter's has 47 chromosomes with an extra X chromosome. This person is male because he has a Y chromosome and is most likely sterile.

21. **(A)** The greater the distance between two genes on the same chromosome, the greater the opportunity that they will separate from each other during meiosis I.

22. **(C)** DNA does not contain sulfur; proteins do. DNA consists of the sugar deoxyribose, which contains carbon, hydrogen, and oxygen. DNA also contains phosphates, which contain phosphorous and oxygen. DNA also contains nitrogenous bases.

23. **(D)** Hydrogen bonds join the two strands of the double helix at the nitrogenous bases.

24. **(C)** Acid rain results from by-products of industry and the internal combustion engine, nitrogen and sulfur. The common acids in acid rain are nitrous acid, nitric acid, and sulfuric acid.

25. **(E)** MHC molecules, also known as HLA (human leukocyte antigens), are a collection of cell surface markers that identify cells as self. IgG is the most abundant circulating antibody, but that has nothing to do with the question. The other choices do not make any sense.

26. **(C)** Prions are infectious proteins that cause several brain diseases: scrapie in sheep, mad cow disease in cattle, and Creutzfeldt-Jakob disease in humans. A prion is a misfolded version of a protein normally found in the brain. If a prion gets into a normal brain, it causes all the normal proteins to misfold in the same way. Prions are not cells and not viruses. However, because they are seriously infectious entities, they are classified with viruses.

27. **(D)** Transformation is the process where one bacterium absorbs genes from another strain of bacteria. This was discovered by Griffith, who performed experiments with several different strains of the bacterium *Diplococcus pneumoniae.*

28. **(C)** Contractile vacuoles are structures within freshwater protists that pump out excess water. These organisms live in an environment that is hypotonic, so water tends to diffuse through the membrane and into the cell. Saltwater protists are isotonic to their environment and do not have the same problem as do freshwater organisms. Bacteria do not have any internal membranes, so they cannot have a contractile vacuole. The rigid cell wall of freshwater plants prevents the plant cells from exploding.

29. **(D)** Wallace published his theory of natural selection before Darwin. However, Darwin's thesis, when he did publish it, was in much greater detail with more data; so Darwin is credited with the theory of natural selection and Wallace is rarely mentioned.

30. **(A)** A plasmid is a foreign, small, circular, self-replicating DNA molecule that inhabits a bacterium. The presence of a plasmid in a bacterium is often indicated by the phenotypic expression of the genes carried by the plasmid. The R plasmid confers resistance to specific antibiotics, such as ampicillin or tetracycline, on the host bacteria. An oncogene is an altered version of a normal gene, called a proto-oncogene, which stimulates cell division. If oncogenes become altered, they become cancer-producing genes. Jumping genes or transposons were discovered by Barbara McClintock. Until their discovery, scientists believed that genes were fixed within the genome. Bacteriophage viruses are viruses that attack bacteria.

31. **(D)** The allantois is analogous to the placenta in mammals because it is the conduit for respiratory gases and the repository of nitrogenous waste between the environment and the embryo. The chorion lies directly under the shell and allows for diffusion of respiratory gases between the air and the embryo. The amnion encloses the embryo in a protective fluid, amniotic fluid. The eggshell is not a membrane. The yolk sac encloses the food for the growing embryo.

32. **(B)** Genetic drift is the change in a gene pool due to chance. If the population is small, the smallest change in the gene pool will have a major effect in allelic frequencies. In a large population, a small change in the gene pool will be diluted by the sheer number of individuals and there will be no change in the frequency of alleles.

33. **(D)** A person with blood type O is homozygous recessive, $I^o I^o$, and must have inherited an O gene from both parents. A parent with blood type AB has an I^A gene and an I^B gene, and lacks the I^o gene. Therefore, the parent with blood type AB cannot be the parent of a child with type O blood.

34. **(E)** An example of convergent evolution is the similar appearance of the shark and the whale, which are not related to each other; the whale is a mammal and the shark is a fish. They have a similar appearance because they evolved in the same environment.

35. **(E)** Use the factor label method to do metric conversions.

$$\frac{3.5 \text{ mm} \times 100 \text{ μm}}{1 \text{ mm}} = 3,500 \text{ μm}$$

36. **(B)** A species is defined in terms of reproductive isolation, meaning that one group of genes becomes isolated from another to begin a separate evolutionary history. If enough time elapses and differing selective forces are sufficiently great, the two populations may become so different that even if they were brought back together, interbreeding would not naturally occur. At this point, speciation is said to have taken place. *Anything that fragments a population and isolates small groups of individuals may cause speciation.*

37. **(D)** Convergent evolution tends to limit variation. The shark and the whale exhibit convergent evolution. They look alike because they evolved in the same environment, not because they share a recent common ancestor. Mutation, crossing-over, sexual reproduction, and outbreeding all are sources of variation in a population.

38. **(C)** The cell walls of fungi are made of chitin, not cellulose. This is one reason that fungi are not classified with plants but are placed in a separate kingdom.

39. **(A)** Choices B–E are all characteristics of dicots. The vascular bundles in a monocot stem are scattered throughout the stem.

40. **(B)** In cell respiration, an oxidative process, electrons flow from glycolysis to the Krebs cycle to the electron transport chain.

41. **(D)** The trait for albinism is autosomal recessive. Since neither parent exhibits the trait but they already have a child with the condition, they must be hybrids. Here is the cross.

    The genotype *a/a* is albino. *AA* and *Aa* are normal.

	A	*a*
A	*AA*	*Aa*
a	*Aa*	*aa*

42. **(D)** The entire chromosome consists of DNA, which consists of genes. There are thought to be about 25,000 genes in humans.

43. **(A)** DNA replicates by semiconservative replication. Each strand of the double helix forms a new helix. The new helix consists of one old strand and one new strand. This was hypothesized by Watson and Crick and proved by Meselson and Stahl. F. Griffith discovered the transformation factor in bacteria in 1927.

44. **(B)** Less than 3% of human DNA codes for proteins. Most of the DNA in eukaryotic chromosomes consists of repetitive sequences that never get transcribed. Operons are found only in prokaryotic cells. Introns are intervening sequences. Exons are expressed sequences; these are the genes.

45. **(D)** The kingdom Protista consists of the greatest variety of organisms of any kingdom. For example, it contains organisms that are photosynthetic and nonphotosynthetic. Also, it contains single-celled and multi-celled organisms. If scientists added another kingdom, Protista would certainly be divided. Bacteria is a domain, not a kingdom.

46. **(A)** Mosses are photosynthetic. They are seedless plants with no vascular tissue. Archaea is a domain, not a kingdom.

47. **(C)** Bacteria recycle nutrients and are therefore critical to the maintenance of any ecosystem. There are bacteria of decay, nitrogen-fixing bacteria, denitrifying bacteria, and nitrifying bacteria.

48. **(D)** All statements are correct except choice D. Both mitochondria and chloroplasts produce ATP. Mitochondria produce energy during cell respiration, and chloroplasts produce energy during the light-dependent reactions of photosynthesis.

49. **(B)** Polymerase chain reaction, PCR, is an automated technique by which a piece of DNA can be rapidly copied in large quantities.

50. **(C)** The Protista are classified by mode of locomotion. Paramecia move by cilia, amoeba move by pseudopods, and euglena move by means of a flagellum.

51. **(A)** The kingdom Protista consist of the most varied organisms of any kingdom. However, one thing all Protista have in common is that they are all eukaryotes.

52. **(E)** Choices A, B, C, and D are modifications for life on land. All plants have cell walls for support and protection, including plants that live in the oceans.

53. **(B)** The phrase "free-living gametophytes" describes a gametophyte generation that is dominant, such as seen in the primitive plants like mosses. All the other choices in the question are characteristics of advanced flowering plants, the angiosperms, which have a dominant sporophyte generation.

54. **(C)** In the operon, the operator is the binding site for the repressor and the promoter is the binding site for RNA polymerase. When RNA polymerase binds to the promoter, the structural genes can begin to transcribe. When the repressor binds to the operator, RNA polymerase is prevented from binding to the promoter, and therefore, transcription is blocked.

55. **(B)** The water potential of pure water is zero. The water inside a cell is less than zero. Since water flows from higher water potential to lower water potential, water will flow into the cell and cause it to swell.

56. **(A)** Cephalization is the development of a head end. It is characteristic of advanced animals with a brain and nerve cord, a nervous system, and sensory apparatus. This helps the animal move easily and rapidly to capture food, find mates, and escape predation. A sessile life is characteristic of simple animals like the sponge.

57. **(B)** Enzymes speed up reactions by lowering the energy of activation, not by adding energy.

58. **(C)** Auxins are responsible for apical dominance. See the section about plant hormones in the chapter "Plants" for details of this important topic.

59. **(D)** Very little ATP is released from glycolysis, only 4 ATP. The important product is pyruvic acid or pyruvate, which is the raw material for the Krebs cycle and still contains a large amount of stored energy.

60–63.   Structure A is a lipid, B is an amino acid, C is cholesterol, D is an unsaturated fatty acid, and E is ATP.

60. **(B)** Channels within the cell membrane are composed of proteins. Choice B is an amino acid, the building block of proteins.

61. **(C)** Cholesterol makes a plasma membrane more stabile, less fluid.

62. **(E)** ATP is produced during cell respiration.

63. **(B)** Proteins are hydrophilic and make up channels within the membrane. Fatty acids are hydrophobic.

64. **(E)** Since the bags are impermeable to sucrose, only water can diffuse across the membrane. If the molarity of sucrose in the bag is higher than the surrounding solution, water will flow into the bag and the bag will gain in mass. If the molarity of sucrose in the bag is lower than the surrounding solution, water will flow out of the bag and the bag will lose mass. Bag A gained the most mass, and bag E lost the most mass. Therefore, bag E had the lowest initial concentration of sucrose.

65. **(C)** Bag C did not lose or gain mass, so it must be isotonic with its surroundings.

66. **(B)** Elements containing the same number of protons but a different number of neutrons are called isotopes, not isomers. All the other statements are correct.

67. **(C)** There are 47 chromosomes instead of the normal 46 and there is an extra 21st chromosome, trisomy 21. This condition causes Down's syndrome. Hemophilia is caused by a mutation in the gene and cannot be seen by looking at the chromosomes. Klinefelter's syndrome is characterized by XXY, Turner's is XO (missing one sex chromosome).

68. **(A)** The person is male. There is an X and a Y (very small in the lower right) chromosome.

69. **(B)** Having an extra chromosome or too few results from nondisjunction. Polyploidy means having an entire extra set of chromosomes, 3*n* or 4*n*. When a piece of one chromosome breaks off and attaches itself to another chromosome that is called translocation. When one gene affects more than one trait, the phenomenon is called pleiotropy. There are 47 chromosomes, one extra. There has not been a deletion.

70. **(A)** Auxins like IAA (indolacetic acid) are responsible for tropisms.

71. **(D)** Abscisic acid counteracts dormancy and generally inhibits growth. It closes stomates in times of stress and promotes seed dormancy. It works in opposition to growth hormones.

72. **(E)** Ethylene is a gas given off by fruit as it ripens.

73. **(A)** Auxins enhance growth. They are used in rooting powder and control apical dominance and phototropisms.

74. **(A)** Oxygen is released as a result of photolysis when water gets broken apart to supply electrons to photosystem II and photons to the light and dark reactions.

75. **(B)** Carbon fixation is the process by which PGAL (sugar) is made. It occurs during the dark reactions and is powered by the ATP produced during the light reactions. Carbon fixation occurs during the Calvin cycle.

76. **(A)** Oxidative phosphorylation is the process by which ATP is produced as protons flow through ATP synthetase channels in the thylakoid or cristae membrane. It occurs during the light-dependent reactions.

77. **(A)** Photolysis is the breaking apart of water during the light reactions. This process provides electrons for the electron transport chain in the thylakoid membrane and protons for the light and dark reactions. Oxygen is released as a result of photolysis.

78–79.
- Picture 1 shows metaphase I. Homologous pairs are lined up on the metaphase plate in preparation for separation.
- Picture 2 shows telophase I as homologous pairs are quite far apart.
- Picture 3 shows interphase. The nuclear membrane is intact and the nucleolus is visible.
- Picture 4 is prophase. It shows homologous chromosomes pairing up into bivalent or tetrads. The process is called synapsis. It is most important because it provides the opportunity for crossing-over to occur, which increases the variation in the species.
- Picture 5 shows metaphase I, where chromosomes are lined up on the metaphase plate, single file, preparing for sister chromatids to separate.

78. **(E)**

79. **(D)**

80. (**D**) The pattern of inheritance of sickle cell disease is autosomal recessive. In order to have the condition, as individual 6 does, one must have two mutated alleles. If you did not know the inheritance pattern of sickle cell, you could have figured it out because neither parent (1 and 2) has the disease.

81. (**D**) All the children are carriers because the mother (6) gave one mutated gene to each offspring.

82. (**C**) Parents 1 and 2 are both carriers. Here is the cross.

	S	*s*
S	*SS*	*Ss*
s	*Ss*	*ss*	← sickle cell disease

There is a 25% chance they would have a child with sickle cell.

83. (**D**) The children of 6 and 7 did not inherit sickle cell because they inherited one healthy gene from their father. The condition requires two diseased genes to express the disease.

84. (**C**) The dashed line shows the catalyzed reaction. Enzymes lower the energy of activation. The letter that shows the uncatalyzed reaction is B.

85. (**D**) The potential energy of the products is the same for both the catalyzed and uncatalyzed reactions.

86. (**A**) The forward reaction is exergonic (exothermic). That means the potential energy of the products is less than the potential energy of the reactants because energy was given off. Enzymes catalyze reactions in both directions. The energy given off by the exothermic (forward) reaction must be the same as the energy absorbed by the endothermic (reverse) reaction.

87. (**D**) Notice the questions states that there is only one species. This is an example of balanced polymorphism; two phenotypically distinct forms of a trait in a single population of one species. Each type of snail represents a different variety of the same species. By analyzing the graph, it is obvious that at some times one variety succeeds better than the other. At other times, the reverse is true. A difference in death rates certainly exists, but it is not consistent. No overall trend shows one variety more successful than another. An example of a cline is varied ear length or coloration within one species of rabbit that lives in very different surroundings. In this question, both varieties lived in one locale.

88. (**C**) Since the two types of snails belong to the same species, they cannot be that different from each other. You can draw no other meaningful conclusion.

89. (**D**) The earliest cleavage of the embryo is spiral. Spiral cleavage is characteristic of protostomes, all animals except the echinoderms and chordates. The fish has a backbone and is a chordate.

90. **(B)** Spiral cleavage is characteristic of protostomes. The fate of the cell is determinant, that is, it is already decided. The blastopore becomes the mouth. The archenteron is the primitive gut and develops inside the gastrula. The gametophyte is the haploid generation of plants that exhibits alternation of generations.

91. **(A)** This is the glycocalyx, sugars attached to membrane proteins that are responsible for identifying the cell. When cells grow too crowded, they stop dividing. This type of inhibition is known as contact inhibition.

92. **(C)** The tail of the balloon is hydrophobic and consists of two hydrocarbon tails. B is the phosphate head of the phospholipid and is hydrophilic. D is an integral protein.

93. **(A)** This is the exponential growth phase of a population newly introduced into an area where there are many resources and little competition.

94. **(C)** The carrying capacity is the maximum amount of life that an ecosystem can support. The population size fluctuates around the carrying capacity.

95. **(E)** The population living in the ecosystem has used up its resources and is poisoning its environment.

96. **(D)** This is a well-established and stable population.

97. **(A)** This curve describes a species that has high survival rate of the young. Most organisms survive into old age when they seem to die at about the same age. This is characteristic of mammals (including humans) and other organisms that have internal fertilization and development of young, protective parenting, and few offspring.

98. **(C)** These animals have very high mortality immediately. The ones that do survive live a long life. This is characteristic of organisms that fertilize externally, producing thousands of eggs that will mostly be destroyed by predation. Examples are fish and amphibians.

99. **(A)** A single action potential in a muscle will cause the muscle to contract locally and minutely for a few milliseconds and then to relax. This is called a twitch and is shown at A. If a second action potential arrives before the first response is over, there will be a summation effect and the contraction will be larger, as seen at B.

100. **(C)** If the rate of stimulation of a muscle is fast enough, the twitches will blur into one smooth, sustained contraction called tetanus, as seen at C.

What Topics Do You Need to Work On?

This table shows an analysis by topic for each question on the test you just took. Some questions are repeated in different areas.

Table 2.1

Topic Analysis			
Biochemistry and Enzymes (Ch. 3)	**Cells and Cell Division (Ch. 4, 7)**	**Cell Respiration and Photosynthesis (Ch. 5, 6)**	**Heredity and Molecular Genetics (Ch. 8, 9)**
1, 38, 57, 60, 61, 62, 63, 84, 85, 86	15, 26, 28, 35, 38, 48, 55, 64, 65, 66, 78, 79, 91, 92	14, 40, 59, 74, 75, 76, 77	6, 20, 21, 22, 23, 26, 27, 30, 41, 42, 43, 44, 49, 54, 67, 68, 69, 80, 81, 82, 83
Classification and Evolution (Ch. 10, 11)	**Animals (Ch. 13, 14, 15, 17)**	**Plants (Ch. 12)**	**Ecology (Ch. 16)**
3, 9, 13, 29, 32, 34, 36, 37, 38, 45, 46, 50, 51, 56	4, 7, 8, 10, 12, 18, 19, 25, 31, 33, 58, 89, 90, 99, 100	2, 5, 39, 52, 53, 70, 71, 72, 73	11, 16, 17, 24, 47, 87, 88, 93, 94, 95, 96, 97, 98

How to Score Your Essay

After you have written the best essay you can write in about 20 minutes, you are ready to grade it. First, though, take a short break to clear your head. When you are ready, reread your essay. Put a 1 to the left or right of any line where you explain any point that is listed below in the **scoring standard**. If you simply list examples, like "induced fit" or "sodium-potassium pump," you get NO points. *You must explain each answer.* Also, if you try to explain something without using the scientific term, you get NO credit. Here is an example. If you say, "There are protein channels in the plasma membrane" without saying what they do or how they function, you get no credit.

After reading and analyzing each part of the question, add up all the points you placed at the end of the line. If a question has three parts, the maximum number of points you can earn in any one part is 3–4. If, in your essay, you happen to include every point listed, you will get a maximum of 10 points. Be honest. You must use and explain scientific terminology and explain all concepts clearly so that anyone would understand what you are trying to say.

Scoring Standard for Free-Response Questions

Question 1
Total—10 points

Plant kingdom

1 pt. **Plasmodesmata**, the openings in the cell walls of plant cells, allow for passage of water and cell products. For example, malic acid in C-4 plants is transported from mesophyll to bundle sheath cells.

1 pt. **Plant hormones** produced in one cell type help coordinate growth, development, and response to environmental stimuli in nearby cells. Ripening fruit gives off ethylene gas, which causes other cells in other fruit to ripen. You may discuss any plant hormone, but you must include specific details about the effect of each hormone: what type of cell it stimulates and what the resulting response is.

1 pt. **Signal transduction pathways** amplify the hormonal signal and connect it to specific cell responses.

1 pt. **Stomates** open and close as guard cells respond to changes in the local environment. Ions and molecules move in and out between guard cells and surrounding cells to effect change. An increase in potassium ions, an increase in turgor pressure, and active transport of H^+ ions out of the guard cells all cause stomates to open. A factor that causes stomates to close is an increase of CO_2 levels within the air space of the leaf.

Immune system

1 pt.   **Histamine** released by basophils and mast cells triggers a multicellular response to increase blood supply to a particular area.

1 pt.   **Chemokines** secreted by monocytes attract phagocytes to an area.

1 pt.   **Natural killer cells** are part of the nonspecific defense of the body but target and kill virus-infected body cells.

1 pt.   **Cytotoxic T cells** kill infected body cells based on the presence of certain receptors on the cell surface of the target cell.

1 pt.   **Helper T cells**, when activated by macrophages, secrete interleukin-2, which stimulate killer T cells and B cells.

1 pt.   **Macrophages** activate helper T cells when they act as **antigen-presenting cells** and present a specific antigen on their cell surface.

Human physiology

1 pt.   **Neurons** stimulate an effector cell, a muscle, gland, or another neuron, by secreting neurotransmitter into a synapse. Neurotransmitters such as acetylcholine stimulate skeletal muscles and cause an action potential at the sarcolemma of the muscle cell.

1 pt.   **Cholecystokinin**, a hormone secreted by the duodenum wall, stimulates the pancreas.

1 pt.   **Secretin**, a hormone secreted by the duodenum wall, stimulates the pancreas.

1 pt.   The **medulla** of the brain monitors CO_2 levels in the blood by monitoring pH levels. If the CO_2 levels are too high (and pH levels too low), the medulla sends a message to increase the breathing rate to remove excess CO_2.

1 pt.   The cells of the **hypothalamus** stimulate the cells of the anterior and posterior pituitary.

1 pt.   The sinoatrial node (SA node), the **pacemaker of the heart**, sends an impulse to the atrioventricular node (AV node), which stimulates the bundle of His to cause the ventricles of the heart to contract.

1 pt.   Since a hormone is a substance produced in one part of the body (by one set of cells) that has its effect in another part of the body (on another set of cells), any gland in the endocrine system can be discussed in depth in answering this question.

1 pt.   The kidney is under hormonal control, including antidiuretic hormone and the RAAS, renin-angiotensin-aldosterone system.

---

10 pts.   Maximum

Question 2
Total—10 points

At the molecular level

1 pt.   Enzymes require cofactors (minerals) and coenzymes (vitamins) to help them work effectively.

1 pt.   The protein-folding problem—Proteins require the assistance of chaperone proteins to fold into the correct shape or conformation.

1 pt.   Cooperativity—The hemoglobin molecule undergoes a conformational change once one oxygen atom bonds to it. Once it undergoes this conformational change, it picks up the next three molecules of oxygen much more easily.

1 pt.   Allosteric inhibition—When an allosteric inhibitor bonds to an allosteric enzyme, the enzyme changes its conformation and cannot bond to its substrate. Phosphofructokinase is an allosteric enzyme important to glycolysis that is inhibited by ATP.

1 pt.   Enzymes enable reactions to occur at a lower energy of activation because the enzymes facilitate the occurrence of more effective collisions.

1 pt.   Induced fit—In order for the substrate to bond to the enzyme, the enzyme undergoes some changes in shape in order to fit the substrate.

1 pt.   Feedback inhibition or negative feedback is a mechanism that regulates chemical reactions in a cell where a buildup of end product inhibits the pathway.

1 pt.   The processes of photosynthesis and cellular respiration complement each other. One uses oxygen and gives off carbon dioxide (cellular respiration). The other uses carbon dioxide and gives off oxygen (photosynthesis).

3 pts.   Maximum

At the cellular level

Cell-to-cell attachment

1 pt.   Tight junctions are cells that work together to produce a barrier.

1 pt.   Desmosomes are cells that are connected by strong welds so they can resist physical stress. These are found in the neck of the uterus, called the cervix.

1 pt.   Gap junctions are large openings between adjacent cells that allow for the rapid passage of necessary information. These are found in heart tissue.

Cell-to-cell communication

1 pt.   Plasmodesmata are openings in the walls of plant cells that allow for the rapid passage of molecules between cells. Malic acid is transported from mesophyll cells to bundle sheath cells in C-4 plants through plasmodesmata.

1 pt. The signal transduction pathway assists in the rapid spread of a chemical signal from one cell to another.

1 pt. Impulses are transmitted from one neuron to another neuron or to an effector (a muscle or gland) via neurotransmitters.

1 pt. The glycocalyx consists of oligosaccharides attached to integral proteins within plasma membranes and is responsible for such phenomena as contact inhibition, the normal trait of cells to stop dividing when they become crowded.

3 pts. Maximum

At the organ/organism level

1 pt. The endocrine system regulates and integrates the functioning of all the organs. For example, the pituitary gland sends an impulse to the thyroid to release thyroxin into the blood. You can give examples of any hormone and its effects for this section.

1 pt. The nervous system integrates the parts of the body through chemical messengers called neurotransmitters.

1 pt. Symbiotic relationships are examples of organisms living in close proximity that impact on each other. Types of symbiotic relationships include: parasitism, mutualism, and commensalism.

1 pt. Predator-prey relationships are interdependent relationships.

1 pt. Mammals are characterized by intense parenting of young.

3 pts. Maximum

At the population level

1 pt. Interdependence can be seen in the food chain and food web, where all animals depend on the ability of plants to convert solar energy into chemical bond energy. In addition, animals at each tropic level depend on the organisms at the level beneath them for nutrition.

1 pt. All life depends on decomposers, which break down and recycle organic matter.

1 pt. Social behavior is any kind of interaction between two or more animals, usually within one population. Types of social behaviors are cooperation, agonistic, dominance hierarchies, territoriality, and altruism.

1 pt. Imprinting is learning that occurs during a sensitive or critical period in the early life of an individual and is irreversible for the length of that period. If the baby does not imprint on the mother, the parent will not care for the offspring and the offspring will die.

3 pts. Maximum

Question 3
Total—10 points

Structure of the membrane

1 pt. The eukaryotic plasma membrane consists of a phospholipid bilayer with proteins dispersed throughout the layers. A phospholipid is amphipathic, meaning it has both a hydrophobic and hydrophilic region.

1 pt. Integral proteins have nonpolar regions that completely span the hydrophobic interior of the membrane. Peripheral proteins are loosely bound to the surface of the membrane. Some proteins are kept in place by attachment to the cytoskeleton, while others drift slowly.

1 pt. Cholesterol molecules are embedded in the interior of the bilayer to stabilize the membrane. The average membrane has the consistency of olive oil and is about 40% lipid and 60% protein.

1 pt. The external surface of the plasma membrane also has carbohydrates attached to it, forming the glycocalyx. All the functions of the glycocalyx are not known, but they are important for cell-to-cell recognition.

1 pt. Proteins within the plasma membrane provide a wide range of functions including transport, as enzymes, as receptor molecules, and as a means of attaching one cell to another.

_____

5 pts. Maximum

Transport through the membrane

1 pt. Transport through a membrane can occur by passive transport or by active transport. Discuss the difference.

1 pt. Small, nonpolar molecules can dissolve in the lipid bilayer of the membrane and pass directly into the cell. Steroid hormones, such as estrogen, which are **lipids**, dissolve through the lipid bilayer and bind with a receptor inside the cell, triggering the cell response.

1 pt. **Calcium ions** flow through single ion-gated channels in the terminal branch of a neuron when the channel is stimulated electrically. This influx of calcium ions causes neurotransmitter to be released into the synapse.

1 pt. **Sodium and potassium ions** are pumped across the membrane of an axon of a nerve cell by the sodium-potassium pump. One ATP molecule is required to pump three sodium ions and two potassium ions across the membrane to repolarize it. This is a necessary part of normal nerve function.

1 pt. Channel proteins form water-filled pores that extend across the lipid bilayer and allow specific solutes to pass through. Some **glucose** passes passively through glucose carriers that operate as uniports (only one type of particle can pass through the membrane). In the kidneys, glucose is carried across the membrane along with sodium ions in what is called symport (two particles flowing in the same direction).

1 pt. **Large amounts of water** are transported passively through aquaporins, special water channels that facilitate diffusion of water. Aquaporins may also function as a gated channel that open and close in response to variables such as turgor pressure.

1 pt. **Cholesterol** is transported into cells from the bloodstream by receptor-mediated endocytosis with the expenditure of ATP.

5 pts. Maximum

Question 4
Total—10 points

Choice of subject animal
1 pt. The animal must be easy to collect, handle, and maintain.
1 pt. All animals used must be of the same species and as close in size and mass to each other as possible.
1 pt. Since any study requires a large sample of animals, the animals should be small.
1 pt. Examples: daphnia, pill bugs, or fruit flies

3 pts. Maximum

Exploring habitat preference
1 pt. Describe how to fashion a choice chamber
  • One possibility is to connect two petri dishes together with a hole between to allow the animals to choose their location.
  • Use flexible tubing (about ½ inch [1 cm] in diameter) and clamps to isolate the animals.
1 pt. You must consider how you will control the stimuli such as light, heat, or pH and discuss it here. Remember, you must change only one variable at a time.

2 pts. Maximum

Discuss the procedure—be explicit
1 pt. Describe the procedure in specific detail. Give the sizes and quantities of things, and show that there is only one variable.
1 pt. How often will you take measurements?
1 pt. Discuss how you determined what is an appropriate interval.
1 pt. How many animals will you use?
1 pt. Discuss why the experiment must be repeated.

1 pt.	If one of your variables is light, you must ensure that the light does not heat up the set-up, thus introducing heat as an uncontrolled variable. One way to do this is to put a container of water between the animals and the light to act as a heat sink.
6 pts.	Maximum

Hypothesis

1 pt.	Discuss what you expect to occur and why.
1 pt.	You may make up data to clarify your hypothesis.
1 pt.	Construct a bar graph from your hypothetical data to clarify.
1 pt.	Use scientific terms to describe what you expect to happen. Words such as **taxis**, **kinesis**, and **orientation behavior** can be useful.
2 pts.	Maximum

SUBJECT AREA
REVIEW

Biochemistry

CHAPTER 3

- Atomic structure
- Bonding
- Polar and nonpolar molecules
- Hydrophobic and hydrophilic
- Characteristics of water
- pH
- Isomers
- Organic compounds
- Enzymes and metabolism

INTRODUCTION

You may think, "This is a biology course, not a chemistry course. Why am I studying chemistry!" Well, the reason you have to know some chemistry is because cells are sacs of chemicals. The fact that sweating cools the skin is a function of the strong forces of attraction between water molecules. Your body maintains your blood at one critical pH because of the bicarbonate buffering system. To prevent heart attacks, you must begin with an understanding about the structure of fatty acids. Mad cow disease is caused by a misfolded protein. Biochemistry affects every aspect of our lives. Here is a review of basic biochemistry.

ATOMIC STRUCTURE

Atoms are the building blocks of all matter. Atoms consist of subatomic particles: **protons**, **neutrons**, and **electrons**.

An atom in the elemental state always has a neutral charge because the number of protons (+) equals the number of electrons (−). Electron configuration is important because it determines how a particular atom will react with atoms of other elements. If all the electrons in an atom are in the lowest available energy levels, the atom is said to be in the **ground state**. When an atom absorbs energy, its electrons move to a higher energy level, and the atom is said to be in the **excited state**. For example, when a chlorophyll molecule in a photosynthetic plant cell absorbs light energy, the molecule becomes excited and electrons get boosted to a higher energy level.

Isotopes are atoms of one element that vary only in the number of neutrons in the nucleus. *Chemically, all isotopes of the same element are identical because they have* *the same number of electrons in the same configuration.* For example: carbon-12 and carbon-14 are isotopes of each other and are chemically identical. Some isotopes, like carbon-14, are radioactive and decay at a known rate called the **half-life**. Knowing the half-life enables us to measure the age of fossils or to estimate the age of the earth. **Radioisotopes** (radioactive isotopes) are useful in many other ways. For example, **radioactive iodine** (I-131) can be used both to diagnose and to treat certain diseases of the thyroid gland. Additionally, a **tracer** such as radioactive carbon can be incorporated into a molecule and used to trace the path of carbon dioxide in a metabolic pathway.

BONDING

A bond is formed when two atomic nuclei attract the same electron/s. *Energy is released when a bond is formed, and energy must be supplied to break a bond.* Atoms bond to acquire a stable configuration, a completed outer shell. There are two main types of bonds, **ionic** and **covalent**.

Ionic bonds result from the **transfer** of electrons. An atom that gains electrons becomes an **anion**, **a n**egative **ion**. An atom that loses an electron becomes a **cation**, a positive ion. The Na^+, Ca^{++}, and Cl^- ions are all necessary for normal nerve function.

Covalent bonds form when atoms **share** electrons. The resulting structure is called a **molecule.** A single covalent bond (–) results when two atoms share a pair of electrons. A double covalent bond (=) results when two atoms share two pairs of electrons. A triple covalent bond (≡) results when two atoms share three pairs of electrons. If electrons are shared *equally* between two identical atoms, the bond is a **nonpolar** bond. This is the type of bond found in **diatomic molecules**, such as H_2 (H–H) and O_2 (O=O). If electrons are shared *unequally*, the bond is referred to as a **polar bond**. This is the case between any two different atoms, such as between atoms of carbon and oxygen in CO_2.

REMEMBER

Bond formation releases energy.

| This bond is nonpolar or balanced: | H–H ↑ | This bond is polar or unbalanced: | C–H ↑ |

POLAR AND NONPOLAR MOLECULES

When two or more atoms form a bond, the entire resulting molecule is either **non-polar** (symmetrical) or **polar** (unbalanced). CO_2 forms a linear molecule that is symmetrical or balanced and therefore nonpolar. It looks like this: O = C = O. H_2O is asymmetrical and a highly polar molecule. See Figure 3.1.

Weak attractions exist between nonpolar molecules, while strong attractions exist between polar molecules. These attractions are responsible for the physical characteristics of the substance such as solubility. Oils and fats, which are nonpolar molecules, will generally dissolve only in nonpolar substances. HCl, a polar substance, and NaCl, an ionic one, readily dissolve in a polar substance like water.

HYDROPHOBIC AND HYDROPHILIC

Hydrophobic means "water hating" or "repelled by water." Hydrophilic means "water loving" or "attracted to water." Substances that are polar will dissolve in water, while substances that are nonpolar will not dissolve in water. We say, "Like dissolves like." You are very familiar with one example. Carbon dioxide (nonpolar), the molecule that gives soda pop its fizziness, does not dissolve in water (polar). This is why the gas escapes when you open a can of soda pop, and it "goes flat."

Lipids, which are nonpolar, are hydrophobic and do not dissolve in water. This is why oil and vinegar salad dressing (a solution of acetic acid and water) separates upon standing. Since the plasma membrane is a phospholipid bilayer, only nonpolar substances can readily dissolve through the plasma membrane. Large polar molecules must travel across a membrane in special hydrophilic (protein) channels.

PROPERTIES OF WATER

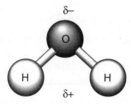

Figure 3.1 H_2O: A Polar Molecule

Because the oxygen atom in a molecule of **water** exerts a greater pull on the shared electrons than do the hydrogen atoms, one side of the molecule has a negative charge and the other side has a positive charge. The molecule is therefore asymmetrical and **highly polar**. See Figure 3.1. In addition, the positive hydrogen of one molecule is attracted to the negative oxygen of an adjacent molecule. The two molecules are held together by this **hydrogen bonding**. These strong hydrogen attractions that water molecules have for each other are responsible for the special characteristics of water that are important for life on earth. See Figure 3.2, which shows the hydrogen of one molecule of water attracted to the oxygen of an adjacent molecule.

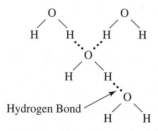

Figure 3.2   Water Molecules with Hydrogen Bonds

- **Water has a high specific heat**. Specific heat is the amount of heat a substance must absorb to increase 1 gram of the substance by 1°C. Because water has a high specific heat, bodies of water resist changes in temperature and provide a stable environmental temperature for the organisms that live in them. Also, because large bodies of water exhibit relatively little temperature change, they moderate the climate of the nearby land. High specific heat is also responsible for the fact that the marine biome has the most stable temperatures of any biome.

- **Water has a high heat of vaporization**. Evaporating water requires a relatively great amount of heat, so evaporation of sweat significantly cools the body surface.

- **Water is the universal solvent**. Because water is a highly polar molecule, it dissolves all polar and ionic substances.

- **Water exhibits strong cohesion tension**. This means that molecules of water tend to attract one another, which results in several biological phenomena.

  1. Water moves up a tall tree from the roots to the leaves without the expenditure of energy by what is referred to as **transpirational-pull cohesion tension**. As one molecule of water is lost from the leaf by transpiration, another molecule is drawn in at the roots.
  2. **Capillary action** results from the combined forces of cohesion and adhesion.
  3. **Surface tension** allows insects to walk on water without breaking the surface.

- **Ice floats because it is less dense than water**. In a deep body of water, floating ice insulates the liquid water below it, allowing fish and other organisms to live beneath the frozen surface during winter. In the spring, the ice melts, becomes denser water, and sinks to the bottom of the lake, causing water to circulate throughout the lake. Oxygen from the surface is returned to the depths, and nutrients released by the activities of bottom-dwelling bacteria during winter are carried to the upper layers of the lake. This cycling of the nutrients in the lake is known as the **spring overturn** and is a necessary part of the life cycle of a lake.

pH

pH is a measure of the acidity and alkalinity of a solution. Anything with a pH of less than 7 is an acid, and anything with a pH value greater than 7 is alkaline or basic. A pH of 7 is neutral. *The value of the pH is the negative log of the hydrogen ion concentration in moles per liter.* See Table 3.1, which shows pH values compared with molarity.

TABLE 3.1

pH Compared with Molarity

pH	Concentration of H⁺ ions in Moles per Liter		
1	1×10^{-1}	=	0.1 molar
2	1×10^{-2}	=	0.01 molar
3	1×10^{-3}	=	0.001 molar
4	1×10^{-4}	=	0.000 1 molar
7	1×10^{-7}	=	0.000 000 1 molar
13	1×10^{-13}	=	0.000 000 000 000 1 molar

A substance with a pH of 3 has 1.0×10^{-3} or 0.001 moles per liter of hydrogen ions in solution, while a substance of pH 4 has a H^+ concentration of 1.0×10^{-4} or 0.0001 moles per liter of hydrogen ions in solution. Therefore, a solution of pH 3 is 10 times more acidic than a solution with a pH of 4. A solution with a pH of 6 is 1,000 times more acidic than a solution with a pH of 9; see Figure 3.3.

IT'S TRICKY

As the H^+ concentration increases, the pH decreases.

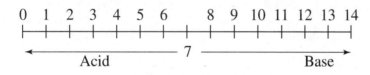

$$0 \quad 1 \quad 2 \quad 3 \quad 4 \quad 5 \quad 6 \qquad 8 \quad 9 \quad 10 \quad 11 \quad 12 \quad 13 \quad 14$$

Acid ⟵ ———— 7 ———— ⟶ Base

Figure 3.3   pH Scale

The pH of some common substances is as follows:

Stomach acid   2
Human blood   7.4
Acid rain   1.5–5.4

The internal pH of most living cells is close to 7. Even a slight change can be harmful. Biological systems regulate their pH through the presence of **buffers**, substances that resist changes in pH. A buffer works by absorbing excess hydrogen ions or donating hydrogen ions when there are too few. The most important buffer in human blood is the **bicarbonate ion**.

H⁺ donor (acid)	Response to rise in pH	H⁺ acceptor (base)		Hydrogen ion
H_2CO_3	⟶ ⟵	HCO_3^-	+	H^+
	Response to drop in pH			

ISOMERS

Isomers are organic compounds that have the same molecular formula but different structures. Therefore, they have different properties. There are three types of isomers: **structural isomers**, **geometric isomers**, and **optical isomers** or **enantiomers**.

Structural isomers differ in the arrangement of their atoms. **Geometric isomers** differ only in spatial arrangement around double bonds, which are not flexible like single bonds are. **Optical isomers** are molecules that are mirror images of each other. The mirror images are called the L- (left-handed) and D- (right-handed) versions. Optical isomers are important in the pharmaceutical industry because the two mirror images may not be equally effective. For example, L-dopa is a drug used in the effective treatment of Parkinson's disease. However, D-dopa, its optical isomer, is biologically inactive and useless in the treatment of the disease. For some reason that we do not understand, all the amino acids in cells are left-handed. See Figure 3.4.

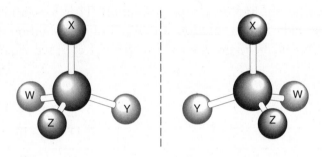

Figure 3.4   Mirror Images

ORGANIC COMPOUNDS

Organic compounds are compounds that contain carbon. The number of different carbon compounds is vast. There are four classes of organic compounds: **carbohydrates**, **lipids**, **proteins**, and **nucleic acids**.

Carbohydrates

Carbohydrates consist of three elements: carbon, hydrogen, and oxygen. The ratio of the number of hydrogen atoms to the number of oxygen atoms in all carbohydrates is always 2 to 1. The empirical formula for all carbohydrates is C_nH_2O. The body uses carbohydrates for quick energy, and 1 gram of any carbohydrate will release 4 calories when burned in a calorimeter. Dietary sources include rice, pasta, bread, cookies, and candy. There are three classes of carbohydrates you should know: monosaccharides, disaccharides, and polysaccharides.

Monosaccharides have a chemical formula of $C_6H_{12}O_6$. Three examples are **glucose**, **galactose**, and **fructose**, which are isomers of each other. The structural formula of glucose is shown in Figure 3.5. Notice the conventional numbering of the carbons in the rings. The numbering begins to the right of the oxygen.

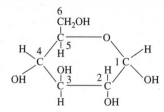

Figure 3.5   Glucose

Disaccharides have the chemical formula $C_{12}H_{22}O_{11}$. They consist of two monosaccharides joined together, with one molecule of water released, by the process known as **dehydration synthesis** or **condensation**. Here are the three condensation reactions of monosaccharides that produce the three disaccharides.

monosaccharide	+	**monosaccharide**	→	**disaccharide**	+	**water**
$C_6H_{12}O_6$	+	$C_6H_{12}O_6$	→	$C_{12}H_{22}O_{11}$	+	H_2O
glucose	+	glucose	→	**maltose**	+	water
glucose	+	galactose	→	**lactose**	+	water
glucose	+	fructose	→	**sucrose**	+	water

Hydrolysis is the breakdown of a compound by adding water. It is the reverse of condensation synthesis.

$$\text{sucrose} + \text{water} \rightarrow \text{glucose} + \text{fructose}$$

Polysaccharides are polymers of carbohydrates and are formed as many monosaccharides join together by dehydration synthesis. There are four important polysaccharides, as shown in Table 3.2.

TABLE 3.2

Polysaccharides		
	Structural	**Storage**
Found in plants:	**Cellulose** Makes up plant cell walls	**Starch** Two forms are **amylose** and **amylopectin**
Found in animals:	**Chitin** Makes up the exoskeleton in arthropods (and cell walls in mushrooms)	**Glycogen** "Animal starch." In humans, this is stored in liver and skeletal muscle

Lipids

Lipids are a diverse class of organic compounds that include **fats**, **oils**, **waxes**, and **steroids**. They are grouped together because they are all hydrophobic, meaning that they are not soluble in water. Structurally, most lipids consist of 1 glycerol and 3 fatty acids; see Figure 3.6.

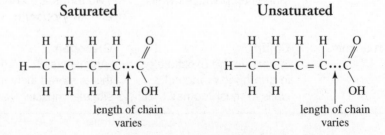

REMEMBER

The cell membrane consists of phospholipids—lipid molecules with a phosphate attached.

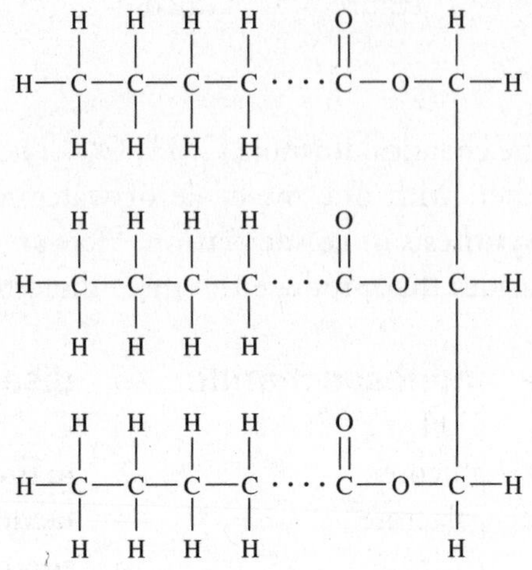

Figure 3.6   Lipid

Glycerol is an alcohol and only exists as shown in Figure 3.7.

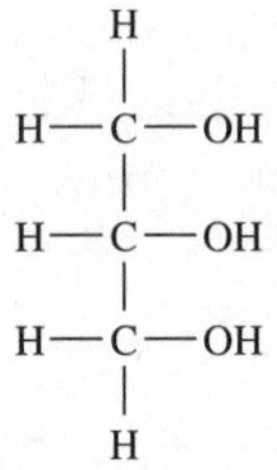

Figure 3.7   Glycerol

A **fatty acid** is a hydrocarbon chain with a carboxyl group at one end. Fatty acids exist in two varieties, **saturated** and **unsaturated**, as shown in Figure 3.8.

Saturated   Unsaturated

Figure 3.8   Fatty Acids

In general, saturated fats come from animals, are solid at room temperature, and when ingested in large quantities, are linked to heart disease. An example of a saturated fat is butter. Saturated fatty acids contain only **single bonds** between carbon atoms. Unsaturated fatty acids, in general, are extracted from plants, are liquid at room temperature, and are considered to be good dietary fats. Unsaturated fatty acids have at least one **double bond** formed by the removal of hydrogen atoms in the carbon skeleton. As a result, they hold fewer hydrogen atoms than saturated fatty acids. One exception is the group of tropical oils such as coconut and palm oil that are saturated, somewhat solid at room temperature, and are as unhealthy as are fats extracted from animals.

Steroids are lipids that do not have the same general structure as other lipids. Instead, they consist of four fused rings. Figure 3.9 shows the steroid cholesterol. Other steroids are testosterone and estradiol.

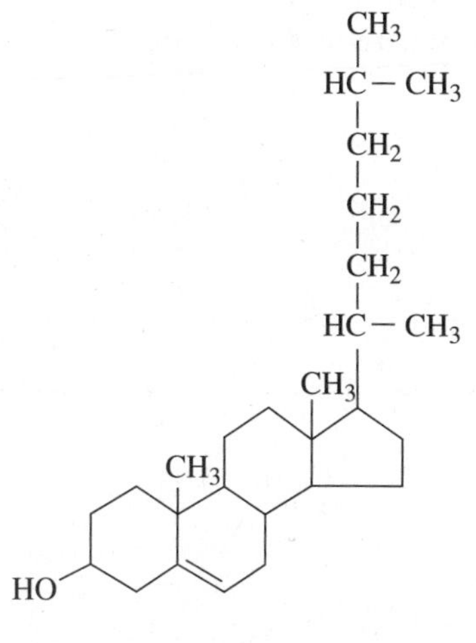

Figure 3.9   The Steroid: Cholesterol

Lipids serve many functions.

- **Energy storage:** One gram of any lipid will release 9 calories per gram when burned in a calorimeter.
- **Structural:** Phospholipids (a lipid where a phosphate group replaces one fatty acid) are a major component of the cell membrane. One steroid, cholesterol, serves as an important component of the plasma membrane of animal cells.
- **Endocrine:** Some steroids are hormones.

Proteins

Proteins are complex macromolecules that carry out many functions in the body including:

- Growth and repair
- Signaling from one cell to another
- Defense against invaders
- Catalyzing chemical reactions

Dietary sources of proteins include fish, poultry, meat, and certain plants like beans and peanuts. One gram of protein burned in a calorimeter releases 4 calories. Proteins consist of the elements S, P, C, O, H, and N. They are **polymers** or **polypeptides** consisting of units called **amino acids**, which are joined by **peptide bonds**.

See Figure 3.10, showing two amino acids combining to form a dipeptide. A **dipeptide** is a molecule consisting of two amino acids connected by one peptide bond.

Amino acids consist of a **carboxyl group**, an **amine group**, and a **variable (R)** all attached to a central asymmetric carbon atom. The R group, also called the side chain or variable, differs with each amino acid. With only 20 different amino acids, cells can build thousands of different proteins.

Each protein has a unique shape or **conformation** that determines what job it performs and how it functions. There are four levels of protein structure that are responsible for a protein's unique conformation. They are **primary**, **secondary**, **tertiary**, and **quaternary structures**; see Figure 3.11.

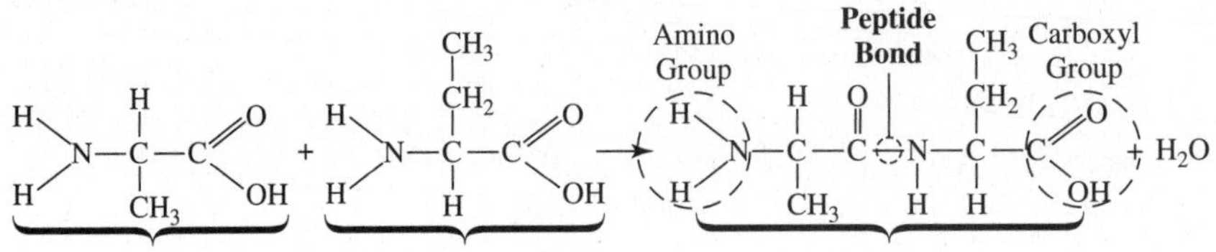

Figure 3.10

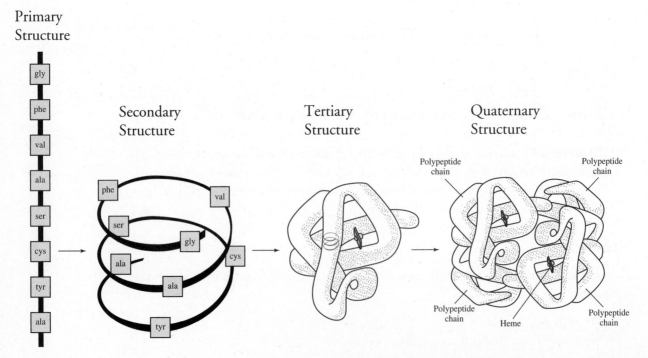

Figure 3.11   Four Levels of Protein Structure

The **primary structure** of a protein refers to the unique linear sequence of amino acids. The slightest change in the amino acid sequence of a protein can have dire consequences. Such is the case with sickle cell anemia, a life-threatening condition that results from a substitution of one amino acid for another in a molecule of hemoglobin.

While working in the 1940s and 1950s **Fred Sanger** was the first to sequence a protein. That protein was **insulin**, and he received the Noble Prize for his work.

The **secondary structure** of a protein results from **hydrogen bonding** within the polypeptide molecule. It refers to how the polypeptide coils or folds into two distinct shapes: an **alpha helix** or a **beta pleated sheet**.

Proteins that exhibit either alpha helix or pleated sheet or both are called **fibrous proteins**. Examples of fibrous proteins are wool, claws, beaks, reptile scales, collagen, and ligaments. The protein that makes up human hair, **keratin**, is composed mostly of alpha helixes, while silk and spider webs consist of proteins made of beta pleated sheets.

Tertiary structure is the intricate three-dimensional shape or conformation of a protein that is superimposed on its secondary structure. Tertiary structure determines the protein's **specificity**. The following factors contribute to the tertiary structure:

- Hydrogen bonding between R groups of amino acids
- Ionic bonding between R groups
- Hydrophobic interactions
- Van der Waals interactions
- Disulfide bonds between cysteine amino acids

Quaternary structure refers to proteins that consist of more than one polypeptide chain. Hemoglobin exhibits quaternary structure because it consists of four polypeptide chains, each one forming a **heme group**.

The Protein-Folding Problem

The concept that *the form of a molecule determines the function of the molecule* is a very important theme in biology. Although scientists have deciphered the three-dimensional shapes of thousands of proteins using x-ray crystallography, they still do not understand all the rules of how proteins spontaneously fold into their unique shapes. This interesting and challenging area of study is referred to as **the protein folding problem**. One important fact that researchers have learned is that molecules called **chaperone proteins** assist in folding other proteins.

Nucleic Acids

The two nucleic acids are either **ribonucleic acid (RNA)** or **deoxyribonucleic acid (DNA)**. They are polymers and carry all heredity information. They consist of repeating units called **nucleotides**. A nucleotide consists of a **phosphate**, a **5-carbon sugar—deoxyribose** or **ribose**, and a **nitrogen base—adenine**, **cytosine**, **guanine**, or either **thymine** (in DNA) or **uracil** (in RNA). The carbon atoms of deoxyribose are numbered from 1 to 5. Nucleic acids are discussed in the genetics section later on. In Figure 3.12, P stands for phosphate.

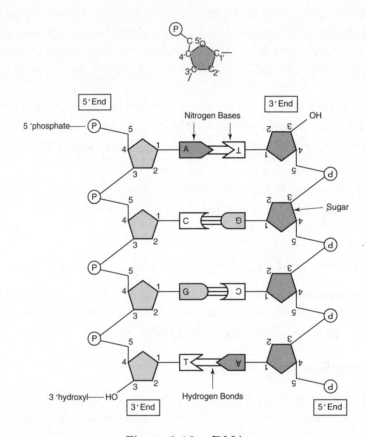

Figure 3.12   DNA

Functional Groups

The components of organic molecules that are most often involved in chemical reactions are known as **functional groups**. These groups are attached to the carbon skeleton, replacing one or more hydrogen atoms that would be present in a hydrocarbon. The differences between testosterone and estradiole are the functional groups attached to the carbon skeleton. Each functional group behaves in a consistent fashion from one organic molecule to another. Table 3.3 shows some common functional groups.

TABLE 3.3

Functional Groups in Organic Compounds

Group	Formula	Name of Compound
Amino	$R-N\begin{smallmatrix}H\\\\H\end{smallmatrix}$	Amine
Carboxyl	$R-C\begin{smallmatrix}O\\\\OH\end{smallmatrix}$	Carboxyl (acid)
Hydroxyl	$R-OH$	Alcohol
Phosphate	$R-O-\overset{O}{\underset{O^-}{\overset{\|\|}{\underset{\|}{P}}}}-O^-$	Organic phosphate

ENZYMES AND METABOLISM

Living systems transform one form of energy to another in order to carry out essential life functions. The **laws of thermodynamics** govern these energy transformations. The **first law of thermodynamics** states that energy cannot be created or destroyed, only transferred. This is the **law of conservation of energy**. The **second law of thermodynamics** states that in the course of energy conversions, the universe becomes more disordered (greater **entropy**). The measure of this disorder or randomness is determined by Gibb's free energy equation: $\Delta G = \Delta H - T\Delta S$, where ΔG represents free energy change, ΔH represents change in heat content, T represents absolute temperature, and ΔS represents entropy. Reactions with a negative ΔG are energy releasing (**exergonic** or **exothermic**), and reactions with a positive ΔG are energy absorbing (**endergonic** or **endothermic**). In cellular reactions, chemical reactions are coupled; *exergonic reactions power the endergonic ones.*

Metabolism is the sum of all the chemical reactions that take place in cells. Some reactions break down molecules (**catabolism**); other reactions build up molecules (**anabolism**). Metabolic reactions take place in a series, called **pathways**, each of which serves a specific function. These multistep pathways are controlled by enzymes and enable cells to carry out their chemical activities with remarkable efficiency.

Enzymes do not provide energy for a reaction. They serve as **catalytic proteins** that speed up reactions by lowering the **energy of activation**, the amount of energy needed to begin a reaction. In the potential energy diagram in Figure 3.13, the potential energy of the products is less than the potential energy of the reactants, so energy is released and the reaction is exothermic. The dotted line shows the same reaction when an enzyme is introduced. The enzyme serves to lower the energy of activation, and the reaction can proceed more quickly.

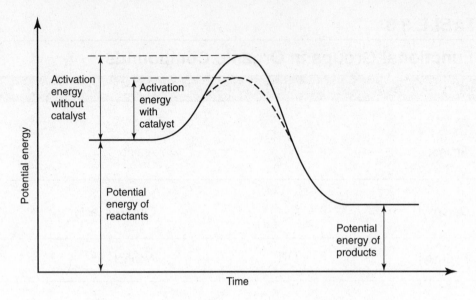

Figure 3.13　Progress of an Exothermic Reaction

In the potential energy diagram in Figure 3.14, the potential energy of the products is greater than the potential energy of the reactants, so energy is absorbed and the reaction is endothermic. The dotted line shows the same reaction when an enzyme is introduced. The enzyme serves to lower the energy of activation, and the reaction can proceed more quickly.

Notice that in both reactions, the only factor altered by the presence of an enzyme is the energy of activation and the activated complex.

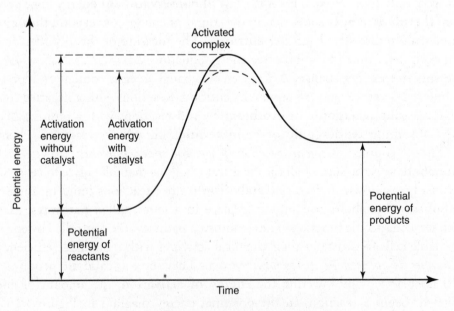

Figure 3.14　Endothermic Reaction

Characteristics of Enzymes

• Enzymes are **globular proteins** that exhibit **tertiary structure**.

• Enzymes are substrate specific. In Figure 3.15 only substrate A will bind to the enzyme.

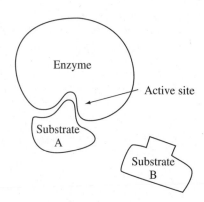

Figure 3.15   Enzyme-Substrate Complex

• The **induced-fit model** describes how enzymes work. As the substrate enters the active site, it induces the enzyme to alter its shape slightly so the substrate fits better. The old **lock and key** model was abandoned because it falsely implied that the lock and the key were unchanging.

• Enzymes remain unchanged in a reaction and are reused.

• Enzymes are named after their substrate, and the name ends in the suffix "**ase**." For example, sucrase is the name of the enzyme that hydrolyzes sucrose, and lactase is the name of the enzyme that hydrolyzes lactose.

• Enzymes catalyze reactions in both directions:

$$\text{lactose} \overset{\text{Enzyme}}{\longleftrightarrow} \text{glucose} + \text{galactose}$$

• Enzymes often require assistance from **cofactors** (inorganic) or **coenzymes** (vitamins).

• The efficiency of the enzyme is affected by temperature and pH. Average human body temperature is 37°C, near optimal for human enzymes. When body temperature is too high, enzymes will begin to denature and lose both their unique conformation and their ability to function. Gastric enzymes become active at low pH, when mixed with stomach acid, while intestinal amylase works best in an alkaline environment; see Figures 3.16 and 3.17.

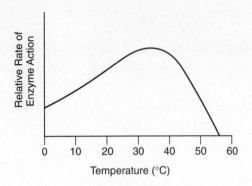

Figure 3.16

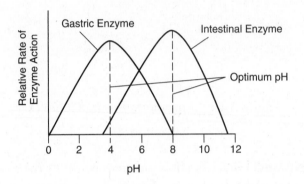

Figure 3.17

Inhibition of Enzyme Activity

Enzymes control reactions in a number of different ways: competitive inhibition, noncompetitive inhibition, allosteric inhibition, and cooperativity.

1. Competitive Inhibition

In **competitive inhibition**, some compounds resemble the normal substrate molecule and compete for the same active site on the enzyme. These mimics or **competitive inhibitors** reduce the productivity of enzymes by preventing the substrate from combining with the enzyme. This inhibition can be reversible or irreversible. See Figure 3.18.

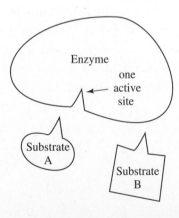

Figure 3.18   Competitive Inhibition

2. Noncompetitive Inhibition

In **noncompetitive inhibition**, the enzyme contains more than one active site and the substrates do not resemble each other. However, the binding of either substrate prevents the other one from binding to the enzyme. See Figure 3.19. Which substrate binds to the enzyme is random and a function of the concentration of each substrate. An example of noncompetitive inhibition is found in the **operon**, where the binding of the repressor to the operator on the DNA strand blocks the binding site for RNA polymerase and no transcription can occur. See Figure 3.20.

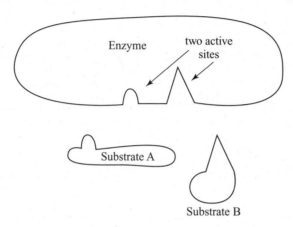

Figure 3.19   Noncompetitive Inhibition

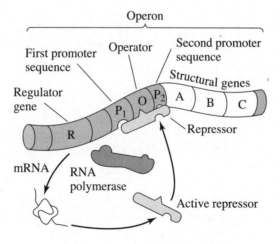

Figure 3.20   The Operon

3. Allosteric Inhibition

Allosteric inhibition involves two active sites, one for a substrate and the other for an **inhibitor**. The enzyme oscillates between two conformations, one active, one inactive. When the inhibitor binds to the allosteric site, the enzyme undergoes a conformational change, the active site for the substrate is altered, and the enzyme cannot catalyze the reaction (see Figure 3.21). A clear example of this type of regulation is found in glycolysis where the enzyme **phosphofructokinase (PFK)**, which catalyzes step 3 in the production of pyruvic acid, is inhibited by ATP (see Figure 5.3 on page 113). This is an example of a **feedback inhibition** where a metabolic pathway is switched off by its end-product. Feedback inhibition prevents a cell from wasting energy and resources.

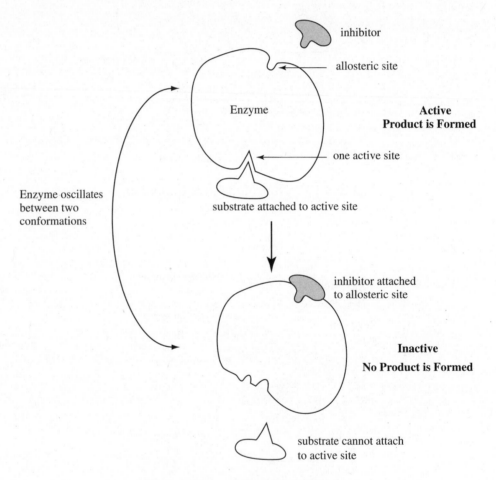

Figure 3.21   Allosteric Inhibition

4. Cooperativity

Sometimes substrates can stimulate an enzyme with quaternary structure (two or more subunits) to be more effective. A classic example is hemoglobin (although not an enzyme, it functions as one). Once hemoglobin binds to one oxygen atom, it can very rapidly bind to three more oxygen atoms.

Multiple-Choice Questions

1. Which of the following is correct about isotopes of carbon?

   (A) They are all radioactive.
   (B) They contain the same number of neutrons but a different number of protons.
   (C) They contain the same number of electrons but are chemically different because the number of neutrons is different.
   (D) They are chemically identical because they have the same number of electrons.
   (E) None of the above is correct.

2. All of the following are characteristics of water EXCEPT?

   (A) water has a relatively high boiling point
   (B) water molecules have little attraction for each other
   (C) water is a universal solvent
   (D) ice is less dense than water
   (E) water has a high specific heat

3. The pH of blood

   (A) is strongly basic
   (B) is strongly acidic
   (C) varies with the needs of the cells
   (D) is about 6.9
   (E) is normally close to 7.4.

4. Which of the following is NOT a base?

   (A) NaOH
   (B) KOH
   (C) $Mg(OH)_2$
   (D) C_2H_5OH
   (E) BaOH

Questions 5–11

In a lab experiment, one enzyme is combined with its substrate at time 0. The product is measured in micrograms at 20-second intervals and recorded on the data table below.

Time (s)	0	20	40	60	80	100	120
Product (µg)	0.0	0.25	0.50	0.70	0.80	0.85	0.85

5. What is the initial rate of the enzyme reaction?

6. What is the rate after 100 seconds?

7. Why is there no increase in product after 100 seconds?

8. What would happen if you added only more enzyme after 100 seconds?

9. What would happen if you added only more substrate after 100 seconds?

10. What would happen if you boiled the enzyme for 10 minutes before you did the experiment?

11. What would happen if you added strong acid to the enzyme 1 hour before you did the experiment?

12. Which of the following is stored in the human liver for energy?

    (A) glucose
    (B) glycogen
    (C) glycerol
    (D) glucagon
    (E) glycine

13. Which of the following is an example of a hydrogen bond?

    (A) the bond between Na^+ and Cl^- ions in salt
    (B) the bond between hydrogen and carbon in glucose or any sugar
    (C) the peptide bond between amino acids
    (D) the intramolecular bond between hydrogen and oxygen within a molecule of water
    (E) the attraction between the oxygen of one water molecule and the hydrogen of an adjacent molecule

14. All of the following statements are correct about enzymes EXCEPT?

    (A) they raise the energy of activation of all reactions
    (B) they enable reactions to occur at a relatively low temperature
    (C) they require coenzymes (vitamins) to work
    (D) they remain unchanged during a reaction
    (E) they are often located within the plasma membrane of the cell

15. Which statement is correct about pH?

    (A) There are no hydrogen ions in a strong basic solution.
    (B) Pure water has a neutral pH of 7 because the concentration of hydrogen ions equals the concentration of hydroxyl ions.
    (C) The concentration of a solution with a pH of 5 is 5 times more acidic than a solution with pH 1.
    (D) The concentration of hydrogen ions in a solution with a pH of 2 is 2,000 times more acidic than a solution with pH of 4.
    (E) A solution with a pH of 5 means there are 5×10^{-1} moles of hydrogen ions in solution.

16. Which of the following can be used to determine the rate of an enzyme-catalyzed reaction?

    (A) the rate of substrate formed
    (B) the decrease in temperature in the system
    (C) the rate of enzyme formed
    (D) the rate of enzyme used up
    (E) the rate of substrate used up

17. Which of the following best describes the reaction shown below?

    $$A + B \rightarrow AB + energy$$

    (A) hydrolysis
    (B) an exergonic reaction
    (C) an endothermic reaction
    (D) catabolism
    (E) an endergonic reaction

<u>Questions 18–19</u>

The graph below demonstrates two chemical reactions. One is catalyzed by an enzyme, one is not.

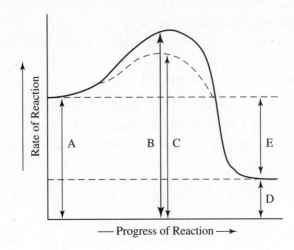

18. Which letter shows the energy of activation for the enzyme-catalyzed reaction?

    (A) A
    (B) B
    (C) C
    (D) D
    (E) E

19. Which letter shows the potential energy of the product?

    (A) A
    (B) B
    (C) C
    (D) D
    (E) E

20. Which level of protein structure is most related to specificity?

    (A) tertiary
    (B) primary
    (C) secondary
    (D) quaternary
    (E) allosterism

Answers to Multiple-Choice Questions

1. **(D)** Since they have the same number of electrons, isotopes of the same element are chemically identical. Only some of the isotopes of carbon (like C-14) are radioactive. Isotopes vary only in the number of neutrons.

2. **(B)** Water molecules have strong attraction for each other because they are polar and because of strong hydrogen bonding between molecules.

3. **(E)** The pH of blood remains very close to 7.4 at all times and is maintained by the bicarbonate buffering system. This is an example of how an organism maintains homeostasis or internal stability.

4. **(D)** This is ethyl alcohol and is not a base. The other compounds contain hydroxide (OH^-) and are all bases.

5. 0.25 µg per 20 seconds or 0.0125 µg/s. Rate is derived by taking the change in amount divided by the time. (0.25 µg − 0.00 µg)/20 s = 0.25 µg/20 s.

6. Zero. This is true because no new product is formed. The calculation is 0.85 − 0.85 = 0.

7. There is no increase in product after 100 seconds because all the enzymes are saturated and are already catalyzing reactions as fast as they can.

8. Assuming there is excess substrate that enzyme is not colliding with, an addition of enzyme after 100 seconds would increase the rate of reaction until the enzyme, once again, became saturated.

9. If you add more substrate after 100 seconds, there would be no change because the enzyme is saturated and reacting with as much substrate as it can. The enzyme cannot handle any more substrate. If you added both substrate and enzyme after 100 seconds, the reaction rate would increase temporarily until the enzyme became saturated once again.

10. If you boiled the enzyme prior to doing the experiment, you would get no product because the enzyme would probably have been denatured by the high heat and would not catalyze the reaction.

11. If you expose the enzyme to strong acid prior to doing the experiment, there would probably be no product because the enzyme would have been denatured by the acid.

12. **(B)** Glycogen is a polysaccharide that stores sugars in the liver and skeletal muscle. Glucose is not stored; it is produced and used up constantly. Glycerol and fatty acids make up lipids. Glucagon is a hormone that is responsible for breaking down glycogen into glucose. Glycine is the simplest amino acid.

13. **(E)** Hydrogen bonding is actually an intermolecular attraction between two molecules, not a covalent bond *within* one molecule. Choice B describes a covalent bond within a molecule. The bond between Na^+ and Cl^- is ionic. A peptide bond exists between the amino group of one amino acid and the carboxyl group of an adjacent amino acid.

14. **(A)** Enzymes lower the energy of activation, thus speeding up the reaction.

15. **(B)** The pH of a solution is neutral when the hydrogen ion (H^+) concentration equals the hydroxyl ion (OH^-) concentration. Pure water has a pH of 7 because it consists of equal amounts of H^+ and OH^-. Strong basic solutions contain mostly hydroxyl ions, but there are some hydrogen ions in it. The concentration of a solution of pH 5 is 10,000 times more acidic than a solution with pH 1. The concentration of hydrogen ions in a solution with a pH of 2 is 100 times more acidic than a solution with pH of 4. A solution with a pH of 5 means there are 1×10^{-5} moles of hydrogen ions in solution.

16. **(E)** Substrate is used up as the product is formed. Since enzymes are never used up, they are reused. So, enzyme levels cannot be used to monitor the progress of a reaction. Enzymes lower the energy needed to begin the reaction (the E_a), but they do not affect the temperature of the system.

17. **(B)** The reaction is exergonic because it releases energy. Endergonic and endothermic are synonyms for a reaction that requires energy. Catabolism is the breaking down of a substance; this reaction is anabolic, a building-up process.

18. **(C)** Enzymes lower the energy of activation.

19. **(D)** The potential energy (PE) of the product is the same for both reactions.

20. **(A)** Tertiary structure dictates the three-dimensional shape and function of a protein.

Free-Response Questions

Directions: Answer all questions. You must answer the question in essay—**not** outline—form. You may use labeled diagrams to supplement your essay, but diagrams alone are *not* sufficient. Before you start to write, read each question carefully so that you understand what the question is asking.

1. Describe the structure and function of enzymes.

2. The unique properties of water make life on earth possible. Select four characteristics of water and:

   a. Identify one characteristic of water, and explain how the structure of water relates to this property.
   b. Describe one example of how this characteristic affects living organisms.

Typical Free-Response Answers

Note: Key words have been written in bold merely to help you focus on what you should be discussing in an essay on enzyme function. You may use sketches to help explain your ideas, but they must be labeled and drawn near the text they explain.

1. Enzymes act as **catalysts**, lowering the **energy of activation**, thus increasing the rate at which reactions occur. Enzymes are large protein molecules that exhibit **tertiary structure** and are folded into a particular shape or **conformation** that results from many intramolecular interactions of the amino acids that make up the molecule. Enzymes are specific and act only on certain **substrates**. The shape of an enzyme determines which substrate an enzyme will act upon. Enzymes fold in such a way that there is one or more active sites where a substrate can bind. Special proteins called **chaperone proteins** assist enzymes in folding in their unique way. Many enzymes require the assistance of **coenzymes** (**vitamins**) or **cofactors** (**minerals**) to function properly. The shape of an enzyme can be altered or **denatured** and the enzyme will no longer function as it is supposed to. High heat and extremes in pH can denature an enzyme. The way in which an enzyme and a substrate form an enzyme-substrate is known as **induced fit**.

   Enzyme activity can be regulated in a number of different ways. One is by **competitive inhibition**, where two substrates compete for the same active site. Another is called **noncompetitive inhibition**, where there are two active sites and two substrates that do not compete for the same active site. However, by binding to one active site, a substrate prevents another substrate from binding to a second active site. A third way in which enzyme activity is controlled is known as **allosteric inhibition**, where an **inhibitor** alters the shape of the allosteric site and prevents the substrate from binding to its active site.

2a. Water is a highly **polar** molecule with strong **hydrogen bonding** between adjacent water molecules. Because there are such strong attractions between molecules, water has certain properties that make life on earth possible. These properties are high cohesion tension, **high specific heat**, **high heat** of **vaporization**, and the fact that water is an **excellent solvent**. Another property of water is that when it is frozen, it is less dense than water. The reason for this is that the bonding between the water molecules in ice holds the molecules rigidly and farther apart than in liquid water. Since the water molecules are farther apart in ice than in water, ice is less dense than water.

2b. Because ice is less dense than water, it floats. This has important consequences for living things. In a deep body of water, such as a lake, floating ice insulates the liquid water below it, allowing life to exist beneath the frozen surface during cold seasons. The fact that ice covers the surface of water in the cold months and melts in the spring results in a stratification of the lake in winter and considerable mixing in the spring when the ice melts. In the spring, the ice melts, becomes denser water, and sinks to the bottom of the lake, causing water to circulate throughout the lake. Oxygen from the surface is returned to the depths, and nutrients released by the activities of bottom-dwelling bacteria during the winter are carried to the upper layers of the lake. This cycling of the nutrients in the lake is known as the spring overturn and is a necessary part of the life cycle of a lake.

The Cell

CHAPTER 4

- Cell theory
- How we study cells
- Structure and function of the cell
- Transport into and out of the cell
- Cell communication

INTRODUCTION

All organisms on earth are believed to have descended from a common **prokaryotic** ancestral cell about three and a half billion years ago. According to the **theory of endosymbiosis**, eukaryotic cells emerged when mitochondria and chloroplasts, once free-living prokaryotes, took up permanent residence inside other larger cells, about one and a half billion years ago. Here was the advent of the radically more complex **eukaryotic cell** with internal membranes that compartmentalized the cell and led to the evolution of multicelled organisms.

Modern cell theory states that all organisms are composed of cells (**Schleiden and Schwann**, 1838) and that all cells arise from preexisting cells (**Virchow**, 1855). Most animal and plant cells have diameters between 10–100 μm (microns or micrometers), although many, like human red blood cells with a diameter of only 8 μm, are smaller.

CELL THEORY

Modern cell theory states:

- All living things are composed of cells
- Cells are the basic unit of all organisms
- All cells arise from preexisting cells

All cells are enclosed by a membrane that regulates the passage of materials between the cell and its surroundings. They also contain nucleic acid, which directs the cell's activities and controls inheritance.

Cells are divided into two varieties: prokaryotes and eukaryotes. **Prokaryotes** have no nucleus or other internal membranes. All bacteria are prokaryotes. **Eukaryotes** have a nucleus and are more complex cells. They make up every other form of life. Human cells are eukaryotic cells. Figure 4.1 shows a typical prokaryotic bacterial cell that lacks a nucleus and all membrane-bound internal structures. Table 4.1 compares prokaryotic and eukaryotic cells.

Table 4.1

Comparison of Cell Types

Prokaryotes	Eukaryotes
No internal membranes: no nuclear membrane, E.R., mitochondria, vacuoles, or other organelles	Contain distinct organelles
Circular, naked DNA	DNA wrapped with histone proteins into chromosomes
Ribosomes are very small	Ribosomes are larger
Metabolism is anaerobic or aerobic	Metabolism is aerobic
Cytoskeleton absent	Cytoskeleton present
Mainly unicellular	Mainly multicellular with differentiation of cell types
Cells are very small: 1–10 μm	Cells are larger: 10–100 μm

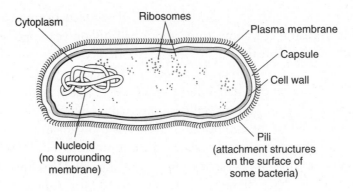

Figure 4.1   Typical prokaryotic cell

HOW WE STUDY CELLS

Microscopes

The main tool for studying cells is the **microscope**, which today can magnify an image over 100,000×. Besides the ability to magnify an image, another important characteristic of a good microscope is the measure of clarity of the image. This is known as **resolution**. The finest microscopes have both **high magnification** and **high resolution**. A toy microscope, which will enlarge an image 400×, has little resolving power, so images appear blurred.

The first microscope was developed by **Antoine van Leeuwenhoek** in the seventeenth century. Further modifications and advances were made by **Robert Hooke**, who developed a microscope that enabled him to study and name **cells** in cork. Today, there are many types of microscopes. Here is an overview.

Light microscopes use light, passing through a living or dead specimen, to form an image. Cells and tissue can be stained to make the organelles easier to see. However, most stains kill cells.

Electron microscopes use electrons passing through a specimen to form an image. These microscopes have superior resolving power as well as magnification over 100,000×. One drawback of electron microscopes is that they *cannot be used to view live specimens* because extensive specimen preparation kills the cells.

- **Transmission electron microscopes** are usually used for studying the *interior of cells*. Images taken with the T.E.M. appear flat and two-dimensional. To prepare tissue, it must first be cut into very small pieces (about 1 mm³) then exposed to a fixative like gluteraldehyde, which stops all biochemical activity. The tissue is then dehydrated, embedded in a polymer, cured overnight, and sliced on an ultramicrotome. This process can take several days.

- **Scanning electron microscopes** are used for studying the *surface of cells* and the images have a three-dimensional appearance. Specimens are coated with a heavy metal, such as gold, and are placed directly in the microscope for observing. The process can take a matter of hours. However, specimens are not alive.

> **REMEMBER**
>
> Tissue studied under an electron microscope is not alive.

Phase-contrast microscopes are used to examine *unstained, living cells*. This microscope is often used to examine cells growing in tissue culture.

Other Tools for Studying Cells

1. **Cell fractionation** uses an **ultracentrifuge** to spin liquid samples at high speed, separating them into layers based on **differences in density**. Tissue or cells are first mashed up in a blender to form a **homogenate** and then spun in a centrifuge. The densest particles settle to the bottom of the centrifuge tube. The supernatant, which is the liquid layer above the pellet, can be poured off and respun. This can be repeated until the desired layer is isolated. Figure 4.2 shows the separation of different organelles from a tissue homogenate. Nuclei are forced to the bottom first, followed by mitochondria and ribosomes.

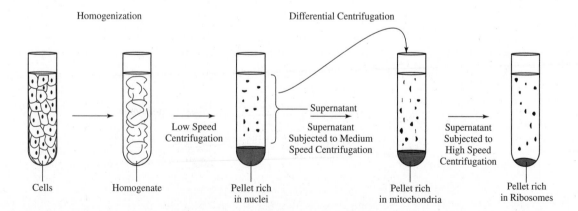

Figure 4.2   Cell Fractionation

2. **Freeze fracture** and **freeze-etching** are multistep techniques used to prepare a detailed cast of the membrane. The tissue is then digested away, leaving only the cast of the tissue. This cast is then examined under the electron microscope.
3. **Tissue culture** is a technique used to study the properties of specific cells *in vitro* (in the laboratory). Cell lines can be grown in culture for years, provided great care is taken with them. While the cells are growing, they can be studied under the **phase-contrast microscope** without staining.

STRUCTURE AND FUNCTION OF THE CELL

A major theme in biology (and therefore a common essay question) is that *function dictates form* and vice versa. As a result, one would expect that all cells do not look alike. And they do not. The nerve cell, whose purpose is to send electrical impulses, is long and spindly. Cells that store fat are rounded, large, and distended. Cells that make up a tough peach pit resemble square building blocks. Figure 4.3. shows a sketch of different cell types, each with a different overall appearance because each has a different function.

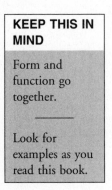

KEEP THIS IN MIND

Form and function go together.

———

Look for examples as you read this book.

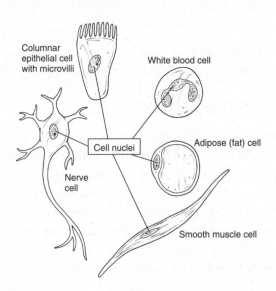

Figure 4.3

All eukaryotic cells have many organelles in common, including ribosomes and mitochondria. However, they also have organelles that are unique to the cell type, such as cell walls in plant cells. Figure 4.4 shows a typical plant cell.

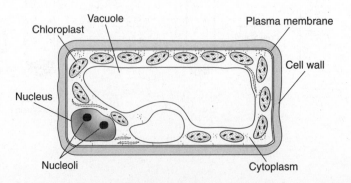

Figure 4.4   Plant cell

The human body is made of approximately two hundred different cell types, each with a different function and, therefore, a different form. Although different cell types have different appearances, they all contain the same organelles. See Figure 4.5, a sketch of a typical animal cell.

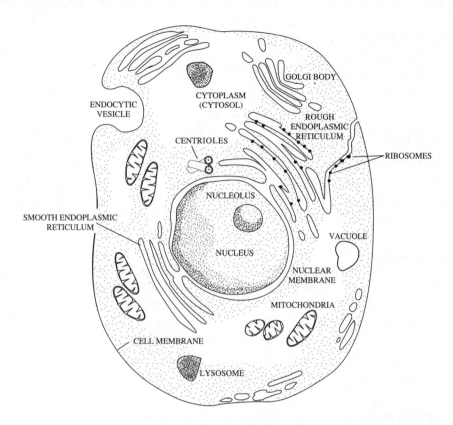

Figure 4.5   Diagram of a Eukaryotic Animal Cell

Nucleus

The nucleus contains chromosomes, which are wrapped with special proteins into a **chromatin network**. It is surrounded by a selectively permeable **nuclear membrane** or **envelope** that contains **pores** to allow for the transport of molecules, like messenger RNA (mRNA), which are too large to diffuse directly through the envelope.

Nucleolus

The nucleus of a cell in interphase contains a prominent region called a **nucleolus**, where components of **ribosomes** are synthesized. One or two nucleoli may be visible. Nucleoli are not membrane-bound structures but are actually a tangle of chromatin and unfinished ribosome precursors.

Ribosomes

Ribosomes are the site of **protein synthesis**. They can be found free in the cytoplasm or attached to endoplasmic reticulum.

Endoplasmic Reticulum

The endoplasmic reticulum (E.R.) is a membranous system of channels and flattened sacs that traverse the cytoplasm. There are two varieties.

- **Rough E.R.** is the site of protein synthesis resulting from the attached ribosomes.
- **Smooth E.R.** has three functions:
  1. Assists in the synthesis of steroid hormones and other lipids
  2. Connects rough E.R. to the Golgi apparatus
  3. Carries out various detoxification processes

Golgi Apparatus

The **Golgi apparatus** lies near the nucleus and consists of flattened membranous sacs stacked next to one another and surrounded by vesicles. They **package** substances produced in the rough endoplasmic reticulum and **secrete** them to other cell parts or to the cell surface for export.

Lysosomes

Lysosomes are sacs of **hydrolytic** (digestive) **enzymes** surrounded by a single membrane. They are the principal site of **intracellular digestion**. With the help of the lysosome, the cell continually renews itself by breaking down and recycling cell parts. Programmed destruction of cells (**apoptosis**) by their own hydrolytic enzymes is a critical part of the development of multicelled organisms. They are generally not found in plant cells.

Peroxisomes

Peroxisomes are found in both plant and animal cells. They contain **catalase**, which converts hydrogen peroxide (H_2O_2), a waste product of respiration in the cell, into water with the release of oxygen atoms. They also detoxify alcohol in liver cells.

Mitochondria

The mitochondria are the site of cellular respiration. All cells have many mitochondria; a very active cell could have 2,500 of them. Mitochondria have an outer double membrane and an inner series of membranes called **cristae**. They also contain their own DNA.

Vacuoles

Vacuoles are single, membrane-bound structures for storage. **Vesicles** are tiny vacuoles. Freshwater Protista have **contractile vacuoles** that pump out excess water.

Plastids

Plastids have a double membrane and are found only in plants and algae. They are of three types.

1. **Chloroplasts** are the site of photosynthesis. In addition to the double outer membrane, they have an inner one that forms a series of structures called **grana** (consisting of **thylakoids**). The grana lie in the **stroma**. They contain their own DNA.
2. **Leucoplasts** store starch and are found in roots like turnips or in tubers like the potato.
3. **Chromoplasts** store carotenoid pigments and are responsible for the red-orange-yellow of carrots, tomatoes, and daffodils.

Cytoskeleton

The **cytoskeleton** of the cell is a complex network of **protein filaments** that extends throughout the cytoplasm and gives the cell its shape, enables it to move, and anchors the organelles to the plasma membrane. The cytoskeleton includes micro-tubules and microfilaments.

- **Microtubules** are hollow tubes made of the protein tubulin that make up the **cilia**, **flagella**, and **spindle fibers**. Cilia and flagella, which move cells from one place to another consist of 9 pairs of microtubules organized around 2 singlet microtubules; see Figure 4.6. **Spindle fibers** help separate chromosomes during mitosis and meiosis and consist of microtubules organized into 9 triplets with no microtubules in the center. Flagella, when present in prokaryotes, are not made of microtubules.

> **STUDY TIP**
>
> Cilia and flagella have the 9 + 2 configuration.

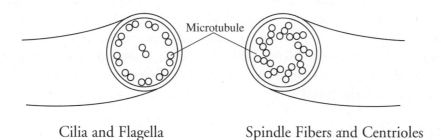

Microtubule

Cilia and Flagella   Spindle Fibers and Centrioles

Figure 4.6

- **Microfilaments**, also called **actin filaments**, help support the shape of the cell. They enable

   1. Animal cells to form a **cleavage furrow** during cell division
   2. Ameoba to move by sending out **pseudopods**
   3. Skeletal muscle to contract as they slide along myosin filaments

Centrioles, Centrosomes, and the Microtubule Organizing Centers

Centrioles, **centrosomes**, or **microtubule organizing centers** (**MTOC**s) are non-membranous structures that lie outside the nuclear membranes. They organize spindle fibers, and give rise to the spindle apparatus required for cell division (see Figure 4.7). Two centrioles oriented at right angles to each other make up one centrosome and consist of 9 triplets of microtubules (just like spindle fibers) arranged in a circle. Plant cells lack centrosomes, but have MTOCs. In animal cells, the MTOC is synonymous with centrosome.

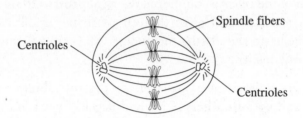

Figure 4.7   Spindle Fibers

Cell Wall

The **cell wall** is one cell structure not found in animal cells. Plants and algae have cell walls made of **cellulose**. The cell walls of fungi are usually made of **chitin**. Those of prokaryotes consist of other polysaccharides and complex polymers. The **primary cell wall** is immediately outside the plasma membrane. Some cells produce a second cell wall underneath the primary cell wall, called the **secondary cell wall**. When a plant cell divides, a thin gluey layer is formed between the two new cells, which becomes the **middle lamella**.

Plasma Membrane

The **cell** or **plasma membrane** is a selectively permeable membrane that regulates the steady traffic that enters and leaves the cell. **S. J. Singer** is famous for his description of the cell membrane in 1972, which he called the **fluid mosaic model**. The eukaryotic plasma membrane consists of a **phospholipid bilayer** with proteins dispersed throughout the layers. A phospholipid is **amphipathic**, meaning it has both a hydrophobic and hydrophilic region. **Integral proteins** have nonpolar regions that completely span the hydrophobic interior of the membrane. **Peripheral proteins** are loosely bound to the surface of the membrane. **Cholesterol molecules** are embedded in the interior of the bilayer to stabilize the membrane. The average membrane has the consistency of olive oil and is about 40 percent lipid and 60 percent protein. Phospholipids move along the plane of the membrane rapidly. Some proteins are kept in place by attachment to the cytoskeleton, while others drift slowly. The external surface of the plasma membrane also has carbohydrates attached to it, forming the **glycocalyx**, which are important for cell-to-cell recognition; see Figure 4.8.

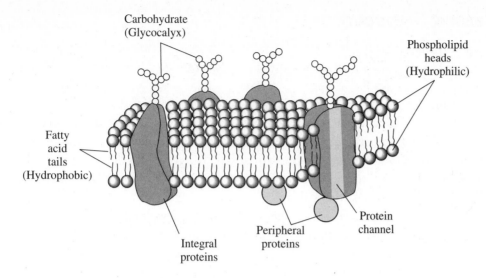

Figure 4.8   Detail of the Plasma Membrane

Proteins in the plasma membrane provide a wide range of functions.

- They **transport** molecules, electrons, and ions through **channels**, **pumps**, **carriers**, and **electron transport chains**.
- They act as enzymes. One **membrane-bound enzyme** located within the cell membrane that synthesizes cyclic AMP (c-AMP) from ATP is **adenylate cyclase**.
- They act as **receptors** for hormones, neurotransmitters, receptor-mediated endocytosis, and for cells of the immune system will be discussed in more detail in other sections.
- **Cell to cell attachments: Desmosomes** serve as anchors for filaments and rivet cells together.

TRANSPORT INTO AND OUT OF THE CELL

Transport is the movement of substances into and out of a cell. Transport can be either active or passive. Active transport requires energy (ATP). Passive transport requires no energy. Figure 4.9 shows an overview of passive and active transport.

STUDY TIP

Facilitated diffusion does not use ATP. It is passive.

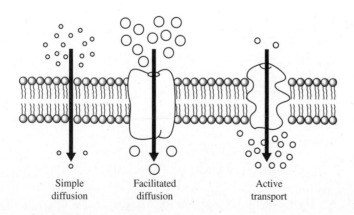

Simple diffusion   Facilitated diffusion   Active transport

Figure 4.9   Overview: Active and Passive Transport

Passive Transport

Passive transport is the movement of molecules *down a concentration gradient from a region of high concentration to a region of low concentration* until equilibrium is reached.

Examples of passive transport are **diffusion** and **osmosis**.

There are two types of **diffusion: simple** and **facilitated**. Simple diffusion does not involve protein channels, but facilitated diffusion does. An example of simple diffusion is found in the glomerulus of the human kidney, where solutes dissolved in the blood diffuse into Bowman's capsule of the nephron. Facilitated diffusion requires a **hydrophilic protein channel** that will passively transport specific substances across the membrane, see Figure 4.10. One type of channel transports single ions such as Na^+, K^+, Ca^{2+}, and Cl^-. (Neither simple nor facilitated diffusion requires energy.)

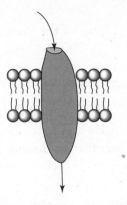

Figure 4.10   Protein Channel

A special case of simple diffusion is called **countercurrent exchange**—the flow of adjacent fluids in opposite directions that maximizes the rate of simple diffusion. One example of countercurrent exchange can be seen in fish gills. Blood flows toward the head in the gills, while water flows over the gills in the opposite direciton. This process maximizes the diffusion of respiratory gases and wastes between the water and the fish; see Figure 4.11.

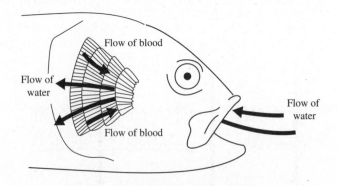

Countercurrent Exchange

Figure 4.11   Countercurrent Exchange

Here is some basic vocabulary to aid in your understanding of transport.

Osmosis is the term used for a specific type of diffusion of *water across a membrane.*

Solvent the substance that does the dissolving

Solute the substance that dissolves

Hypertonic having greater concentration of *solute* than another solution

Hypotonic having lesser concentration of *solute* than another solution

Isotonic two solutions containing equal concentration of *solutes*

Osmotic potential the tendency of water to move across a permeable membrane into a solution

Water potential Scientists look at movement of water in terms of water potential. **Water potential**, symbolized by the Greek letter psi, (Ψ), results from two factors: solute concentration and pressure. The water potential for *pure water is zero*; the addition of solutes lowers water potential to a value less than zero. Therefore, the water potential inside a cell is a negative value. Water will move across a membrane from the solution with the higher water potential to the solution with the lower water potential.

Figure 4.12 shows a sketch of two containers. Solution *A* is hypertonic to solution *B*.

> **REMEMBER**
>
> Water diffuses toward the hypertonic area.

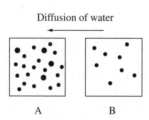

Figure 4.12

Example 1: In Figure 4.13 the cell is in an isotonic solution. Water diffuses in and out, but there is no net change in the size of the cell.

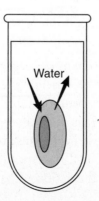

Figure 4.13   Cell in Isotonic Solution

Example 2: The cell in Figure 4.14 is in a **hypotonic** solution. The concentration of solute in the beaker is less than the concentration of solute in the cell. Therefore, *water* will flow into the cell, causing the cell to swell or burst. If the cell is a plant cell, the cell wall will prevent the cell from bursting. The cell will merely swell or become **turgid**. This turgid pressure is what keeps plants like celery crisp. If a plant loses too much water (dehydrates), it loses its **turgor pressure** and wilts.

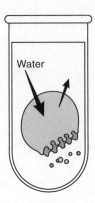

Figure 4.14   Cell in Hypotonic Solution

Example 3: The cell in Figure 4.15 is in a **hypertonic** solution. The concentration of **solute** in the beaker is greater than the concentration of solute in the cell. Therefore, water will flow out of the cell because water flows from high concentration of water to low concentration of water. As a result, the cell shrinks, exhibiting **plasmolysis**.

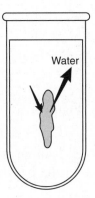

Figure 4.15   Cell in Hypertonic Solution

Aquaporins are special water channel proteins that facilitate the diffusion of massive amounts of water across a cell membrane. These channels do not affect the water potential gradient or the direction of water flow but, rather, the rate at which water diffuses down its gradient. It is possible that aquaporins can also function as **gated channels** that open and close in response to variables such as turgor pressure of a cell. The sudden change in a cell in response to changes in tonicity as seen in Figure 4.15 may be the result of the action of aquaporins.

Active Transport

Active transport is the movement of molecules *against a gradient,* which requires energy, usually in the form of ATP. There are many examples of active transport.

- **Pumps** or **carriers** carry particles across the membrane by active transport.

    1. **Plastoquinone** in the thylakoid membrane of chloroplasts is a mobile electron carrier.
    2. The **sodium-potassium pump** that pumps Na^+ and K^+ ions across a nerve cell membrane to return the nerve to its resting state is another example; see Figure 4.16. The sodium-potassium pump moves Na^+ and K^+ ions against a gradient, pumping two K^+ ions for every three Na^+ ions.

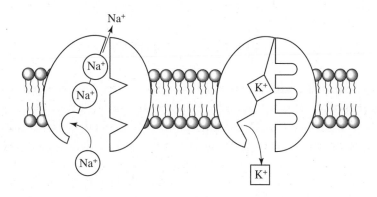

Figure 4.16   Sodium-Potassium Pump

3. The **electron transport chain** in mitochondria consists of proteins that pump protons across the cristae membrane.

- The **contractile vacuole** in freshwater Protista pumps out excess water that has diffused inward because the cell lives in a hypotonic environment.
- **Exocytosis** in nerve cells occurs as vesicles release neurotransmitters into a synapse.
- **Pinocytosis**, *cell drinking,* is the uptake of large, dissolved particles. The plasma membrane invaginates around the particles and encloses them in a vesicle; see Figure 4.17.

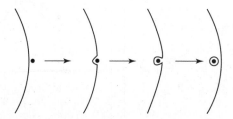

Figure 4.17   Pinocytosis

- **Phagocytosis** is the engulfing of large particles or small cells by pseudopods. The cell membrane wraps around the particle and encloses it into a vacuole. This is the way human white blood cells engulf bacteria and also the way in which ameoba gain nutrition; see Figure 4.18.

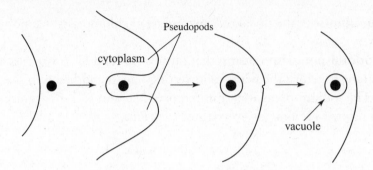

Figure 4.18   Phagocytosis

- **Receptor-mediated endocytosis** enables a cell to take up large quantities of very **specific substances**. It is a process by which extracellular substances bind to receptors on the cell membrane. Once the **ligand** (the general name for any molecule that binds specifically to a receptor site of another molecule) binds to the receptors, endocytosis begins. The receptors, carrying the ligand, migrate and cluster along the membrane, turn inward, and become a **coated vesicle** that enters the cell. This is the way cells take in **cholesterol** from the blood; see Figure 4.19.

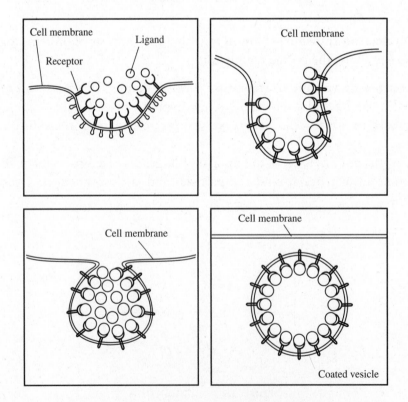

Figure 4.19   Receptor-Mediated Endocytosis

Bulk Flow

Bulk flow is a general term for the overall movement of a fluid in one direction in an organism. In humans, **blood** moves around the body by bulk flow as a result of blood pressure created by the pumping heart. **Sap** in trees moves by bulk flow from the leaves to the roots due to active transport in the phloem. Bulk flow movement is always from **source** (where it originates) **to sink** (where it is used).

CELL COMMUNICATION

In multicelled organisms, individual cells must work together to create a harmonious organism. Examples are cell junctions, signal transduction pathway, and cell-to-cell recognition.

Cell Junctions

Cell junctions can be classified into four types: tight junctions, desmosomes, gap junctions, and plasmodesmata.

TIGHT JUNCTIONS

Tight junctions are belts around the epithelial cells that line organs and serve as a barrier to prevent leakage into or out of those organs. In the **urinary bladder**, they prevent the urine from leaking out of the bladder into the surrounding body cavity; see Figure 4.20.

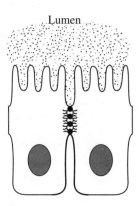

Figure 4.20   Tight Junction

DESMOSOMES

Desmosomes are found in many tissues and have been compared to **spot welds** that rivet cells together. They consist of clusters of **cytoskeletal filaments** from adjacent cells that are looped together. They occur in tissues that are subjected to severe mechanical stress, such as skin epithelium or the neck of the uterus, which must expand greatly during childbirth; see Figure 4.21.

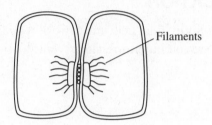

Figure 4.21   Desmosome

GAP JUNCTIONS

Gap junctions permit the passage of materials directly from the cytoplasm of one cell to the cytoplasm of an adjacent cell. In the muscle tissue of the **heart**, the flow of ions through the gap junctions coordinates the contractions of the cardiac cells; see Figure 4.22.

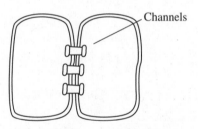

Figure 4.22   Gap Junction

PLASMODESMATA

Plasmodesmata connect one plant cell to the next. They are analogous to gap junctions in animal cells.

Signal Transduction Pathway

The **signal transduction pathway** relies on plasma membrane proteins in a multi-step process in which a small number of extracellular signal molecules produce a major cellular response. Three stages occur in this type of cell signaling: **reception**, **transduction**, and **response**. In reception, the signal molecule, commonly a protein that does not enter the cell, binds to a specific receptor on the cell surface, causing the receptor molecule to undergo a change in conformation. This change in conformation leads to transduction, a change in signal form, where the receptor relays a message to a secondary messenger. This secondary messenger, such as cyclic AMP (cAMP), induces a response within the cell.

Cell-to-cell Recognition

Cell-to-cell recognition is the cell's ability to distinguish one type of neighboring cell from another and is crucial to the functioning of a multicelled organism. A feature of all cells that aids in cell communication is the **glycocalyx** (see Figure 4.8 on page 99), which consists of oligosaccharides (small chains of sugar molecules) attached to integral proteins within the plasma membrane. The glycocalyx is responsible for such phenomena as **contact inhibition**, the normal trait of cells to stop dividing when they become too crowded.

Multiple-Choice Questions

Matching Column

1. Produces ATP   (A) Golgi apparatus

2. Produces proteins   (B) Microtubules

3. Packages and secretes substances (C) Rough endoplasmic reticulum

4. Contains hydrolytic enzymes   (D) Mitochondria

5. Directly assists with cell division (E) Lysosomes

6. Which of the following is *not* normally found in a plant cell?

    (A) mitochondria
    (B) endoplasmic reticulum
    (C) plastids
    (D) plasma membrane
    (E) centrioles

7. Which of the following is *present* in a prokaryote cell?

    (A) mitochondria
    (B) ribosomes
    (C) endoplasmic reticulum
    (D) chloroplasts
    (E) nuclear membrane

8. Membranes are components of all of the following except a

   (A) microtubule
   (B) nucleus
   (C) Golgi apparatus
   (D) mitochondrion
   (E) lysosome

9. When biologists wish to study the internal structure of cells with the best resolution and the highest magnification, they would most likely use a

   (A) good light microscope
   (B) transmission electron microscope
   (C) scanning electron microscope
   (D) phase-contrast microscope
   (E) centrifuge

10. A scientist has made a homogenate of human liver cells in a blender and then spun that mixture in an ultracentrifuge, as shown below. Which layer would include the most mitochondria?

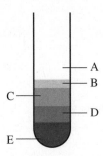

    (A) A
    (B) B
    (C) C
    (D) D
    (E) E

11. What is the approximate size of a human red blood cell?

    (A) 0.01 micrometer
    (B) 8 micrometers
    (C) 80 micrometers
    (D) 8 nanometers
    (E) 8,000 micrometers

12. Smooth E.R. carries out all of the following activities *except*

    (A) lipid production
    (B) detoxification
    (C) connects rough E.R. to the Golgi
    (D) produces RNA
    (E) assists in synthesis of steroid hormones

13. An animal cell in a hypertonic solution would

    (A) swell
    (B) swell and exhibit turgor
    (C) exhibit plasmolysis
    (D) shrink and then swell
    (E) not be affected; only a plant cell would be affected

14. Which one of the following would *not* normally diffuse through the lipid bilayer of a plasma membrane?

    (A) CO_2
    (B) amino acid
    (C) starch
    (D) water
    (E) O_2

15. Which of the following requires ATP?

    (A) the uptake of cholesterol by a cell
    (B) the facilitated diffusion of glucose into a cell
    (C) countercurrent exchange
    (D) the diffusion of oxygen into a fish's gills.
    (E) simple diffusion in the nephron of a kidney

16. All of the following cellular activities require ATP EXCEPT

    (A) sodium-potassium pump
    (B) cells absorbing oxygen
    (C) receptor-mediated endocytosis
    (D) amoeboid movement
    (E) phagocytosis

17. Which of the following best characterizes the structure of the plasma membrane?

    (A) rigid and unchanging
    (B) rigid but varying from cell to cell
    (C) fluid but unorganized
    (D) very active
    (E) rigid and organized

18. The cytoplasmic channels between plant cells are called

    (A) desmosomes
    (B) middle lamellae
    (C) plasmodesmata
    (D) tight junctions
    (E) reticulum

Questions 19–21
The following questions refer to the figure below, which shows the plasma membrane.

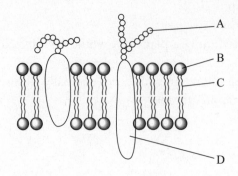

19. Identify the hydrophilic portion of a lipid molecule.

20. Identify the proteins involved in transport.

21. Identify the glycocalyx.

22. Which organelle contains DNA?

    (A) ribosomes
    (B) mitochondria
    (C) Golgi body
    (D) lysosomes
    (E) cilia

Matching Column

23. Smooth endoplasmic reticulum

24. Centrioles

25. Peroxisome

26. Nucleolus

27. Lysosome

(A) Consists of microtubules

(B) Digests damaged cells

(C) Synthesizes ribosomes

(D) System of membranes; detoxifies alcohol in liver cells

(E) Breaks down the by-product of cell respiration

Answers to Multiple-Choice Questions

1. **(D)** Mitochondria release energy from organic molecules and store it in ATP molecules.

2. **(C)** Ribosomes attached to rough endoplasmic reticulum produce proteins.

3. **(A)** The Golgi apparatus receives newly synthesized proteins and lipids from the E.R. and distributes them to the plasma membrane, lysosomes, and secretory vesicles.

4. **(E)** Lysosomes are the principal sites of intracellular digestion.

5. **(B)** Microtubules make up the spindle fibers, which connect to the centromeres of chromosomes and assist in mitosis.

6. **(E)** Plant cells lack centrioles. Instead they have microtubule organizing regions. They also have mitochondria, ribosomes, plastids, and endoplasmic reticulum.

7. **(B)** Prokaryotes have NO internal membranes. Therefore, they lack mitochondria, E.R., chloroplasts, and nuclear membrane. They do have small ribosomes.

8. **(A)** Microtubules are part of the cytoskeletal structure and are made of the protein tubulin. The others all consist of selectively permeable plasma membranes.

9. **(B)** Although all electron microscopes have high magnification and excellent resolution, the transmission electron microscope is used to study the interior of cells while the scanning electron microscope studies the surface of cells.

10. **(D)** Centrifugation causes the densest structures to sink to the bottom and the lightest to remain on top. Nuclei are the densest, and mitochondria are the next most dense. The least dense layer would consist of ribosomes.

11. **(B)** Human red blood cells are small, 8 micrometers (μm) or 80 nanometers (nm). An average cell is about 80 micrometers.

12. **(D)** Smooth ER connects the rough E.R. to the Golgi, carries out detoxification, and produces lipids like steroids. The nucleolus produces RNA.

13. **(C)** An animal cell in a hypertonic solution would shrink because the concentration of water is greater inside the cell than outside the cell. Since water flows down a gradient, it would flow out of the cell. Plasmolysis means cell shrinking.

14. **(C)** Starch, a polysaccharide, is too large to diffuse through the plasma membrane.

15. **(A)** The uptake of cholesterol occurs by receptor-mediated endocytosis, which requires energy. B is an example of facilitated diffusion, and C and D are examples of countercurrent exchange. B, C, D, and E are all examples of passive transport and do not require energy.

16. **(B)** Oxygen is absorbed by diffusion. All the other choices are examples of active transport.

17. **(D)** The plasma membrane is organized and made of many small particles that move about readily. Hence the name, fluid mosaic. A membrane is a very active structure. A cell's activity is limited by how fast plasma membranes can take in and get rid of materials.

18. **(C)** Plasmodesmata are functionally like gap junctions in animal cells and are a means of cytoplasmic communication among plant cells. A system of plasmodesmata is called a symplast.

19. **(B)** The phospholipid head is polar and hydrophilic.

20. **(D)** This is a protein channel.

21. **(A)** The glycocalyx is involved in cell-to-cell communication.

22. **(B)** The nucleus, chloroplasts, and mitochondria all contain DNA.

23. **(D)** The smooth endoplasmic reticulum detoxifies the cell, produces steroids like cholesterol and connects the rough endoplasmic reticulum to the Golgi.

24. **(A)** Centrioles consist of microtubules in a 9 triplet configuration.

25. **(E)** Peroxisomes break down peroxide, which is a poisonous by-product of cell respiration.

26. **(C)** The nucleolus is responsible for synthesizing ribosomal RNA, one of the main components of ribosomes.

27. **(B)** Lysosomes contain hydrolytic enzymes that break down nutrients, waste products, and cell debris.

Free-Response Questions

Directions: Answer all questions. You must answer the question in essay—**not** outline—form. You may use labeled diagrams to supplement your essay, but diagrams alone are *not* sufficient. Before you start to write, read each question carefully so that you understand what the question is asking.

1. Compare and contrast the characteristics of prokaryotic and eukaryotic cells.

2. Living cells are highly organized and regulated.

    a. Describe the structure of the plasma membrane
    b. Explain how the plasma membrane contributes to the regulation of the cell

Typical Free-Response Answers

Note: In the following essay, key words are in bold to make them more visible to you so you can focus on the importance of stating scientific terms when you write an essay.

1. Being the ancestor of all cells, prokaryotic cells have certain things in common with eukaryotic cells. They both have DNA, a cell wall (in the case of plant cells), and cytoplasm. Prokaryotic cells, like eukaryotic cells, carry out a wide range of metabolic processes including glycolysis, photosynthesis, and respiration; but prokaryotes carry them out in a far simpler fashion. Prokaryotes are generally small (1–10 μm in diameter); while eukaryotic cells are 10 times larger, (10–100 μm). Prokaryotic cells have circular DNA while eukaryotes have long, linear DNA molecules with many noncoding regions encircled by a nuclear membrane. Prokaryotes are mainly unicellular, while eukaryotes are mostly multicellular. Prokaryotes have no cytoskeleton and no internal membranes. Metabolism in prokaryotes can be anaerobic or aerobic, but in eukaryotes metabolism is aerobic. Eukaryotes have an elaborate cytoskeletal structure and an array of internal membranes such as Golgi, endoplasmic reticulum, mitochondria, chloroplasts, and vacuoles.

2a. In 1972, S. J. Singer elucidated the structure of the cell membrane, which he called the **fluid mosaic model**. The plasma membrane is a selectively permeable and dynamic, fluid structure consisting of a continuous double phospholipid layer about 5nm thick, with proteins dissolved throughout the bilayer. The phospholipids are ***amphipathic***, meaning they have polar (**hydrophilic**) heads and nonpolar (**hydrophobic**) tails. The polar heads face outward and the nonpolar tails face toward the interior of the membrane. Proteins are dispersed throughout the lipid layers. **Peripheral proteins** are bound to one face or the other and do not extend into the hydrophobic interior of the lipid bilayer. **Integral membrane proteins** are amphipathic and pass through the lipid layers. In eukaryotes, large quantities of **cholesterol** serve to enhance the flexibility and the mechanical stability of the bilayer.

2b. The lipid bilayer of the plasma membrane is a **selectively permeable membrane**. The lipid layer allows for the passage of water and small, nonpolar molecules. The protein component carries out most of the other functions. **Channel proteins** are large pores that extend across the membrane and allow the passage of specific molecules of appropriate size and charge to flow down a gradient into or out of a cell. **Carrier proteins** behave like membrane-bound enzymes and transport specific molecules across the membrane against a gradient. One example of proteins that work by active transport is the **sodium-potassium pump** in nerve cells. The Na-K pump acts as an *antiport*, carrying two different ions in opposite directions as the protein changes its conformation. Some proteins act as proton pumps, which are responsible for ATP production in mitochondria and the thylakoid membrane of chloroplasts. Other proteins are specific markers that are embedded on the surface of the cell and take part in the cell-cell recognition process. Plasma membranes vary from cell type to cell type depending on their particular function.

Cell Respiration

CHAPTER 5

- ATP—adenosine triphosphate
- Glycolysis
- Anaerobic respiration—fermentation
- Structure of the mitochondrion
- Aerobic respiration: The Krebs cycle
- NAD and FAD
- Aerobic respiration: The electron transport chain
- Oxidative phosphorylation and chemiosmosis
- Summary of ATP production

INTRODUCTION

Cell respiration is the means by which cells extract energy stored in food and transfer that energy to molecules of **ATP**. Energy that is temporarily stored in molecules of ATP is instantly available for every cellular activity such as passing an electrical impulse, contracting a muscle, moving cilia, or manufacturing a protein. The equation for the complete aerobic respiration of one molecule of glucose (seen below) is highly exergonic (releases energy). See Figure 5.1.

$$C_6H_{12}O_6 + 6O_2 \rightarrow 6CO_2 + 6H_2O + \text{energy}$$
$$\text{free energy} = \Delta G = -686 \text{ kcal/mole}$$

> **STUDY TIP**
>
> Cell respiration is an oxidative process.

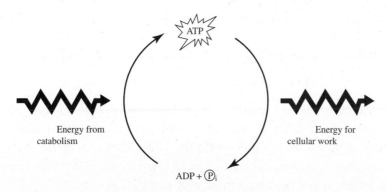

Figure 5.1

There are two types of cell respiration: anaerobic and aerobic. If oxygen is not present (anaerobic respiration), **glycolysis** is followed by either **alcoholic fermentation** or **lactic acid fermentation**. If oxygen is present (aerobic respiration), glycolysis is only the first phase of aerobic respiration. It is followed by the **Krebs cycle**, the **electron transport chain**, and **oxidative phosphorylation**.

ATP—ADENOSINE TRIPHOSPHATE

A molecule of ATP (adenosine triphosphate) consists of **adenosine** (the nucleotide adenine plus ribose) plus three **phosphates**. ATP is an unstable molecule because the three phosphates in ATP are all negatively charged and repel one another. When one phosphate group is removed from ATP by hydrolysis, a more stable molecule, ADP (adenosine diphosphate), results. *The change, from a less stable molecule to a more stable molecule, always releases energy.* ATP provides energy for all cells activities by transferring phosphates from ATP to another molecule, as seen in Figure 5.1. Figure 5.2 shows the structure of a molecule of ATP.

<table>
<tr><td>

STUDY TIP

Be able to identify this molecule. It sometimes shows up on AP tests.

</td></tr>
</table>

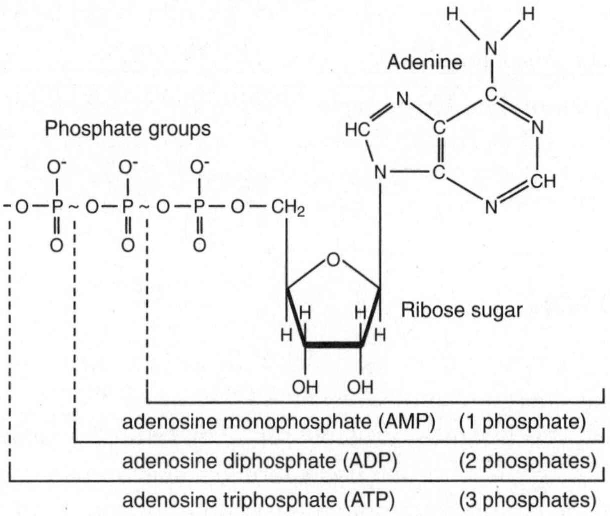

Figure 5.2   The Structure of Adenosine Triphosphate (ATP)

GLYCOLYSIS

Glycolysis is a ten-step process that breaks down 1 molecule of glucose (a six-carbon molecule) into 2 three-carbon molecules of **pyruvate** or **pyruvic acid** and releases **4** molecules of **ATP**. The energy of activation for glycolysis is 2 ATP. After subtracting 2 ATPs from the 4 ATPs released from the reaction, the glycolysis of 1 molecule of glucose results in a *net gain of 2 ATP*. Here is the simplified equation.

$$2 \text{ ATP} + 1 \text{ Glucose} \rightarrow 2 \text{ Pyruvate} + 4 \text{ ATP}$$

Glycolysis occurs in the cytoplasm and produces ATP *without using oxygen*. Each step is catalyzed by a different enzyme. Although this process releases only one-fourth of the energy stored in glucose (most of the energy remains locked in pyruvate), the reaction is critical. The end product, pyruvate, is the *raw material for the Krebs cycle*, which is the next step in aerobic respiration.

During glycolysis, ATP is produced by **substrate level phosphorylation**—by direct enzymatic transfer of a phosphate to ADP. Only a small amount of ATP is produced this way.

There is one other important thing about glycolysis. The enzyme that catalyzes the third step, phosphofructokinase (**PFK**), is an **allosteric** enzyme. It inhibits glycolysis when the cell has enough ATP and does not need to produce any more.

If ATP is present in the cell in large quantities, it inhibits PFK by altering the conformation of that enzyme, thus stopping glycolysis. As the cell's activities use up ATP, less ATP is available to inhibit PFK and glycolysis continues, ultimately to produce more ATP. *This is an important example of how a cell regulates ATP production through allosteric inhibition;* see Figure 5.3.

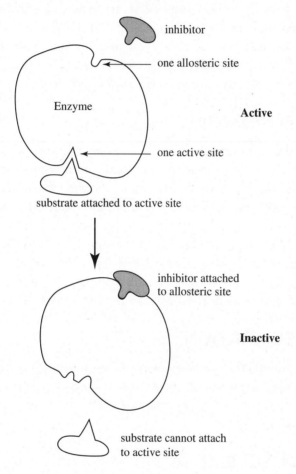

inhibitor

one allosteric site

Enzyme

Active

one active site

substrate attached to active site

inhibitor attached
to allosteric site

Inactive

substrate cannot attach
to active site

Figure 5.3   Allosteric Inhibition

ANAEROBIC RESPIRATION—FERMENTATION

Anaerobic respiration or **fermentation** is not a synonym for glycolysis. It is an anaerobic, catabolic process that consists of glycolysis *plus* alcohol or lactic acid fermentation. Anaerobic respiration originated millions of years ago when there was no free oxygen in the earth's atmosphere. Even today it is the sole means by which anaerobic bacteria such as **botulinum** (the bacterium that causes a form of food poisoning, botulism) release energy from food. There are two types of anaerobes: facultative and obligate. **Facultative anaerobes** can tolerate the presence of oxygen; they simply do not use it. **Obligate anaerobes** cannot live in an environment containing oxygen.

Fermentation can generate ATP during anaerobic respiration *as long as there is an adequate supply of NAD⁺ to accept electrons* during glycolysis. Without some mechanism to convert NADH back to NAD⁺, glycolysis would shut down. Fermentation consists of glycolysis plus the reactions that regenerate NAD⁺. Two types of fermentation are **alcohol fermentation** and **lactic acid fermentation**.

Alcohol Fermentation

Alcohol fermentation or simply **fermentation** is the process by which certain cells convert pyruvate from glycolysis into **ethyl alcohol** and **carbon dioxide** in the **absence of oxygen** and, *in the process, oxidize NADH back to NAD⁺*. The bread-baking industry depends on the ability of yeast to carry out fermentation and produce carbon dioxide, which causes bread to rise. The beer, liquor, and wine industry depends on yeast to ferment sugar into ethyl alcohol.

Lactic Acid Fermentation

During lactic acid fermentation, pyruvate from glycolysis is reduced to form **lactic acid** or **lactate**. This is the process that the dairy industry uses to produce yogurt and cheese. Also in the process, *NADH gets oxidized back to NAD⁺*. **Human skeletal muscles** also carry out lactic acid fermentation when the blood cannot supply adequate oxygen to muscles during strenuous exercise. Lactic acid in the muscle causes fatigue and burning. The lactic acid continues to build up until the blood can supply the muscles with adequate oxygen to repay the oxygen debt. With normal oxygen levels, the muscle cells will revert to the more efficient aerobic respiration and the lactic acid is then converted back to pyruvate in the liver.

AEROBIC RESPIRATION

When oxygen is present, cells carry out aerobic respiration. It is highly efficient and produces a lot of ATP. This process consists of an anaerobic phase—glycolysis—and an aerobic phase. The aerobic phase has two parts: the Krebs cycle and oxidative phosphorylation.

THE KREBS CYCLE

The **Krebs cycle** is a cyclical series of enzyme-catalyzed reactions also known as the **citric acid cycle**. It takes place in the **matrix of mitochondria** and requires pyruvate, the product of glycolysis. Figure 5.4 illustrates the Krebs cycle in detail. Here are some other important points:

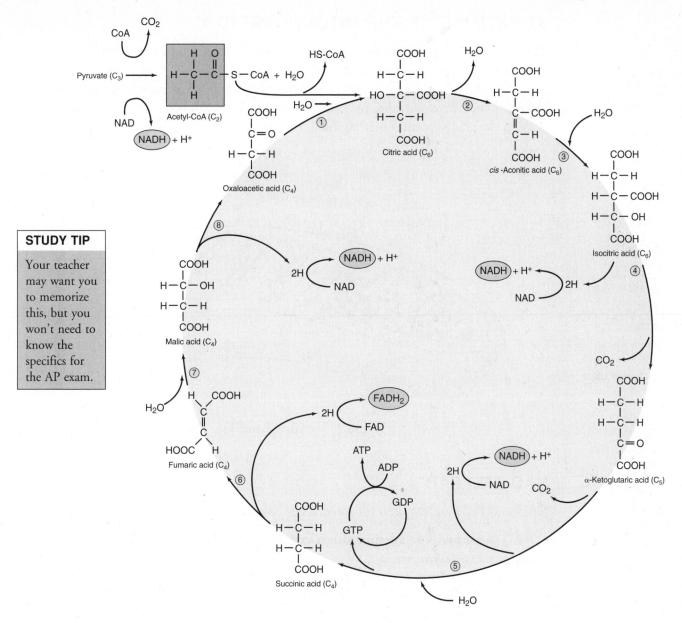

Figure 5.4   The Krebs Cycle

- In the first step, acetyl co-A combines with **oxaloacetic acid** (**OAA** or oxaloacetate) to produce citric acid. Hence the cycle also has the name citric acid cycle.
- Remember that each molecule of glucose is broken down to 2 molecules of pyruvate during glycolysis. The respiration of each molecule of glucose causes the Krebs cycle to turn two times.
- Before it enters the Krebs cycle, pyruvate must first combine with coenzyme A (a vitamin) to form **acetyl co-A**, which does enter the Krebs cycle. The conversion of pyruvate to acetyl co-A produces **2 molecules of NADH**, 1 NADH for each pyruvate.
- Each turn of the Krebs cycle releases **3 NADH, 1 ATP, 1 FADH**, and the waste product **CO_2**, which is exhaled. (Remember, two turns of the Krebs cycle occur per glucose molecule.)
- During the Krebs cycle, ATP is produced by **substrate level phosphorylation**—by direct enzymatic transfer of a phosphate to ADP. Very little energy is produced this way compared with the amount produced by oxidative phosphorylation.

STRUCTURE OF THE MITOCHONDRION

The mitochondrion is enclosed by a double membrane. The outer membrane is smooth, but the inner, or **cristae**, membrane is folded. This inner membrane divides the mitochondrion into two internal compartments, the **outer compartment** and the **matrix**. The Krebs cycle takes place in the matrix; the electron transport chain takes place in the cristae membrane. Figure 5.5 shows a diagram of the mitochondrion.

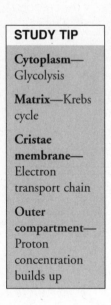

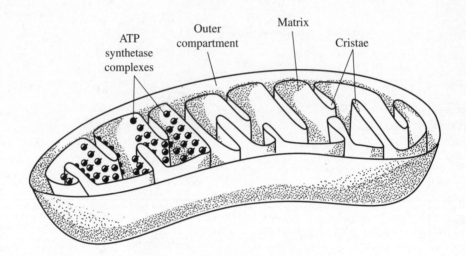

Figure 5.5   Mitochondrion

NAD AND FAD

NAD and FAD are required for normal cell respiration.

* NAD (**nicotinamide adenine dinucleotide**) and FAD (**flavin adenine dinucleotide**) are **coenzymes** that carry protons or electrons from glycolysis and the Krebs cycle to the electron transport chain.
* The enzyme NAD or FAD dehydrogenase facilitates the transfer of hydrogen atoms from a substrate, such as glucose, to its coenzyme NAD^+.
* Without NAD^+ to accept protons and electrons from glycolysis and the Krebs cycle, both processes would cease and the cell would die.
* NAD and FAD are vitamin derivatives.
* NAD^+ is the oxidized form. NAD_{re} or NADH is the reduced form. NADH carries 1 proton and 2 electrons.
* FAD is the oxidized form. FAD_{re} or $FADH_2$ is the reduced form.

AEROBIC RESPIRATION

The **electron transport chain (ETC)** is a **proton pump** in the mitochondria that uses the energy released from the exergonic flow of electrons to pump protons from the matrix to the outer compartment. This results in the establishment of a **proton gradient** inside the mitochondrion. The electron transport chain makes no ATP directly but sets the stage for ATP production during **chemiosmosis**; see Figure 5.6.

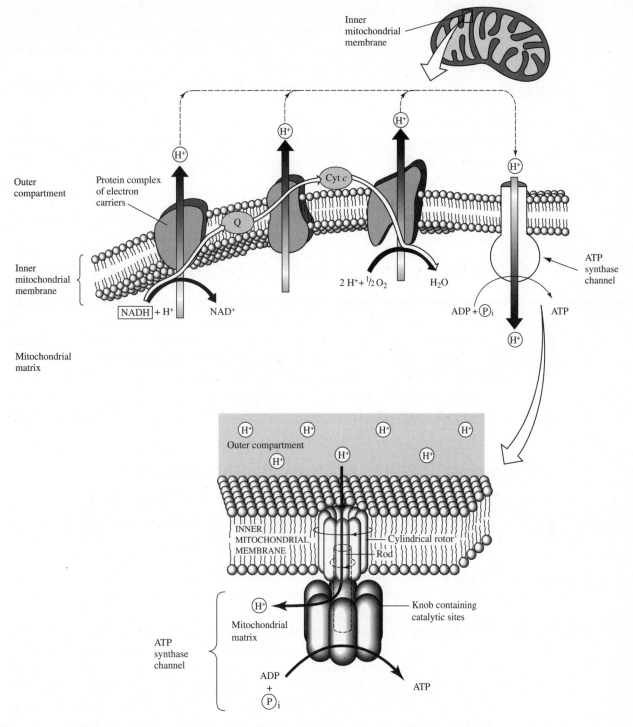

Figure 5.6   The Electron Transport Chain, Proton Gradient, and ATP Synthase

Important Points

- The ETC is a collection of molecules embedded in the **cristae membrane** of the mitochondrion.
- There are thousands of copies of the ETC in every mitochondrion due to the extensive folding of cristae membrane.

STUDY TIP

Here is where most of the ATP is made in the cell.

- The ETC carries electrons delivered by NAD and FAD from glycolysis and the Krebs cycle to oxygen, the **final electron acceptor**, through a series of **redox reactions**. In a redox reaction, one atom gains electrons or protons (**reduction**), and one atom loses electrons (**oxidation**).
- The highly electronegative **oxygen** acts to pull electrons through the electron transport chain.
- **NAD** delivers its electrons to a higher energy level in the chain than does **FAD**. As a result, NAD will provide more energy for ATP synthesis than does FAD. Each NAD produces 3 ATP molecules, while each FAD produces 2 ATP molecules.
- The ETC consists mostly of **cytochromes**. These are proteins structurally similar to hemoglobin. Cytochromes are present in all aerobes and are used to trace evolutionary relationships.

OXIDATIVE PHOSPHORYLATION AND CHEMIOSMOSIS

Most of the energy produced during cell respiration occurs in the mitochondria by a process known by the general name of **oxidative phosphorylation**. This term means the phosphorylation of ADP into ATP by the oxidation of the carrier molecules NADH and $FADH_2$. This energy-coupling mechanism was elucidated in 1961 by Peter Mitchell, who named it the **chemiosmotic theory**. According to the Mitchell hypothesis, chemiosmosis uses potential energy stored in the form of a **proton (H^+) gradient** to phosphorylate ADP and produce ATP (ADP + P $\rightarrow$ ATP).

Important Points of Oxidative Phosphorylation

- It is powered by the redox reactions of the electron transport chain.
- Protons are pumped from the **matrix** to the **outer compartment** by the electron transport chain. The electron transport chain is an energy converter that couples the exergonic flow of electrons with the endergonic pumping of protons across the cristae membrane and into the outer compartment.
- A proton gradient is created between the outer compartment and the inner matrix.
- Protons cannot diffuse through the cristae membrane; they can flow only down the gradient into the matrix through **ATP synthase** (or **ATP synthetase**) **channels**. This is **chemiosmosis**, the key to the production of ATP. *As protons flow through the ATP-synthase channels, they generate energy to phosphorylate ADP into ATP.* This process is similar to how a hydroelectric plant converts the enormous potential energy of water flowing through a dam to turn turbines and generate electricity.
- **Oxygen** is the **final hydrogen acceptor**, combining $\frac{1}{2}$ an oxygen molecule with 2 electrons and 2 protons, thus forming **water**. This water is a waste product of cell respiration and is excreted.

SUMMARY OF ATP PRODUCTION

ATP is produced in two ways.

- **Substrate level phosphorylation** occurs when an enzyme, a **kinase**, transfers a phosphate from a substrate directly to ADP. Only a small amount of ATP is produced this way. This is the way energy is produced during glycolysis and the Krebs cycle.
- **Oxidative phosphorylation** occurs during chemiosmosis. This is the way 90 percent of all ATP is produced from cell respiration. During oxidative phosphorylation, NAD and FAD lose protons (become oxidized) to the electron transport chain, which pumps them to the outer compartment of the mitochondrion, creating a steep proton gradient. This electrochemical or proton gradient powers the phosphorylation of ADP into ATP.

During respiration, most energy flows in this sequence:

Glucose → NAD$_{re}$ and FAD → electron transport chain → chemiosmosis → ATP

Table 5.1 and Figure 5.7 summarize the maximum energy that can be produced from the aerobic respiration of 1 molecule of ATP. This is hypothetical because some cells are more efficient than others and cells vary in their efficiency at different times. In general, 1 NADH entering the ETC produces 3 ATPs. One FADH entering the ETC produces 2 ATPs. Here is the tally of ATPs produced from *1 molecule of glucose*. Remember that during glycolysis, 2 pyruvates are formed and each enters the Krebs cycle separately. So after glycolysis, all numbers for FADH$_2$, NADH and ATP are doubled.

Table 5.1 shows all the ATP produced by substrate level phosphorylation and oxidative phosphorylation of 1 molecule of glucose.

TABLE 5.1

Maximum Energy Produced from the Aerobic Respiration of 1 Molecule of Glucose				
	FADH$_2$	NADH	ATP	Total ATP
Glycolysis (Substrate Level Phosphorylation)		2	2 net	4 (2 net) 4
From Pyruvate to Acetyl CoA		2		6
Krebs cycle	2	6	2	4 18 2
Total ATP from 1 molecule of glucose				38 (36 net)

Figure 5.7 shows an overview of ATP production from cell respiration.

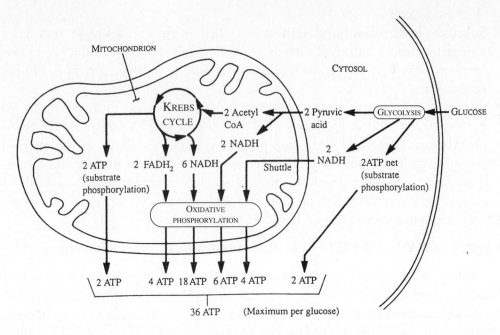

Figure 5.7

Multiple-Choice Questions

1. The role of oxygen in aerobic respiration is

   (A) to transport CO_2
   (B) most important in the Krebs cycle
   (C) to transport electrons in glycolysis
   (D) to provide electrons for the electron transport chain
   (E) as the final hydrogen acceptor in the electron transport chain

2. The loss of protons and electrons is known as

   (A) dehydration
   (B) hydrogenation
   (C) reduction
   (D) oxidation
   (E) Both B and D are correct.

3. $C_6H_{12}O_6 + 6O_2 \rightarrow 6H_2O + 6CO_2 + 38$ ATP

   The process shown is

   (A) reduction and is endergonic
   (B) reduction and is exergonic
   (C) oxidation and is endergonic
   (D) oxidation and is exergonic
   (E) neither oxidation nor reduction but must be endergonic because energy must be added to the reaction to get it to begin

4. Most energy during cell respiration is harvested during

    (A) the Krebs cycle
    (B) oxidative phosphorylation
    (C) glycolysis
    (D) anaerobic respiration
    (E) fermentation

5. Plants carry out photosynthesis during the day and respiration only during the night.

    (A) true
    (B) false

6. All of the following processes produce ATP except

    (A) lactic acid formation
    (B) oxidative phosphorylation
    (C) glycolysis
    (D) the Krebs cycle
    (E) both A and B

7. After strenuous exercise, a muscle cell would contain decreased amounts of _____ and increased amounts of _____.

    (A) glucose; ATP
    (B) ATP; glucose
    (C) ATP; lactic acid
    (D) lactic acid; ATP
    (E) CO_2; pyruvic acid

Match each process with its correct location

8. Glycolysis   (A) The cristae membrane

9. Electron transport chain   (B) Cytoplasm

10. Krebs cycle   (C) Inner matrix of the mitochondria

The three circles represent three major processes in aerobic respiration.

glucose $\rightarrow$ (process A) $\rightarrow$ (process B) $\rightarrow$ (process C) $\rightarrow$ $CO_2 + H_2O$

11. Process C represents

    (A) glycolysis
    (B) the Krebs cycle
    (C) the electron transport chain
    (D) substrate level phosphorylation
    (E) cannot be determined

Questions 12–13
The following questions refer to the sketch of a mitochondrion, shown below.

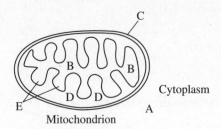

12. Identify the site of the Krebs cycle in the sketch of the mitochondrion.

13. Identify the site of the ATP synthase.

14. Each NAD molecule carrying hydrogen to the electron transport chain can produce a maximum of _____ molecules of ATP?

    (A) 1
    (B) 2
    (C) 3
    (D) 4
    (E) Cannot be determined

15. Which is true of aerobic respiration but not true of anaerobic respiration?

    (A) CO_2 is produced.
    (B) ATP is produced.
    (C) Water is produced.
    (D) Alcohol is produced.
    (E) Pyruvate is produced.

16. Which of the following is the most important thing that happens during aerobic respiration?

    (A) Electrons move down the electron transport chain in a series of redox reactions.
    (B) Acetyl CoA enters the Krebs cycle.
    (C) NAD carries hydrogen to the electron transport chain.
    (D) ATP is produced.
    (E) CO_2 is produced in the Krebs cycle.

17. The ATP produced during fermentation is generated by which of the following?

    (A) the electron transport chain
    (B) substrate level phosphorylation
    (C) the Krebs cycle
    (D) chemiosmosis
    (E) the citric acid cycle

18. In addition to ATP, what is produced during glycolysis?

    (A) pyruvate and NADH
    (B) CO_2 and H_2O
    (C) CO_2 and ethyl alcohol
    (D) CO_2 and NADH
    (E) H_2O and ethyl alcohol

19. Which of the following probably evolved first?

    (A) the Krebs cycle
    (B) oxidative phosphorylation
    (C) glycolysis
    (D) the electron transport chain
    (E) the citric acid cycle

20. Which is an example of a feedback mechanism?

    (A) Phosphofructokinase, an allosteric enzyme that catalyzes step 3 of glycolysis, is inhibited by ATP.
    (B) NAD carries hydrogen to the ETC.
    (C) ATP is produced in mitochondria as protons flow through the ATP-synthase channel.
    (D) Energy is released from glucose as it decomposes into CO_2 and H_2O.
    (E) Lactic acid gets converted back to pyruvic acid in the human liver.

21. Which process of cell respiration is most closely associated with intracellular membranes?

    (A) fermentation
    (B) the Krebs cycle
    (C) glycolysis
    (D) substrate level phosphorylation
    (E) oxidative phosphorylation

22. During cell respiration, most ATP is formed as a direct result of the net movement of

    (A) sodium ions diffusing across a membrane
    (B) electrons flowing against a gradient
    (C) electrons flowing through a channel
    (D) protons flowing through a channel
    (E) protons flowing against a gradient

Answers to Multiple-Choice Questions

1. **(E)** Oxygen is the final hydrogen and electron acceptor in the electron transport chain. Oxygen does not take part in glycolysis or in the Krebs cycle.

2. **(D)** The loss of hydrogen or loss of electrons is known as oxidation. The gain of hydrogen or of electrons is known as reduction because when electrons are added to an atom, the charge on the atom is reduced.

3. **(D)** This process, cell respiration, is exergonic because energy is being released. It is an oxidation reaction because carbon loses hydrogen atoms and oxidation is defined as the loss of hydrogen atoms.

4. **(B)** A small amount of ATP is produced by substrate level phosphorylation during glycolysis and the Krebs cycle. However, most of the ATP produced during cell respiration occurs by chemiosmosis or oxidative phosphorylation during the electron transport chain.

5. **(B)** All organisms carry out respiration all the time.

6. **(A)** Lactic acid is produced from pyruvic acid during fermentation but does not result in the production of energy. All the other choices produce ATP. Oxidative phosphorylation is the process by which ATP is produced during the ETC.

7. **(C)** During sustained strenuous exercise, muscle cells use up ATP and oxygen as they carry out anaerobic respiration. Anaerobic respiration produces lactic acid, which causes weakness and fatigue in the muscle.

8. **(B)** Glycolysis occurs in the cytoplasm.

9. **(A)** The electron transport chain is located within in the cristae membrane of the mitochondria.

10. **(C)** The Krebs cycle occurs in the matrix of the mitochondria.

11. **(C)** The order of energy in producing processes in cell respiration is glycolysis, the Krebs cycle, then the electron transport chain.

12. **(B)** The Krebs cycle takes place in the matrix of the mitochondrion.

13. **(E)** The ATP synthase molecules lie within the cristae membrane.

14. **(C)** Each NADH can produce up to 3 ATP, and each FAD produces 2 ATP.

15. **(C)** Water is produced as oxygen combines with hydrogen flowing through the ATP synthase during chemiosmosis of aerobic respiration only. This does not occur in anaerobic respiration. CO_2 is produced in both the aerobic (Krebs cycle) and in anaerobic fermentation. ATP is produced in anaerobic and aerobic respiration. Alcohol is produced in only anaerobic respiration.

16. **(D)** Choices A, B, C, and E all describe events that occur during cell respiration and lead up to the production of ATP. However, the most important event in respiration is the production of ATP.

17. **(B)** Substrate level phosphorylation is responsible for the production of ATP during glycolysis and the Krebs cycle (also known as the citric acid cycle). Most ATP, however, is produced by chemiosmosis or oxidative phosphorylation.

18. **(A)** ATP, NADH, and pyruvate are the end products of glycolysis. CO_2 is a by-product of the Krebs cycle. H_2O is a waste product of chemiosmosis and is formed when protons combine with electrons and oxygen. Ethyl alcohol is a by-product of alcoholic fermentation.

19. **(C)** Glycolysis occurs in the first phase of cell respiration and does not require oxygen. The first organisms on earth were probably anaerobes because no free oxygen was available in the atmosphere. Choices A, B, D, and E all require free oxygen.

20. **(A)** Choices B–E are all correct statements about respiration, but they are not examples of a feedback mechanism. Choice A is an example of negative feedback because when plenty of ATP is available in the cell to meet demand, respiration slows down, conserving valuable molecules and energy for other functions.

21. **(E)** Oxidative phosphorylation occurs as protons flow through the ATP synthetase channel within the cristae membrane of the mitochondria. The Krebs cycle occurs within the matrix of the mitochondria. Glycolysis occurs in the cytoplasm of the cell. Substrate level phosphorylation explains how ATP is produced from the Krebs cycle and glycolysis.

22. **(D)** Most of the ATP produced during cell respiration comes from oxidative phosphorylation as protons flow through the ATP synthetase channel in the cristae membrane.

Free-Response Question

Membranes are important structural features of cells.

a. Describe the structure of a membrane.

b. Discuss the role of membranes in ATP synthesis.

Typical Free-Response Answer

Note: Key words are in bold to remind you that you must focus on scientific terminology.

a. The structure of the plasma membrane was elucidated by **S. J. Singer** in 1972, who described the membrane as a **fluid mosaic**, meaning it is made of small pieces that move. The plasma membrane consists of a **phospholipid bilayer** with the **hydrophilic head** of the phospholipid end facing outward and the **hydrophobic tail** facing inward. Protein molecules are dispersed throughout the membrane. Some proteins, **integral proteins**, completely span the membrane. Others, **peripheral proteins**, are loosely bound to the surface of the membrane. **Cholesterol molecules** are embedded in the interior to stabilize the membrane. The average membrane has the consistency of olive oil and is about 40 percent lipid and 60 percent protein. The membrane is **selectively permeable** and therefore controls what enters and leaves the cell. In general only small, uncharged, and hydrophobic molecules can diffuse freely through the membrane. Large, polar molecules cannot diffuse through the membrane; they must pass through special **protein channels**.

b. Cristae membranes in mitochondria play a special role in ATP synthesis. The cristae membrane contains **electron transport chains (ETC)**, collections of molecules embedded in the membrane. Most of these molecules are proteins that carry electrons from higher to lower energy levels. The ETC uses the exergonic flow of electrons to pump protons across the cristae membrane to the outer compartment of the mitochondria to create an **electrochemical** or **proton gradient**. The key here is that the cristae membrane does not allow protons to diffuse through the membrane. Protons can pass through only special protein channels, large enzyme complexes, called **ATP synthase channels**. As protons flow through the ATP synthase channels, like water flowing through a dam, energy is generated to produce ATP by a process known as **chemiosmosis**.

Photosynthesis

CHAPTER 6

- Photosynthetic pigments
- The chloroplast
- Photosystems
- Light-dependent reactions—
  The light reactions
- Calvin cycle
- Photorespiration
- C-4 photosynthesis
- CAM plants

INTRODUCTION

Photosynthesis is the process by which light energy is converted to chemical bond energy and carbon is fixed into organic compounds. The general formula is:

$$6CO_2 + 12H_2O \quad \xrightarrow{\text{light}} \quad C_6H_{12}O_6 + 6H_2O + 6O_2$$

There are two main processes of photosynthesis: the **light-dependent** or simply **light reactions** and the **light-independent reactions**. The light reactions use light energy directly to produce ATP that powers the light-independent reactions. The light-independent reactions consist of the Calvin cycle, which actually produces sugar. To power the production of sugar, the Calvin cycle uses the ATP formed during the light reactions. Both reactions occur only when light is present.

STUDY TIP

Photo—Light
Synthesis—Sugar
production: think
of the Calvin
cycle

PHOTOSYNTHETIC PIGMENTS

Photosynthetic pigments absorb light energy and use it to provide energy to carry out photosynthesis. Plants contain two major groups of pigments, the chlorophylls and carotenoids. **Chlorophyll *a*** and **chlorophyll *b*** are green and absorb all wavelengths of light in the red, blue, and violet range. The **carotenoids** are yellow, orange, and red. They absorb light in the blue, green, and violet range. **Xanthophyll**, another photosynthetic pigment, is a carotenoid with a slight chemical variation. Pigments found in red algae, the **phycobilins**, are reddish and absorb light in the blue and green range. Chlorophyll *b*, the carotenoids, and the phycobilins are known as **antenna pigments** because they capture light in wavelengths other than those captured by chlorophyll *a*. Antenna pigments absorb photons of light and pass the energy along to chlorophyll *a*, which is directly involved in the transformation of light energy to sugars. Figure 6.1 is a graph showing the absorption spectrum for the photosynthetic pigments.

Chlorophyll *a* is the pigment that participates directly in the light reactions of photosynthesis. It is a large molecule with a single **magnesium** atom in the head surrounded by alternating double and single bonds. The head, called the porphyrin ring, is attached to a long hydrocarbon tail. The double bonds play a critical role in the light reactions. They are the source of the electrons that flow through the electron transport chains during photosynthesis. Figure 6.2 is a drawing of chlorophyll *a*.

CAN YOU INTERPRET THIS GRAPH?

Chlorophyll *a* absorbs light in the violet, blue, and red ranges. It reflects green, yellow, and orange light.

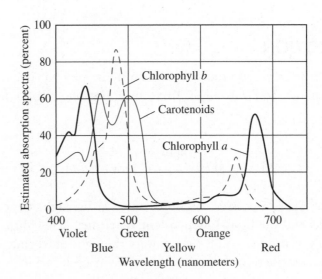

Figure 6.1

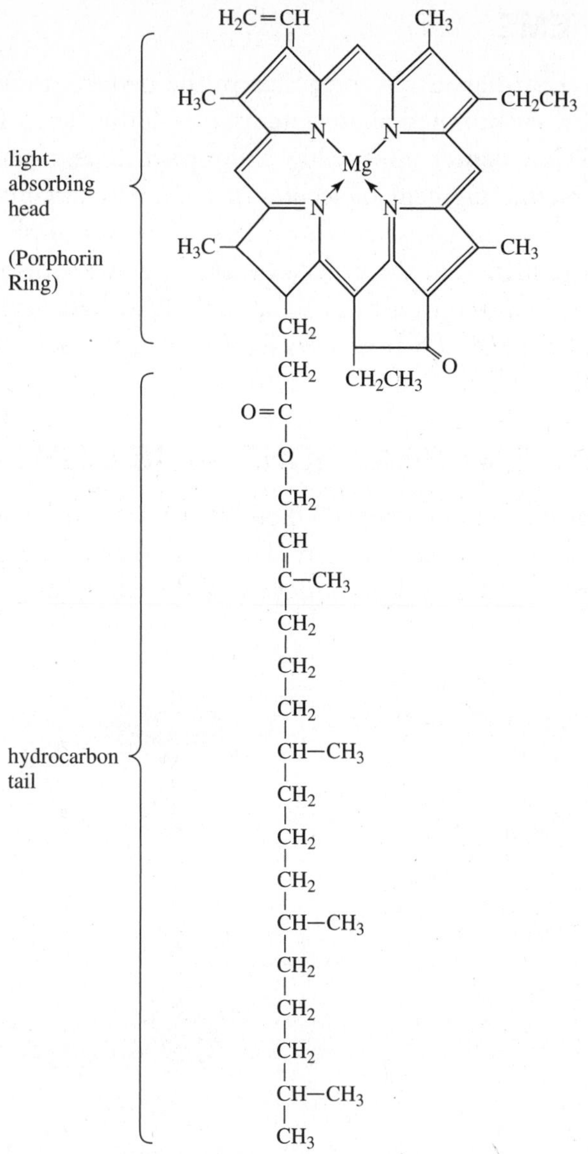

$H_2C=CH$ CH_3

H_3C ─   ─ CH_2CH_3

N   N

Mg

N   N

H_3C ─   ─ CH_3

CH_2 CH_2CH_3 O

light-
absorbing
head

(Porphorin
Ring)

CH_2

$O=C$

O

CH_2

CH

$C-CH_3$

CH_2

CH_2

CH_2

$CH-CH_3$

CH_2

CH_2

CH_2

$CH-CH_3$

CH_2

CH_2

CH_2

$CH-CH_3$

CH_3

hydrocarbon
tail

Figure 6.2   Chlorophyll *a*

THE CHLOROPLAST

The chloroplast contains **grana**, where the light reactions occur, and **stroma**, where the light-independent reactions occur. The grana consist of layers of membranes called **thylakoids**, the site of photosystems I and II. The chloroplast is enclosed by a double membrane. Figure 6.3 is a sketch of a chloroplast.

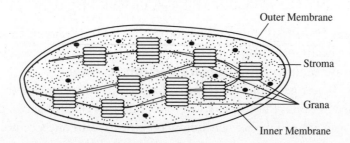

Outer Membrane

Stroma

Grana

Inner Membrane

Figure 6.3   Chloroplast

PHOTOSYSTEMS

Photosystems are light-harvesting complexes in the thylakoid membranes of chloroplasts. There are a few hundred photosystems in each thylakoid. Each photosystem consists of a **reaction center** containing chlorophyll *a* and a region containing several hundred antenna pigment molecules that funnel energy into chlorophyll *a*. Two types of photosystems cooperate in the light reactions of photosynthesis, **PS I** and **PS II**. They are named in the order in which they were discovered. However, PS II operates first, followed by PS I. PS I absorbs light best in the 700 nm range; hence it is also called **P700**. PS II absorbs light best in the 680 nm range; hence it is also called **P680**.

LIGHT-DEPENDENT REACTIONS—THE LIGHT REACTIONS

Light is absorbed by the photosystems (PS II and PS I) in the thylakoid membranes and electrons flow through electron transport chains. As shown in Figure 6.4 there are two possible routes for electron flow: **noncyclic** and **cyclic photophosphorylation**.

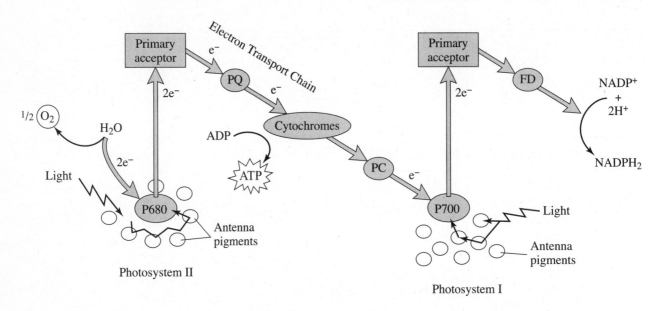

Figure 6.4   Noncyclic Photophosphorylation

Noncyclic Photophosphorylation

<table>
<tr><td>

REMEMBER

Electrons flow from water to P680 to P700 to NADP, which carries them to the Calvin cycle.

</td></tr>
</table>

During **noncyclic photophosphorylation**, electrons enter two electron transport chains, and *ATP and NADPH (nicotinamide dinucleotide phosphate) are formed.* The process begins in PS II and proceeds through the following steps.

- **Photosystem II—P680.** Energy is absorbed by P680. Electrons from the double bonds in the head of chlorophyll *a* become energized and move to a higher energy level. They are captured by a **primary electron acceptor**.
- **Photolysis.** Water gets split apart, providing electrons to replace those lost from chlorophyll *a* in P680. Photolysis splits water into two electrons, two protons (H^+), and one oxygen atom. Two oxygen atoms combine to form one O_2 molecule, which is released into the air as a waste product of photosynthesis.

- **Electron transport chain**. Electrons from P680 pass along an electron transport chain consisting of **plastoquinone (PQ)**, a complex of two cytochromes and several other proteins, and ultimately end up in P700 (PSI). This flow of electrons is exergonic and provides energy to produce ATP by **chemiosmosis**, just as it does in mitochondria. Because this ATP synthesis is powered by light, it is called **photophosphorylation**. See Figure 6.5.
- **Chemiosmosis**. This is the process by which ATP is formed during the light reactions of photosynthesis. Protons that were released from water during photolysis are pumped by the thylakoid membrane from the stroma into the **thylakoid space (lumen)**. ATP is formed as these protons diffuse down the gradient from the thylakoid space, through the **ATP-synthase channels**, and into the stroma. The ATP produced here provides the energy that powers the Calvin cycle. (This is similar to the way energy is produced in mitochondria.)
- **NADP** becomes reduced when it picks up the two protons that were released from water in P680. Newly formed **NADPH** carries hydrogen to the Calvin cycle to make sugar in the light-independent reactions.
- **Photosystem I—P700**. Energy is absorbed by P700. Electrons from the head of chlorophyll *a* become energized and are captured by a primary electron receptor. This process is similar to the way it happens in P680. One difference is that the electrons that escape from chlorophyll *a* are replaced with electrons from photosystem II, P680, instead of from water. Another difference is that this electron transport chain contains ferrodoxin and ends with the production of NADPH, not ATP.

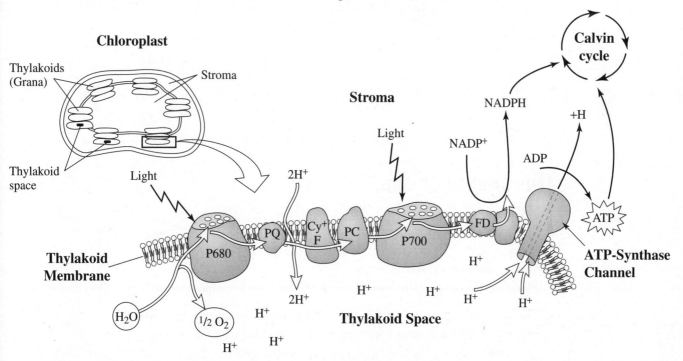

Figure 6.5 The light-dependent reactions

Here is the overview of noncyclic photophosphorylation.

light → P680 → ATP produced → P700 → NADPH produced
        oxygen   (NADPH carries
        released   H⁺ to the Calvin
                                     cycle)

Cyclic Photophosphorylation

The sole purpose of cyclic photophosphorylation is to produce ATP. No NADPH is produced, and no oxygen is released.

DETAILS

The production of sugar that occurs during the Calvin cycle consumes enormous amounts of ATP, so periodically, the chloroplast runs low on ATP. When it does, the chloroplast carries out **cyclic photophosphorylation** to replenish the ATP levels. Cyclic electron flow takes photoexcited electrons on a short-circuit pathway. Electrons travel from the P680 electron transport chain to P700, to a primary electron acceptor, and then back to the cytochrome complex in the P680 electron transport chain. Cyclic photophosphorylation is shown in Figure 6.6.

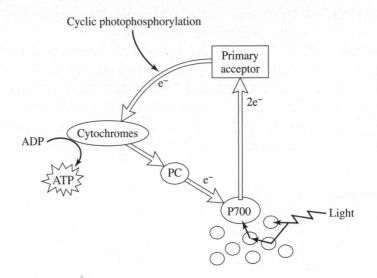

Figure 6.6   Cyclic Photophosphorylation

THE CALVIN CYCLE

The **Calvin cycle** is the main business of the light-independent reactions. It is a cyclical process that produces the 3-carbon sugar **PGAL (phosphoglyceraldehyde)**. Carbon enters the stomates of a leaf in the form of CO_2 and becomes *fixed* or incorporated into PGAL.

Here are the important aspects of the Calvin cycle, as shown in Figure 6.7.

- The process that occurs during the Calvin cycle is **carbon fixation**.
- It is a **reduction reaction** since carbon is gaining hydrogen.
- CO_2 enters the Calvin cycle and becomes attached to a 5-carbon sugar, **ribulose biphosphate (RuBP)**, forming a 6-carbon molecule. The 6-carbon molecule is unstable and immediately breaks down into two 3-carbon molecules of **3-phosphoglycerate (3-PGA)**. The enzyme that catalyzes this first step is ribulose biphosphate carboxylase **(rubisco)**.
- The Calvin cycle does not directly depend on light. Instead, it uses the products of the light reactions: ATP and NADPH.
- The Calvin cycle, like the light-dependent reactions, **occurs only in the light**.

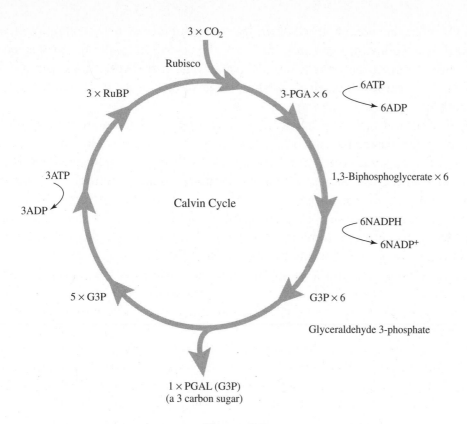

Figure 6.7

PHOTORESPIRATION

In most plants, CO_2 enters the Calvin cycle and is fixed into **3-phosphoglycerate (3-PGA)** by the enzyme rubisco. These plants are called **C-3 plants** because the first step produces the compound 3-PGA, which contains three carbons. This first step is not very efficient because rubisco binds with O_2 as well as with CO_2. When rubisco binds with O_2 instead of CO_2, this process, **photorespiration**, diverts the process of photosynthesis in two ways.

1. Unlike normal respiration, no ATP is produced
2. Unlike normal photosynthesis, no sugar is formed.

Instead, **peroxisomes** break down the products of photorespiration. If photorespiration does not produce any useful products, why do plants carry it out at all? This process is probably a vestige from the ancient earth when the atmosphere had little or no free oxygen to divert rubisco and sugar production.

C-4 PHOTOSYNTHESIS

C-4 photosynthesis is a modification for dry environments. C-4 plants exhibit modified anatomy and biochemical pathways that enable them to minimize excess water loss and maximize sugar production. As a result, C-4 plants thrive in hot and sunny environments where C-3 plants would wilt and die. Examples of C-4 plants are **corn**, **sugar cane**, and **crabgrass**. In C-4 plants, a series of steps precedes the Calvin cycle. The steps pump CO_2 that entered the leaf away from the air spaces near the stomates. The details are described:

- CO$_2$ enters the mesophyll cell of the leaf and combines with a 3-carbon molecule, **PEP** (**phosphoenolpyruvate**), to form the 4-carbon molecule **oxaloacetate**. Hence, the plants have the name C-4 plants. The enzyme that catalyzes this reaction, **PEP carboxylase**, does not bind with oxygen and can therefore fix CO$_2$ more efficiently than rubisco.

- From oxaloacetate the mesophyll cell produces **malic acid** or **malate**, which it pumps through the **plasmodesmata** into the adjacent **bundle sheath cell**. Once the malate is in the bundle sheath cell, it releases its CO$_2$, which gets incorporated into PGAL by the Calvin cycle. Because the bundle-sheath cell is deep within the leaf and little oxygen is present, rubisco can fix CO$_2$ efficiently without being diverted to the dead end of photorespiration. This biochemical pathway is called the **Hatch-Slack pathway**. Its purpose is to remove CO$_2$ from the air space near the stomate.

- With CO$_2$ sequestered inside bundle-sheath cells, there is a steep CO$_2$ gradient between the airspace in the mesophyll of the leaf near the stomates and the atmosphere around the leaf. Thus, C-4 plants can maximize the amount of CO$_2$ that diffuses into the air space in the leaf and minimize the length of time the stomates must remain open.

- **Kranz anatomy** refers to the structure of C-4 leaves that differs from C-3 leaves. In C-4 leaves, the bundle sheath cells lie under the mesophyll cells, deep within the leaf where CO$_2$ is sequestered. The light reactions occur in the mesophyll cells and the dark reactions occur in the bundle-sheath cells. In C-3 leaves, all photosynthetic cells have direct access to CO$_2$. Figure 6.8 is a sketch of both C-3 and C-4 leaves.

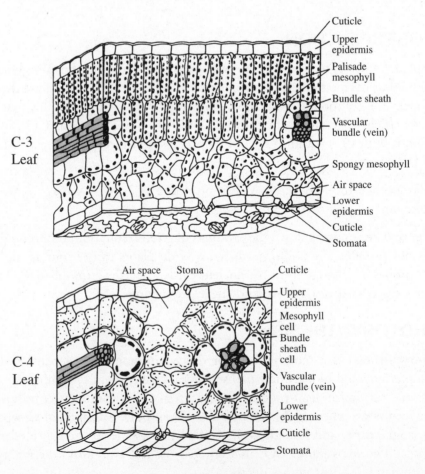

Figure 6.8

CAM PLANTS

CAM plants carry out a form of photosynthesis, **crassulacean acid metabolism**, which is another adaptation to dry conditions. These plants keep their stomates closed during the day and open at night; the reverse of how most plants behave. The mesophyll cells store CO_2 in organic compounds they synthesize at night. During the day, when the light reactions can supply energy for the Calvin cycle, CO_2 is released from the organic acids made the night before to become incorporated into sugar. Figure 6.9 shows three sketches comparing carbon fixation in C-3, C-4, and CAM plants.

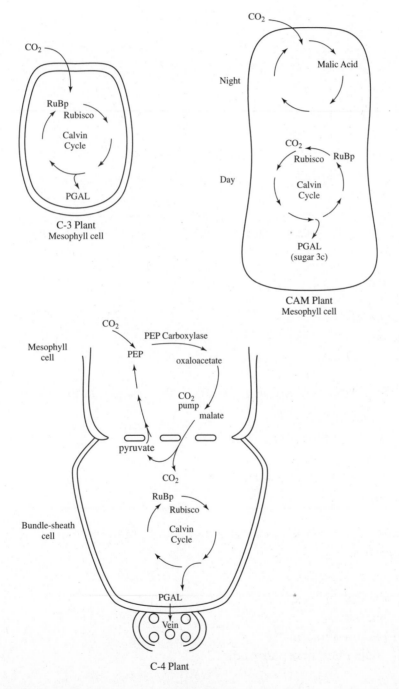

Figure 6.9

Multiple-Choice Questions

1. The graph below shows an absorption spectrum for an unknown pigment molecule. What color would this pigment appear?

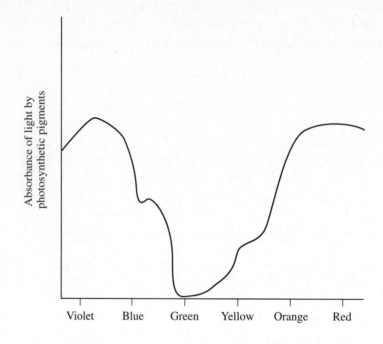

   (A) red
   (B) yellow
   (C) green
   (D) blue
   (E) violet

2. Cyclic photophosphorylation results in the production of

   (A) ATP only
   (B) ATP and NADPH
   (C) NADPH only
   (D) ATP, NADPH, and sugar
   (E) sugar only

3. Plants give off oxygen as a waste product of photosynthesis. This oxygen comes from

   (A) the Krebs cycle
   (B) the Calvin cycle
   (C) photolysis
   (D) photorespiration
   (E) cyclic photophosphorylation

4. How many turns of the Calvin cycle are required to produce one molecule of glucose?

   (A) 1
   (B) 2
   (C) 3
   (D) 6
   (E) 12

Questions 5–11
Indicate which of the following events occurs during

   (A) light-dependent reactions
   (B) light-independent reactions

5. Oxygen is released.

6. Carbon gets reduced.

7. Oxidative photophosphorylation

8. ATP is produced.

9. Electrons flow through an electron transport chain.

10. Oxidation of NADPH

11. Reduction of NADP$^+$

12. CAM plants keep their stomates closed during the daytime to reduce excess water loss. They can do this because they

   (A) can fix CO_2 into sugars in the mesophyll cells.
   (B) can use photosystems I and II at night
   (C) modify rubisco so it does not bind with oxygen
   (D) can incorporate CO_2 into organic acids at night
   (E) have lenticels instead of stomates

13. Which of the following is NOT directly associated with photosystem II?

   (A) P680
   (B) harvesting light energy by chlorophyll
   (C) release of oxygen
   (D) splitting of water
   (E) production of NADPH

14. This graph shows the rate of photosynthesis for two plants under experimental conditions.

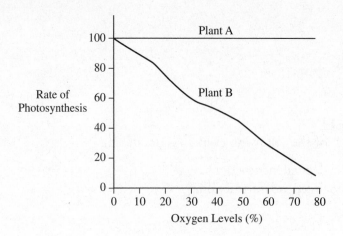

    From this graph, what is the best conclusion about the mechanism of photosynthesis in the two plants?

    (A) A is a C-3 plant, and B is a C-4 plant.
    (B) A is a C-4 plant, and B is a C-3 plant.
    (C) A is a CAM plant, and B is a C-4 plant.
    (D) A and B are both CAM plants.
    (E) There is not enough information to determine.

15. Where in the cell is ATP-synthase located?

    (A) in the plasma membrane
    (B) in the nuclear membrane
    (C) in the thylakoid membrane of the chloroplast
    (D) in the cristae membrane of mitochondria
    (E) both C and D

16. Which one of the following is NOT required for photosynthesis to occur?

    (A) CO_2
    (B) O_2
    (C) ATP
    (D) NADP
    (E) All of the above are required.

17. Which one is NOT correct about where the Calvin cycle occurs?

    (A) in the spongy mesophyll in C-3 plants
    (B) in the mesophyll in C-4 plants
    (C) in the palisade layer in C-3 plants
    (D) in the bundle sheath in C-4 plants
    (E) none is correct

Answers to Multiple-Choice Questions

1. (**C**) If light is absorbed, it is not reflected. Only reflected colors are seen. The graph shows that red and blue are most absorbed and that green is most reflected. Therefore the color of the pigment is green.

2. (**A**) Electrons undergoing cyclic phosphorylation move from P680 to P700 and then cycle back to P680. The sole purpose of cyclic photo-phosphorylation is the production of ATP. No NADPH is produced, and no oxygen is released. This process is necessary when the cell needs more ATP because ATP has been used up by the Calvin cycle. Sugar is produced during the light-independent reactions only, not during the light-dependent ones.

3. (**C**) Photolysis breaks apart water into oxygen, hydrogen ions, and electrons to provide electrons for the electron transport chain. The Krebs cycle is part of cell respiration, not photosynthesis. The Calvin cycle uses CO_2 from the atmosphere and hydrogen from photolysis to make sugar. Cyclic photophosphorylation does not cause photolysis. Photorespiration uses up oxygen; it does not produce oxygen.

4. (**D**) Since glucose is a 6-carbon molecule and each turn of the cycle absorbs 1 molecule of CO_2, 6 turns of the Calvin cycle are needed to produce glucose. Three turns of the Calvin cycle produce 1 PGAL (phospho-glyceraldehyde), a 3-carbon molecule.

5. (**A**) Oxygen is released from the photolysis of water during the light-dependent reactions.

6. (**B**) Carbon is reduced when it enters the Calvin cycle in the dark reactions and combines with hydrogen to yield PGAL.

7. (**A**) Oxidative photophosphorylation is a type of chemiosmosis and explains how ATP is formed during the light-dependent reactions. This is also the way in which energy is produced during aerobic respiration. A steep proton gradient provides the energy for the production of ATP.

8. (**A**) Energy is produced from the process of chemiosmosis, which occurs during the light-dependent reactions of photosynthesis.

9. (**A**) Electrons flow through the electron transport chain during the light reactions, and ATP and NADPH are produced.

10. (**B**) NADP gains a proton (is reduced) during the light-dependent reactions and carries the proton to the Calvin cycle of the light-independent reactions, where it loses the proton (is oxidized).

11. (**A**) $NADP^+$ (also written as NADP) is reduced in the light reactions when it gains hydrogen ions from water.

12. **(D)** During the night, CAM plants fix CO_2 into a variety of organic acids, like malic acid. These acids are stored in vacuoles until the daylight when the CO_2 is released into the Calvin cycle. The Calvin cycle occurs in the bundle-sheath cells, not the mesophyll cells. CAM plants use PEP carboxylase to fix CO_2 into malic acid initially. This is then pumped into the bundle-sheath cells, where CO_2 combines with normal, unmodified rubisco. Lenticels are openings in the stems of woody plants and do not have anything to do with photosynthesis.

13. **(E)** Photosystem II (PS II) is also known as P680. This photosystem absorbs light with an average wavelength of 680nm. When light is absorbed by P680, electrons from chlorophyll *a* become energized and move to a higher energy level and into an electron transport chain within the thylakoid membrane. Water splits apart during photolysis to provide electrons to replace those lost in chlorophyll *a*. Protons that were released from water during photolysis are pumped by the thylakoid membrane from the stroma into the thylakoid space. ATP is formed as protons flow down a steep gradient and through ATP-synthase channels.

14. **(B)** Plant A is not affected by the increase in the concentration of oxygen in the air because in C-4 plants, PEP carboxylase does not react with oxygen. Plant B is a C-3 plant because the increased oxygen levels cause the plant to undergo photorespiration and to carry out less photosynthesis.

15. **(E)** The ATP-synthase is the enzyme located within the membranes of mitochondria and chloroplasts. It produces ATP during the light reaction of photosynthesis and during cell respiration.

16. **(B)** All of the choices are required for photosynthesis to occur except oxygen, which is released as a by-product of photosynthesis.

17. **(B)** In C-3 plants, the Calvin cycle occurs in all photosynthetic cells in both the palisade and mesophyll layers. However, in C-4 plants, the light reactions occur in only the mesophyll cells while the Calvin cycle occurs in the bundle-sheath cells.

Free-Response Question

> **Directions:** Answer all questions. You must answer the question in essay—**not** outline—form. You may use labeled diagrams to supplement your essay, but diagrams alone are *not* sufficient. Before you start to write, read each question carefully so that you understand what the question is asking.

The rate of photosynthesis varies with different environmental conditions such as light intensity, wavelength of light, temperature, and so on.

a. Devise an experiment to demonstrate that ONE environmental condition will alter the rate of photosynthesis.
b. State your hypothesis, the procedure, how you would collect the data, the results you expect.
c. The scientific theory behind your expectations.

> **REMEMBER**
> 1. Make sure you answer the question asked.
> 2. State why you chose the organism.
> 3. The experimental and the control must be identical in all ways except the one you are testing.

Typical Free-Response Answer

This experiment will test the hypothesis that wavelengths of light will alter the rate of photosynthesis. The organism chosen for this experiment is elodea, which is inexpensive, readily available, and easy to take care of. Elodea merely requires an aerated, freshwater tank and sunlight.

Set up four 500 mL beakers of freshwater, each containing the same mass of elodea. Place the beakers into four large boxes that do not allow light to enter. This is required to maintain control over the wavelengths of light the elodea is exposed to. Place a light source in each box that is the same distance from each beaker. Include a heat sink between the beaker and the light source to absorb heat and to make sure that the temperature in all beakers remains the same. All light must be of the same intensity so as not to introduce another variable into the experiment. Set up each light source with appropriate light filters so that each beaker is exposed to only one wavelength of light: green, blue, red, or yellow.

The rate of photosynthesis will be measured by counting the number of bubbles of oxygen released in 30-second intervals for a period of 30 minutes. Repeat the experiment five times to verify the results.

I would expect the following results. The elodea exposed to blue light would produce the most bubbling. The elodea exposed to the red and yellow light would produce less bubbling in that order, and the elodea exposed to the green light would produce no bubbles.

The theory behind this experiment is that photosynthetic organisms release oxygen gas from photolysis during the light reactions of photosynthesis. The rate of photosynthesis can therefore be monitored by measuring the amount of oxygen released. The amount of oxygen released is measured by counting the bubbles released. The rate at which photosynthesis occurs is a function of the energy absorbed. The greater the amount of light absorbed, the greater the rate of photosynthesis. Light that is not absorbed is reflected and cannot be used as an energy source.

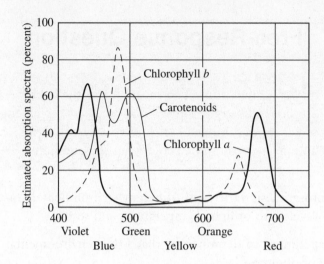

Photosynthetic pigments absorb light of different wavelengths. The figure above shows the absorption spectrum for the photosynthetic pigments chlorophyll *a*, chlorophyll *b*, and the carotenoids. Blue light is absorbed the most, red somewhat less, yellow even less, and green is reflected.

If you wanted the variable to be light intensity, only one thing would vary from the setup above. You must use lightbulbs rated with different lumens, lightbulbs of different intensity. However, the color (wavelength of light) must be the same for all the bulbs. You would not need the filters.

If you wanted temperature to be the variable, everything would be the same as the setup above except for two factors. The lightbulbs would all be identical, but you would control the temperature of each box that contained the light and plants.

Cell Division

CHAPTER 7

- The cell cycle
- Cell division and cancerous cells
- Meiosis
- Meiosis and genetic variation
- Details of the cell cycle

INTRODUCTION

Cell division functions in **growth**, **repair**, and **reproduction**. Two types of cell division occur, **mitosis** and **meiosis**. **Mitosis** produces two genetically identical daughter cells and conserves the chromosome number ($2n$). **Meiosis** occurs in sexually reproducing organisms and results in cells with half the chromosome number of the parent cell (n).

	BE CAREFUL
	Know the difference between mitosis and meiosis.

Any discussion about cell division must first consider the structure of the chromosome.

A chromosome consists of a highly coiled and condensed strand of DNA. A replicated chromosome consists of two **sister chromatids**, where one is an exact copy of the other. The **centromere** is a specialized region that holds the two chromatids together. The **kinetochore** is a disc-shaped protein on the centromere that attaches the chromatid to the **mitotic spindle** during cell division. See Figure 7.1 of a replicated chromosome

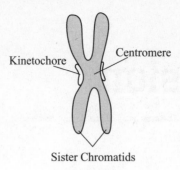

Figure 7.1

THE CELL CYCLE

Living and dividing cells pass through a regular sequence of growth and division called the **cell cycle**. The timing and rate of cell division are crucial to normal growth and development. Two important factors limit cell size and promote cell division, the **ratio of the volume of a cell to the surface area** and the **capacity of the nucleus to control the entire cell**.

Ratio of the Cell Volume to Surface Area

As a cell grows, the area of the cell membrane *increases as the square* of the radius, while the volume of the cell *increases as the cube* of the radius. Therefore, as a cell grows larger, the volume inside the cell increases at a faster rate than does the cell membrane. Since a cell depends on the cell membrane for exchange of nutrients and waste products, the ratio of cell volume to membrane size is a major determinant of when the cell divides.

Capacity of the Nucleus

The nucleus must be able to provide enough information to produce adequate quantities to meet the cell's needs. In general, metabolically active cells are small. However, cells that have evolved a strategy to exist as large, active cells exist in several kingdoms. Large sophisticated cells like the **paramecium** have two nuclei that each control different cell functions. **Human skeletal muscle cells** are giant **multinucleate** cells. The fungus slime mold actually consists of one giant cell containing thousands of nuclei.

Phases of the Cell Cycle

The cell cycle consists of five major phases: **G₁**, **S**, and **G₂** (which together make up **interphase**), **mitosis**, and **cytokinesis**.

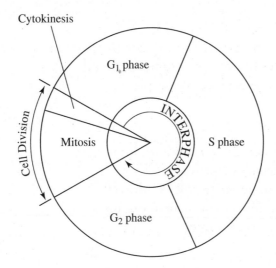

Figure 7.2   The cell cycle

INTERPHASE

Interphase consists of **G₁**, **S**, and **G₂**. The **G₁ phase** is a period of intense growth and biochemical activity. **S** stands for the synthesis or replication of DNA. **G₂** is the phase when the cell continues to grow and to complete preparations for cell division. More than 90 percent of the life of a cell is spent in interphase. When a cell is in interphase and not dividing, the chromatin is threadlike, not condensed. Within the nucleus are one or more **nucleoli**. A single **centrosome**, consisting of two **centrioles**, can be seen in the cytoplasm of an animal cell. Plant cells lack centrosomes but have **microtubule organizing centers**, **MTOC**s. See Figure 7.3, which shows an animal cell during interphase.

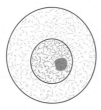

Figure 7.3   Interphase

MITOSIS

Mitosis consists of the actual dividing of the nucleus. It is a continuous process. However, scientists have divided it into four arbitrary divisions: **prophase**, **metaphase**, **anaphase**, and **telophase**. Here are the characteristics of each phase. Figure 7.4 shows each of these phases.

Prophase

- The nuclear membrane begins to disintegrate.
- The strands of chromosomes begin to condense into discrete observable structures, like that in Figure 7.1 on page 144.
- The nucleolus disappears.
- In the cytoplasm, the mitotic spindle begins to form, extending from one centrosome to the other.
- Prophase is the longest phase of mitosis.

Metaphase

- The *chromosomes line up in a single file* located on the equator or metaphase plate.
- Centrosomes are at opposite poles of the cell.
- Spindle fibers run from the **centrosomes** to the **kinetochores** in the **centromeres**.

Anaphase

- Centromeres of each chromosome separate, as spindle fibers pull apart the sister chromosomes.
- This is the shortest phase of mitosis.

Telophase

- Chromosomes cluster at opposite ends of the cell, and the nuclear membrane reforms.
- The supercoiled chromosomes begin to unravel and to return to their normal, pre-cell division condition as long, threadlike strands.
- Once two individual nucleoli form, mitosis is complete.

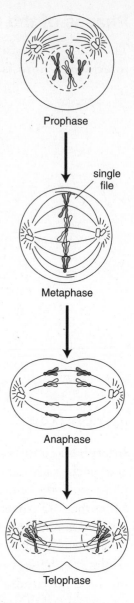

Figure 7.4   Mitosis

CYTOKINESIS

Cytokinesis consists of the dividing of the cytoplasm. It begins during mitosis, often during anaphase. In animal cells, a **cleavage furrow** forms down the middle of the cell as **actin** and **myosin microfilaments** pinch in the cytoplasm. In plant cells, a **cell plate** forms during telophase as vesicles from the Golgi coalesce down the middle of the cell. Daughter plant cells do not separate from each other. A sticky **middle lamella** cements adjacent cells together. See Figure 7.5.

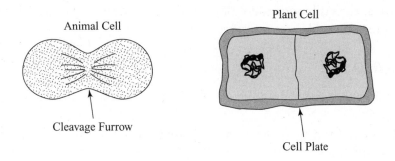

Figure 7.5   Cytokinesis

CELL DIVISION AND CANCEROUS CELLS

Normal cells grow and divide until they become too crowded; then they stop dividing and enter G_0 (G zero). This reaction to overcrowding is called **contact inhibition** or **density-dependent inhibition**. Another characteristic of normal animal cells is **anchorage dependence**. To divide, a cell must be attached or anchored to some surface, such as a Petri dish (in vitro) or an extracellular membrane (in vivo). Cancer cells show neither contact inhibition nor anchorage dependence. They divide uncontrollably and do not have to be anchored to any membrane. That is why cancer cells can migrate or metastasize to other regions of the body.

MEIOSIS

Meiosis is a form of cell division that produces **gametes** (sex cells, or **sperm** and **ova**) with the **haploid chromosome number** (**n**). There are two stages in meiosis. **Meiosis I** (reduction division) is the process by which **homologous chromosomes separate**. **Meiosis II** is like mitosis. In meiosis I, each chromosome pairs up precisely with its homologue into a **synaptonemal complex** by a process called **synapsis** and forms a structure known as a **tetrad** or **bivalent**. Synapsis is important for two reasons. First, it ensures that each daughter cell will receive one homologue from each parent. Second, it makes possible the process of **crossing-over** by which homologous chromatids exchange genetic material. *Crossing-over is a common and highly organized mechanism to ensure greater variation among the gametes.* (See the chapter "Heredity.") In meiosis II, **sister chromatids separate**. The two stages of meiosis are further divided into phases. At the beginning of meiosis, cells have the **diploid chromosome number** (**2n**). By the end of meiosis, cells contain the **haploid chromosome number** (**n**). Each meiotic cell division consists of the same four stages as mitosis: prophase, metaphase, anaphase, and telophase.

Meiosis I

PROPHASE I

- **Synapsis**, the pairing of homologues, occurs.
- **Crossing-over**, the exchange of homologous bits of chromosomes, occurs.
- **Chiasmata**, the visible manifestations of the cross-over events, are visible.
- This is the longest phase.

METAPHASE I

- The homologous pairs of chromosomes are lined up **double file** along the metaphase plate.
- **Spindle fibers** from the poles of the cell are attached to the centromeres of each pair of homologues.

ANAPHASE I

- Homologous chromosomes are separated as they are pulled by spindle fibers and migrate to opposite poles.

TELOPHASE I

- Homologous pairs continue to separate until they reach the poles of the cell. Each pole has the haploid number of chromosomes.

CYTOKINESIS I

- Cytokinesis usually occurs simultaneously with telophase I.

In some species, an interphase occurs between meiosis I and meiosis II. In other species, none occurs. In either case, chromosomes do not replicate between meiosis I and II because chromosomes already exist as double or replicated chromosomes.

Meiosis II

Meiosis II is functionally the same as mitosis and consists of the same phases: prophase, metaphase, anaphase, telophase, and cytokinesis. The chromosome number remains haploid, and daughter cells are genetically identical to the parent cell.

Figure 7.6 is a sketch comparing meiosis and mitosis. Each parent cell contains four chromosomes. The cell undergoing meiosis experiences one crossover.

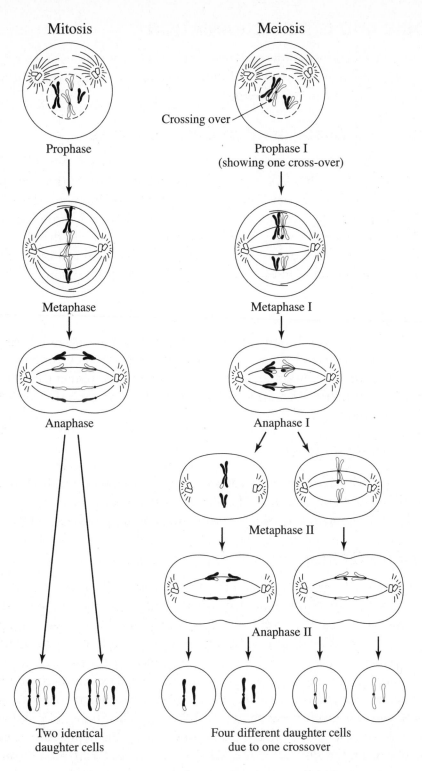

Figure 7.6   Comparison of Mitosis and Meiosis

MEIOSIS AND GENETIC VARIATION

Three types of genetic variation result from the processes of meiosis and fertilization. They are **independent assortment of chromosomes**, **crossing-over**, and **random fertilization of an ovum by a sperm**.

Independent Assortment of Chromosomes

During meiosis, homologous pairs of chromosomes separate depending on the random way in which they line up on the **metaphase plate** during metaphase I. Each pair of chromosomes can line up in two possible orientations. There is a 50 percent chance that a particular gamete will receive a maternal chromosome and a 50 percent chance it will receive a paternal chromosome. Given that there are 23 pairs of chromosomes in humans, the number of possible combinations of maternal and paternal chromosomes in each gamete is 2^{23}, or about 8 million.

Crossover

Crossover produces **recombinant chromosomes** that combine genes inherited from both parents. For humans, an average of two or three crossover events occur in each chromosome pair. In addition, at metaphase II, these recombinant chromosomes line up on the metaphase plate in random fashion. This increases the possible types of gametes even more.

Random Fertilization

One human ovum represents one of approximately 8 million possible chromosome combinations. The same is true for the human sperm. Thus, when one sperm fertilizes one ovum, 8 million × 8 million recombinations are possible.

THE CELL CYCLE

A **cell cycle control system** regulates the rate at which cells divide. Several **checkpoints** act as built-in stop signals that halt the cell unless they are overridden by go-ahead signals. Three checkpoints exist in G_1, G_2, and **M**. The G_1 checkpoint is known as the **restriction point** and is the most important one in mammals. If it receives a go-ahead, the cell will most likely complete cell division. On the other hand, if it does not get the appropriate signal, the cell will exit the cycle and become a nondividing cell arrested in the G_0 **(G zero) phase**. Since the activity of a cell varies, the rate at which it needs to divide also varies. This timing of cell division is controlled by two kinds of molecules: **cyclins** and **cyclin-dependent kinases** or **CDKs**. The first CDK discovered was **MPF**, which stands for **M-phase promoting factor**. In humans, the frequency of cell division varies with the cell type. Bone marrow cells are always dividing in order to produce a constant supply of red and white blood cells. Cells in the human intestine normally divide twice per day to renew tissue that is destroyed during digestion. Nerve and muscle cells are arrested in G_0 and do not divide or regenerate at all, hence the great danger from spinal cord injury. Liver cells are also arrested in G_0 but can be induced to divide when liver tissue is damaged.

Other Details

The processes of **spermatogenesis** and **oogenesis** are discussed in the chapter "Animal Reproduction and Development." For information about **meiosis** and **alternation of generations** in plants, refer to the chapter "Plants."

Multiple-Choice Questions

1. Which of the following does NOT occur by mitosis?

   (A) growth
   (B) production of gametes
   (C) repair
   (D) development in the embryo
   (E) cleavage

The following two questions refer to the sketch below of a cell containing chromosomes.

2. How many chromosomes are in this cell?

   (A) 2
   (B) 4
   (C) 8
   (D) 16
   (E) 32

3. How many chromatids are in this cell?

   (A) 2
   (B) 4
   (C) 8
   (D) 16
   (E) 32

4. Which is a factor that limits cell size?

    (A) how active a cell is
    (B) what kind of activity a cell is engaged in
    (C) the ratio of volume to cell surface area
    (D) whether the cell is a plant cell or animal cell
    (E) whether the cell is a prokaryote or eukaryote

5. In which stage of the life of a cell is the nucleolus always visible?

    (A) prophase
    (B) anaphase
    (C) telophase
    (D) cytokinesis
    (E) interphase

6. Which of the following cells is a giant multinucleate cell?

    (A) neuron
    (B) liver
    (C) tracheids and vessels in plants
    (D) companion cell in plants
    (E) skeletal muscle

7. Which of the following is NOT found in plant cells?

    (A) cell plate
    (B) actin and myosin filaments
    (C) microtubule organizing center
    (D) cleavage furrow
    (E) middle lamella

8. If a cell has 24 chromosomes at the beginning of meiosis, how many chromosomes will it have at the end of meiosis?

    (A) 6
    (B) 12
    (C) 24
    (D) 48
    (E) The number varies with the species.

9. If a cell has 24 chromosomes, how many will it have at the end of mitosis?

    (A) 6
    (B) 12
    (C) 24
    (D) 48
    (E) The number varies with the species.

10. All of the following are true of meiosis EXCEPT

    (A) crossover occurs during prophase I
    (B) there is no replication of chromosomes between meiosis I and meiosis II
    (C) in plants, spindle fibers are attached to the centriole
    (D) synapsis occurs during prophase I
    (E) the longest phase is prophase

Questions 11–13
Match the event of meiosis with the stages shown below.

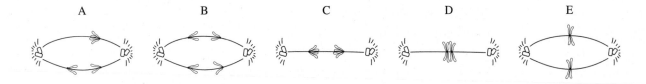

11. Identify metaphase I of meiosis.

    (A) A
    (B) B
    (C) C
    (D) D
    (E) E

12. Identify metaphase II of meiosis.

    (A) A
    (B) B
    (C) C
    (D) D
    (E) E

13. Identify anaphase II.

    (A) A
    (B) B
    (C) C
    (D) D
    (E) E

14. The synaptonemal complex forms during

    (A) anaphase I of meiosis
    (B) interphase of any cell
    (C) anaphase II of meiosis
    (D) telophase I
    (E) prophase I of meiosis

15. Homologous chromosomes separate during

    (A) prophase I
    (B) prophase II
    (C) anaphase of mitosis
    (D) anaphase I
    (E) anaphase II

16. Chiasmata are most closely related to which of the following?

    (A) mitotic cell division
    (B) fertilization
    (C) cytokinesis
    (D) crossing-over
    (E) nondisjunction

17. Which is NOT a source of genetic variation?

    (A) independent assortment of chromosomes
    (B) crossover
    (C) random fertilization
    (D) mitosis
    (E) recombinant chromosomes

18. Which of the following cells are permanently arrested in the G_0 phase?

    (A) bone marrow
    (B) liver cells
    (C) cancer cells
    (D) nerve cells
    (E) skin cells

19. Which is TRUE of the cell cycle?

    (A) The timing of cell division is controlled by cyclins and CDKs.
    (B) A characteristic of cancer cells is density-dependent inhibition.
    (C) The cell cycle is controlled solely by signals external to the cell.
    (D) The cell cycle is controlled soley by internal signals.
    (E) The cell cycle is controlled soley by nerve impulses.

20. A cell that passes the restriction point will most likely

    (A) stop dividing
    (B) divide
    (C) show density-dependent inhibition
    (D) die
    (E) become arrested in the G_0 phase

Answers to Multiple-Choice Questions

1. (**B**) The production of gametes, sperm, and eggs occurs by meiosis, where the chromosome number gets cut in half.

2. (**B**) There are 4 replicated chromosomes in this cell but 8 chromatids. The number of chromosomes is determined by the number of centromeres, 1 per chromosome.

3. (**C**) See #2.

4. (**C**) Two important factors limit cell size and promote cell division: ratio of the volume of a cell to the surface area and capacity of the nucleus.

5. (**E**) Most of the life of the cell is spent in interphase when the chromosomes are threadlike and not visible under a light microscope. When the cell divides, chromosomes must be condensed. When supercoiled or condensed, chromosomes appear like the Xs and Ys we commonly see them as. The nucleolus is not a real structure but threadlike chromosomes organized in a way that form a sphere.

6. (**E**) The skeletal muscle cell is a giant cell with many nuclei. At some time in its evolutionary history, it apparently underwent mitosis many times (which accounts for the many nuclei) without undergoing cytokinesis (which accounts for its large size).

7. (**D**) A cleavage furrow is a shallow groove in the cell surface in animal cells where cytokinesis is taking place. The cell plate is seen in dividing plant cells. The middle lamella is a layer between two adjacent plant cells.

8. (**B**) Meiosis cuts the chromosome number in half, from $2n$ to n. This occurs so that after fertilization, when two gametes fuse, the embryo will have the correct chromosome number, $2n$.

9. (**C**) The cells that result from mitotic cell division have the same number of chromosomes as the parent cell.

10. (**C**) Plants do not have centrioles, they have only microtubule organizing centers.

11. (**D**) During metaphase I of meiosis, homologous pairs line up on the metaphase plate in double file.

12. (**E**) During metaphase II of meiosis, single chromosomes line up on the metaphase plate in preparation for division. During meiosis II, sister chromatids separate.

13. (**B**) During anaphase II, sister chromatids are separating.

14. **(E)** The synaptonemal complex forms during prophase I and holds the two replicated chromosomes tightly together as a bivalent or tetrad so that crossing-over can occur without error.

15. **(D)** Homologous chromosomes separate during anaphase I of meiosis.

16. **(D)** Chiasmata are the microscopically visible regions of homologous chromatids where crossing-over has occurred.

17. **(D)** The daughter cells resulting from mitotic cell division are genetically identical to each other and to the mother cell. Sources of variation are independent assortment of chromosomes, crossover, random fertilization, and recombinant chromosomes.

18. **(D)** A cell arrested in the G_0 phase is not dividing. Most human body cells are not actively dividing. Highly specialized cells such as nerve and muscle cells never divide. Liver cells can be induced to divide when necessary, and human skin and bone marrow cells are always dividing. Also, cancer cells are always rapidly dividing.

19. **(A)** The timing of the cell cycle responds to external and internal cues and to fluctuations in levels of cyclins and cyclin-dependent kinases (CDKs). Normal cells stop dividing when crowded. This phenomenon is called contact inhibition or density-dependent inhibition. Cancer cells are characterized by uncontrolled growth.

20. **(B)** In mammalian cells, the G_1 checkpoint is known as the restriction point. If the cell receives the go-ahead signal at the G_1 checkpoint, it will usually complete the cycle and divide. In contrast, if the cell is not stimulated to pass the restriction point, it will switch into a nondividing mode known as G_0.

Free-Response Question

Directions: Answer all questions. You must answer the question in essay—**not** outline—form. You may use labeled diagrams to supplement your essay, but diagrams alone are *not* sufficient. Before you start to write, read each question carefully so that you understand what the question is asking.

An organism is heterozygous at two gene loci on different chromosomes.

a. Explain how these alleles are transmitted by the process of mitosis to daughter cells.

b. Explain how these alleles are distributed to gametes by the process of meiosis.

c. Explain how the behavior of these two pairs of chromosomes during meiosis provides the physical basis for two of Mendel's laws of heredity.

Typical Free-Response Answer

Author's note: These questions are about Mendelian inheritance and can be used as essays for that topic as well as this one. In this essay, the key words are in bold to show how many possible terms there are to discuss. You might consider underlining as you write to keep track of your key words.

a. Mitotic cell division produces daughter cells that are **genetically identical** to the parent cell. Mitosis is the division of the nucleus and consists of the following stages: **prophase**, **metaphase**, **anaphase**, and **telophase**. In preparation for mitosis, the **DNA replicates** itself. **Cytokinesis**, the actual division of the cytoplasm, almost always follows mitosis but is not part of mitosis. During prophase, the nuclear membrane begins to break apart and the **nucleolus** disappears as the chromosomes condense and become visible under a light microscope. In metaphase, chromosomes line up **single file** on the metaphase plate. **Sister chromatids** begin to separate during anaphase. They are pulled apart by **spindle fibers** connected at one end to the centrioles and at the other end to the **kinetochore** within the **centromere** of the chromosome. In telophase, the chromosomes reform into a circle as the nuclear membrane begins to reform. The two daughter cells that will result from this mitotic cell division each contain the same chromosomes, *T, t, Y,* and *y.*

b. Meiosis is a form of cell division that produces **gametes** (sex cells) with the **haploid chromosome number (*n*)**. Two stages occur in meiosis: **meiosis I (reduction division)** in which homologous chromosomes separate, and **meiosis II**, which is like mitosis, in which sister chromatids separate. The two stages of meiosis are further divided into phases. Each meiotic cell division consists of the same four stages as mitosis: prophase, metaphase, anaphase, and telophase.

In meiosis I, the homologous pairs line up double file on the metaphase plate. One homologue comes from the mother, and one comes from the father. During anaphase I, the homologues are pulled apart by spindle fibers and migrate to the poles; one homologue goes to each daughter cell. How the homologues separate or **segregate** is determined by how they line up on the metaphase plate; and how they line up on the metaphase plate is random.

If the four chromosomes line up like this:

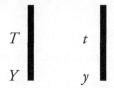

the daughter cells will contain the alleles *TY* and *ty*.

If the four chromosomes line up like this:

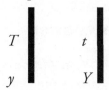

the daughter cells will contain the alleles: *Ty* and *tY*.

Meiosis II is like mitosis. After two meiotic cell divisions, there are **four haploid gametes**. Unlike cells that are formed by mitosis, these gametes are very different from each other.

c. Mendel's **law of segregation** states that during gamete formation, the two alleles for each trait are separated or segregated. This separation is random. In the example given, *T* separates from *t* and *Y* separates from *y* during anaphase of meiosis I.

Mendel's **law of independent assortment** states that during gamete formation, the alleles of a gene for one trait, such as *T* or *t*, segregate independently from the alleles of another gene, such as *Y* or *y*. This is shown in section B above. How they assort or segregate is random. Additionally, the law of independent assortment applies only if the two genes are on separate chromosomes. **Linked genes** will not assort independently.

Heredity

CHAPTER 8

- Basics of probability
- Multiplication and addition
- Law of dominance
- Law of segregation
- Monohybrid cross
- Backcross or testcross
- Law of independent assortment
- Incomplete dominance
- Codominance
- Multiple alleles
- Gene interactions
- Expressivity, penetrance, and the environment
- Sex-influenced inheritance
- Linked genes
- Sex-linkage
- Crossover and gene mapping
- The pedigree
- X Inactivation—The Barr body
- Mutations
- Nondisjunction
- Genomic imprinting
- Extranuclear inheritance

INTRODUCTION

The father of modern genetics is **Gregor Mendel**, an Austrian monk who, in the 1850s, bred garden peas in order to study patterns of inheritance. Mendel was successful because he brought an experimental and quantitative approach to the study of inheritance. First, he studied traits that were clear-cut, with no intermediates between varieties. Second, he collected data from a large sample, hundreds of plants from each of several generations. Altogether, Mendel collected ten thousand plants. Third, he applied statistical analysis to his carefully collected data.

Until the nineteenth century, people thought that inheritance was blended, a mixture of fluids that passed from parents to children. In contrast, Mendel's theory of genetics is one of **particulate inheritance** in which inherited characteristics are carried by discrete units that he called *elementes*. These *elementes* eventually became known as genes.

BASICS OF PROBABILITY

Probability is the likelihood that a particular event will happen. If an event is an absolute certainty, its probability is 1. If the event cannot happen, its probability is 0. The probability of anything else happening is between 0 and 1. Probability cannot predict whether a particular event will actually occur. However, if the sample is large enough, probability can predict an average outcome.

Understanding probability is important to the study of genetics because predicting outcomes is what Punnett squares enable us to do. What is the chance that two brown-eyed people can give birth to a child with blue eyes? That is probability, and that is what this chapter is all about.

When Do You Multiply? Multiplication Rule

To find the probability of **two independent events** happening, **multiply** the chance of one happening by the chance that the other will happen. For example, the chance of a couple having two boys depends on two independent events. The chance of the first child being a boy is $\frac{1}{2}$; and the chance of the next child being a boy is $\frac{1}{2}$. The chance that the couple will have two boys is therefore $\frac{1}{2} \times \frac{1}{2} = \frac{1}{4}$. The chance of having three boys is $\frac{1}{2} \times \frac{1}{2} \times \frac{1}{2} = \frac{1}{8}$.

When Do You Add? Addition Rule

When more than one arrangement of events producing the specified outcome is possible, the probabilities for each outcome are added together. For example, if a couple is planning on having two children, what is the chance that they will have one boy and one girl (in either order)? Here is how you solve this problem. The probability of having a boy then a girl is $\frac{1}{2} \times \frac{1}{2} = \frac{1}{4}$. The probability of having a girl and then a boy is $\frac{1}{2} \times \frac{1}{2} = \frac{1}{4}$. Therefore, the probability of having one boy and one girl is $\frac{1}{4} + \frac{1}{4} = \frac{1}{2}$.

LAW OF DOMINANCE

Mendel's first law is the **law of dominance**, which states that when two organisms, each **homozygous** (pure) for two opposing traits are crossed, the offspring will be **hybrid** (carry two different alleles) but will exhibit only the **dominant trait**. The trait that remains hidden is the **recessive trait**.

	T	T
t	Tt	Tt
t	Tt	Tt

Parent (P): TT × tt
                       Pure tall   Pure dwarf

Offspring (F_1): Tt
                               All hybrid tall

Law of dominance
All offspring are tall

LAW OF SEGREGATION

The **law of segregation** states that during the formation of gametes, the two traits carried by each parent separate. See Figure 8.1.

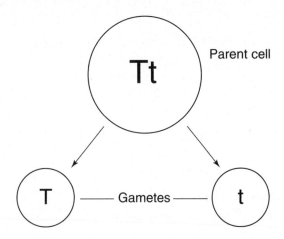

Figure 8.1   Law of Segregation

The cross that best exemplifies this law is the **monohybrid cross**, *Tt* × *Tt*. In the monohybrid cross, a trait that was not evident in either parent appears in the F₁ generation.

MONOHYBRID CROSS

The **monohybrid cross** (*Tt* × *Tt*) is a cross between two organisms that are each hybrid for one trait. The **phenotype** (appearance) ratio from this cross is 3 tall to 1 dwarf plant. The **genotype** (type of genes) **ratio**, 1 to 2 to 1, given as percentages is 25 percent homozygous dominant to 50 percent heterozygous to 25 percent homozygous recessive. These results are always the same for any monohybrid cross.

	T	*t*
T	*TT*	*Tt*
t	*Tt*	*tt*

F₁: *Tt* × *Tt*

F₂: *TT, Tt,* or *tt*

Monohybrid cross

BACKCROSS OR TESTCROSS

The **testcross** or **backcross** is a way to determine the genotype of an individual plant or animal showing the dominant trait. The individual in question (*B*/___) is crossed with a homozygous recessive individual (*b/b*). If the individual being tested is in fact homozygous dominant, all offspring of the testcross will be *B/b* and will show the dominant trait. There can be no offspring showing the recessive trait. If the individual being tested is hybrid (*B/b*), one-half of the offspring can be expected to show the recessive trait. Therefore, if any offspring show the recessive trait, the parent of unknown genotype must be hybrid.

	B	B
b	Bb	Bb
b	Bb	Bb

B = black

b = white

If the parent of unknown genotype is *BB,*
there can be no white offspring.

	B	b
b	Bb	bb
b	Bb	bb

B = black

b = white

If the parent of unknown genotype is hybrid,
there is a 50% chance that any offspring will be white.

LAW OF INDEPENDENT ASSORTMENT

The **law of independent assortment** applies when a cross is carried out between
two individuals hybrid for two or more traits that are **not on the same chromosome**. This cross is called the dihybrid cross. This law states that during gamete
formation, the alleles of a gene for one trait, such as height (*Tt*), segregate independently from the alleles of a gene for another trait such as seed color (*Yy*). Figure 8.2
represents a hybrid individual (*TtYy*) where the traits will assort independently. In
it, *T* = tall, *t* = short, *Y* = yellow seed, and *y* = green seed.

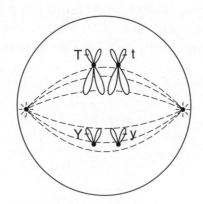

Figure 8.2

The genes for height and seed color are not on the same chromosome and will
assort independently. The only factor that determines how these alleles segregate or
assort is how the homologous pairs line up in metaphase of meiosis I, which is a
random event.

During metaphase I if the homologous pairs
happen to line up like this:

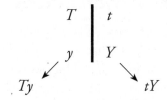

they will produce these gametes:

TY ty

If the homologous pairs happen to line up like this:

$$T \mid t$$
$$y \mid Y$$

they will produce these gametes:

Ty tY

    In contrast, if the gene for tall is **linked** to the gene for yellow seed color and the
gene for short is linked to the gene for green seed color, the genes will **not assort
independently**. If a plant is tall, it will have yellow seeds. If a plant is short, it will
have green seeds. See Figure 8.3.

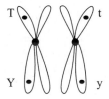

Figure 8.3   Linked Genes

    The following describes a cross that adheres to the law of independent assortment.
Two flowers are crossed that are homozygous for diffrent traits and have the genes
for height and seed color on different homologous chromosomes.

P: *T/T Y/Y*   ×   *t/t y/y*
                Homozygous tall plant   Homozygous short plant
                 with yellow seeds   with green seeds

Gametes: *TY*   *ty*

F₁: *T/t Y/y*
                 Homozygous tall plant
                 with yellow seeds

The *T/T Y/Y* parent can produce gametes carrying only *TY* genes. The *t/t y/y* parent can produce gametes carrying only *ty* genes. A Punnett square is not needed because only one outcome is possible in the F₁: *T/t Y/y*. The phenotype of all members of the F₁ generation is tall plants with yellow seeds. Their genotype is known as **dihybrid**.

The Dihybrid Cross

A cross between two F₁ plants is called a **dihybrid cross** because it is a cross between individuals that are hybrid for two different traits, such as height and seed color. This cross can produce four different types of gametes: *TY, Ty, tY,* and *ty*. The figure below shows how to set up the Punnett square for this cross. There are 16 squares; therefore all the ratios are out of 16.

Dihybrid cross	*TY*	*Ty*	*tY*	*ty*
TY	*TTYY*	*TTYy*	*TtYY*	*TtYy*
Ty	*TTYy*	*TTyy*	*TtYy*	*Ttyy*
tY	*TtYY*	*TtYy*	*ttYY*	*ttYy*
ty	*TtYy*	*Ttyy*	*ttYy*	*ttyy*

Obviously, many different genotypes are possible in the resulting F₂ generation, but you need not pay attention to them. Just pay attention to, the **phenotype ratio** of the dihybrid cross. It is **9:3:3:1**—9 tall, yellow; 3 tall, green; 3 short, yellow; and 1 short, green. To show this as a probability:

F₂: 9/16 tall, yellow   3/16 tall, green   3/16 short, yellow   1/16 short, green

INCOMPLETE DOMINANCE

Incomplete dominance is characterized by **blending**. Here are two examples. A long watermelon (*LL*) crossed with a round watermelon (*RR*) produces all oval watermelon (*RL*). A black animal (*BB*) crossed with a white (*WW*) animal produces all grey (*BW*) animals. Since neither trait is dominant, the convention for writing the genes uses different capital letters.

A red Japanese four o'clock flower (*RR*) crossed with a white Japanese four o'clock flower (*WW*) produces all pink offspring (*RW*).

	R	R
W	RW	RW
W	RW	RW

If two pink four o'clocks are crossed, there is a 25 percent chance that the offspring will be red, a 25 percent chance the offspring will be white, and a 50 percent chance the offspring will be pink.

	R	W
R	RR	RW
W	RW	WW

CODOMINANCE

In **codominance**, *both traits show*. A good example is the *MN* blood groups in humans. (These are not related to *ABO* blood groups). There are three different blood groups: *M, N,* and *MN*. These groups are based on two distinct molecules located on the surface of the red blood cells. There is a **single gene locus** at which **two allelic variants** are possible. A person can be homozygous for one type of molecule (*mm*), be homozygous for the other molecule (*NN*), or be hybrid and have both molecules (*MN*) on their red blood cells. The *MN* genotype is not intermediate between *M* and *N* phenotypes. Both *M* and *N* traits are expressed because both molecules are present on the surface of the red blood cells; see Figure 8.4.

MM   NN   MN

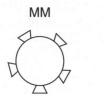

Figure 8.4

MULTIPLE ALLELES

Most genes in a population exist in only two allelic forms. For example, pea plants can be either tall (*T*) or short (*t*). When *there are more than two allelic forms of a gene*, that is referred to as **multiple alleles**. In humans there are four different blood types: **A**, **B**, **AB**, and **O** determined by the presence of specific molecules on the surface of the red blood cells. There are three alleles, *A*, *B*, and *O*, which determine the four different blood types. A and B are codominant and are often written as I^A and I^B. (*I* stands for immunoglobin.) When both alleles are present, they both manifest themselves, and the person has *AB* blood type. In addition, O is a recessive trait and is often written as i. A person can have any one of the six blood genotypes shown in Table 8.1.

TABLE 8.1

Human Blood Types and Genotypes	
Blood Type	**Genotype**
A	Homozygous *A*: *AA*
A	Hybrid *A*: *Ai*
B	Homozygous *B*: *BB*
B	Hybrid *B*: *Bi*
AB	*AB*
O	*ii*

GENE INTERACTIONS

Pleiotropy

Pleiotropy is the ability of *one single gene to affect an organism in several or many ways*. In chickens, the **frizzle trait** is a gene for a malformed feather. This mutation causes the development of feathers that cannot keep the animal warm, resulting in changes in several organ systems. Another example of pleiotropy can be seen in Siamese cats. The allele responsible for the coloration pattern (light body, dark extremities) is the same allele responsible for the fact that many Siamese cats are cross-eyed. An example of pleiotropy in humans is **Marfan syndrome**, in which a single defective gene results in abnormalities of the eyes, the skeleton, and the great blood vessels.

Epistasis

In **epistasis**, two separate genes control one trait, but *one gene masks the expression of the other gene*. The gene that masks the expression of the other gene is **epistatic** to the gene it masks. The best way to explain this gene interaction is by example. In guinea pigs, *a gene for production of melanin is epistatic to one for the deposition of melanin*. The gene related to production of melanin has two alleles, *C*, which causes pigment to be produced, and *c*, which does not. Therefore, *c/c* results in no melanin being produced, resulting in an albino animal. The second gene has an allele, *B*, which causes deposition of melanin, and an allele, *b*, which causes deposition of only a moderate amount of melanin, producing a brown coat. However, neither *B* nor *b* can deposit melanin if *C* is not present to make the melanin in the first place.

P:	*C/C B/B*	×	*c/c b/b*
	Black		Albino

F₁:	*C/c B/b*	×	*C/c B/b*
	Black		Black

F₂:	9 *C/_ B/_*	3 *C/_ b/b*	3 *c/c B/_*	1 *c/c b/b*
	Black	Brown	Albino	Albino

Polygenic Inheritance

Many characteristics such as skin color, hair color, and height result from a blending of several separate genes that vary along a continuum. These traits are **polygenic**. Two parents who are short carry more genes for shortness than for tallness. However, they can have a child who inherits mostly genes for tallness from both parents and who will be taller than the parents. This wide variation in genotypes always results in a bell-shaped curve in an entire population. See Figure 8.5, which shows the distribution of skin pigmentation across a population.

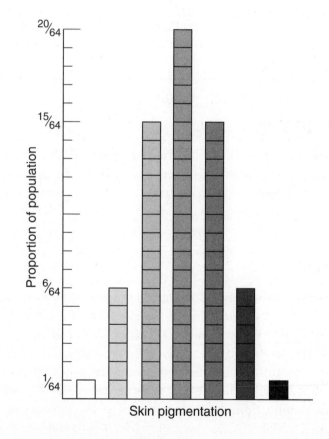

Figure 8.5

Genes and the Environment

The environment can alter the expression of genes. In fruit flies, the expression of the mutation for vestigial wings (short, shriveled wings) can be altered by temperature. When raised in a hot environment, fruit flies homozygous recessive for vestigial wings can grow wings almost as long as normal wild-type wings. In humans, the development of intelligence is the result of an interaction of genetic predisposition and the environment, or **nurture versus nature**.

SEX-INFLUENCED INHERITANCE

Inheritance can be influenced by the sex of the individual carrying the traits. An example can be seen in male-pattern baldness in humans, where hair is very thin on top of the head. This is not a sex-linked trait but, rather, a **sex-influenced trait**. Males and females express the gene for pattern baldness differently; see Table 8.2.

TABLE 8.2

Sex-influenced Inheritance of Male-Pattern Baldness		
Genotype	**Phenotype**	
	Female	*Male*
BB	Bald	Bald
Bb	Not Bald	Bald
bb	Not Bald	Not Bald
B = bald; *b* = not bald		

LINKED GENES

Genes on the same chromosome are called **linked genes**. Since there are many more genes than chromosomes, thousands of genes are linked. Humans have 46 chromosomes in every cell. Therefore, humans have 46 linkage groups. Linked genes are inherited together and do not assort independently (unless they are separated by a crossover event).

SEX-LINKAGE

REMEMBER

Sex-linked traits are located on the X chromosome.

Of the 46 human chromosomes, 44 (22 pairs) are **autosomes** and 2 are **sex chromosomes**, X and Y. Traits *carried on the X chromosome* are called **sex-linked**. Few genes are carried on the Y chromosome. Females (XX) inherit two copies of the sex-linked genes. If a sex-linked trait is due to a **recessive mutation**, a female will express the phenotype only if she carries two mutated genes (X–X–). If she carries only one mutated X-linked gene, she will be a **carrier** (X–X). If a sex-linked trait is due to a **dominant mutation**, a female will express the phenotype with only one mutated gene (X–X). Males (XY) inherit only one X-linked gene. As a result, if the male inherits a mutated X-linked gene (X–Y), he will express the gene. Recessive sex-linked traits are much more common than dominant sex-linked traits; so males suffer with sex-linked conditions more often than females do.

Here are some important facts about sex-linked traits.

- Common examples of recessive sex-linked traits are **color blindness**, **hemophilia**, and Duchenne muscular dystrophy**.**
- All daughters of affected fathers are carriers (shaded squares).

Punnett square

	X–	Y
X	X–X	XY
X	X–X	XY

- Sons cannot inherit a sex-linked trait from the father because the son inherits the Y chromosome from the father.
- A son has a 50 percent chance of inheriting a sex-linked trait from a carrier mother (shaded square).

Punnett square

	X	Y
X–	X–X	X–Y
X	XX	XY

- There is no carrier state for X-linked traits in males. If a male has the gene, he will express it.
- It is uncommon for a female to have a recessive sex-linked condition. In order to be affected, she must inherit a mutant gene from *both* parents.

CROSSOVER AND LINKAGE MAPPING

The farther apart two genes are on one chromosome, the more likely they will be separated from each other during meiosis because a crossover event will occur between them. The site at which a crossover and recombination occurs is called a **chiasma**. The result of a crossover is a **recombination**. *Crossover and recombination are a major source of variation in sexually reproducing organisms.*

Figure 8.6 shows one crossover between homologous chromosomes. Without the crossover, the resulting four gametes would contain the following genes: *AB, AB, ab,* and *ab.* There would be only two different types of gametes, *AB* and *ab.* With one crossover (as shown), the four resulting gametes contain the following genes: *AB, Ab, aB* and *ab.* There would be four different types of gametes. That one crossover results in twice the variation in the type of gametes possible.

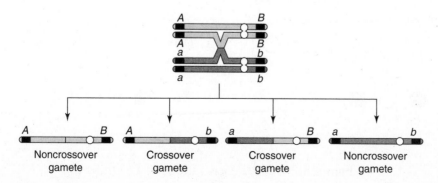

Figure 8.6

By convention, one **map unit** distance on a chromosome is *the distance within which recombination occurs 1 percent of the time.* The rate of crossover gives no information about the actual distance between genes, but it tells us the order of the linked genes on the chromosome.

Here is one example. Genes *A, B,* and *D* are linked. The crossover frequencies for *B* and *D* is 5 percent, *B* and *A* is 30 percent, and for *D* and *A* is 25 percent. The **linkage map** that can be constructed from this data is *BDA* or *ADB*. Whether you read it forward or backward does not matter.

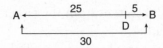

THE PEDIGREE

A **pedigree** is a family tree that indicates the phenotype of one trait being studied for every member of a family. *Geneticists use the pedigree to determine how a particular trait is inherited.* By convention, females are represented by a circle and males by a square. The carrier state is not always shown. If it is, though, it is sometimes represented by a half-shaded-in shape. A shape is completely shaded in if a person exhibits the trait.

The pedigree in Figure 8.7 shows three generations of deafness. Try to determine the pattern of inheritance. First, eliminate any possibilities. Dominance can be ruled out (either sex-linked or autosomal) because in order for a child to have the condition, she or he would have had to have received one mutant gene from one afflicted parent, and nowhere is that the case. (All afflicted children have unaffected parents.) Also, you can rule out sex-linked recessive, because in order for F_3 generation daughter #1 to have the condition, she would have had to inherit two mutant traits (X–X–), one from each parent. However, her father does not have the condition. Therefore, the trait must be autosomal recessive.

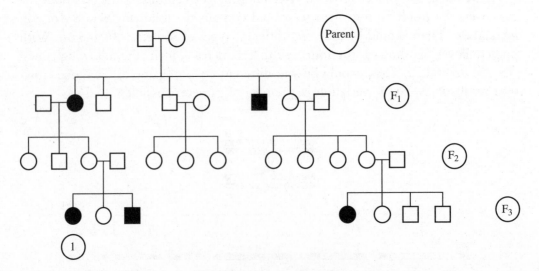

Figure 8.7   Three Generations of Deafness

X INACTIVATION—THE BARR BODY

Early in the development of the embryo of a female mammal, one of the X chromosomes is inactivated in every **somatic** (body) **cell**. This inactivation occurs randomly. The process results in an embryo that is a **genetic mosaic**, some cells have one X inactivated, some cells have the other X inactivated. Therefore, all the cells of female mammals are not identical. The inactivated chromosome condenses into a dark spot of chromatin and can be seen at the outer edge of the nucleus of all somatic cells in the female. This dark spot is called a **Barr body**.

Proof of X chromosome inactivation can be seen in the genetics of the female calico cat where the alleles for black and yellow fur are carried on the X chromosome. Male cats, having only a single X chromosome, can be either yellow ($X^Y Y$) **or** black ($X^B Y$). Calico cats, which are almost always female, have coats with patches of both yellow and black ($X^B X^Y$). These patches of fur developed from embryonic cells with different deactivated X chromosomes. Some fur-producing cells contained the X^B active chromosome and produce black fur. Other fur-producing cells contain the X^Y active chromosome and produce yellow fur. The result is a cat with yellow and black patches of fur, the characteristic calico appearance.

Another example of X chromosome inactivation is evident in humans. A certain X-linked recessive mutation prevents the development of sweat glands. A woman who is heterozygous for this trait is not merely a carrier. Because of X inactivation, she has patches of normal skin and patches of skin lacking sweat glands.

MUTATIONS

Mutations are any changes in the genome. They can occur in the somatic (body) cells and be responsible for the spontaneous development of cancer, or they can occur during gametogenesis and affect future offspring. Even though radiation and certain chemicals cause mutations, when and where mutations occur is random.

There are two types of mutations, gene mutations and chromosome mutations. Gene mutations are caused by a change in the DNA sequence. Some human genetic disorders caused by gene and chromosome mutations are listed and described in Table 8.3. The nature of gene mutations at the DNA level is discussed in the next chapter.

Although gene mutations cannot be seen under a microscope, **chromosome mutations** can. A procedure called a **karyotype** shows the size, number, and shape of chromosomes and can reveal the presence of certain mutations. Karyotypes can be used to scan for chromosomal abnormalities in developing fetuses. Figure 8.8 on page 173 shows a karyotype of a male with Down syndrome due to an extra chromosome 21.

TABLE 8.3

Gene and Chromosome Mutations

Genetic Disorder	Pattern of Inheritance	Description
Phenylketonuria (PKU)	Autosomal recessive	Inability to break down the amino acid phenylalanine. Requires elimination of phenylalanine from diet, otherwise serious mental retardation will result.
Cystic fibrosis	Autosomal recessive	The most common lethal genetic disease in the U.S. 1 out of 25 Caucasians is a carrier. Characterized by buildup of extra-cellular fluid in the lungs, digestive tract, etc.
Tay-Sachs disease	Autosomal recessive	Onset is early in life and is caused by lack of the enzyme necessary to break down lipids needed for normal brain function. It is common in Ashkenazi Jews and results in seizures, blindness, and early death.
Huntington's disease	Autosomal dominant	A degenerate disease of the nervous system resulting in certain and early death. Onset is usually in middle age.
Hemophilia	Sex-linked recessive	Caused by the absence of one or more proteins necessary for normal blood clotting.
Color blindness	Sex-linked recessive	Red-green color blindness is rarely more than an inconvenience.
Duchenne muscular dystrophy	Sex-linked recessive	Progressive weakening of muscle control and loss of coordination
Sickle cell disease	Autosomal recessive	A mutation in the gene for hemoglobin results in deformed red blood cells. Carriers of the sickle cell trait are resistant to malaria.

Chromosomal Disorder	Pattern of Inheritance	Description
Down syndrome	47 chromosomes due to trisomy 21	Characteristic facial features, mental retardation, prone to developing Alzheimer's and leukemia
Turner's syndrome	XO 45 chromosomes due to a missing sex chromosome	Small stature, female
Klinefelter's syndrome	XXY 47 chromosomes due to an extra X chromosome	Have male genitals, but the testes are abnormally small and the men are sterile

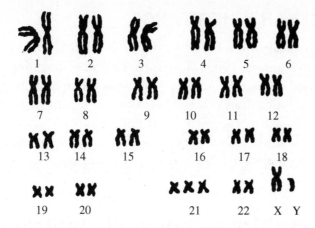

Figure 8.8   Karyotype of a male with trisomy 21

Chromosomal aberrations include:

- **Deletion**—when a fragment lacking a centromere is lost during cell division
- **Inversion**—when a chromosomal fragment reattaches to its original chromosome but in the reverse orientation
- **Translocation**—when a fragment of a chromosome becomes attached to a non-homologous chromosome
- **Polyploidy**—when a cell or organism has extra sets of chromosomes.

NONDISJUNCTION

Nondisjunction is an error that sometimes happens during meiosis in which homologous chromosomes fail to separate as they should. See Figure 8.9. When this happens, one gamete receives two of the same type of chromosome and another gamete receives no copy. The remaining chromosomes may be unaffected and normal. If either aberrant gamete unites with a normal gamete during fertilization, the resulting zygote will have an abnormal number of chromosomes. Any abnormal number of chromosomes is known as **aneuploidy**. If a chromosome is present in triplicate, the condition is known as **trisomy**. People with Down syndrome have an extra chromosome 21. The condition is referred to as **trisomy 21**. Cancer cells grown in culture almost always have extra chromosomes. An organism in which the cells have an extra set of chromosomes is referred to as **triploid** ($3n$). An organism with extra sets of chromosomes is referred to as **polyploid**. Hugo DeVries, the scientist who coined the term **mutation**, was studying plants that were polyploidy. Polyploidy is common in plants and results in plants of abnormally large size. As the word is used today, mutation refers to any **genetic** or **chromosomal abnormality**.

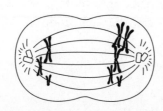

Figure 8.9   Nondisjunction

EXTRANUCLEAR INHERITANCE

Extranuclear genes are those found in the mitochondria and chloroplasts. They were discovered in plants in 1909. Since then, they have been linked to several rare and severe inherited diseases in humans. Defects in the mitochondrial DNA reduce the amount of ATP a cell can make. Therefore, the organs most affected by these mutations are the ones that require the most energy: the nervous system and muscles. Because the mitochondria passed to the zygote all come from the cytoplasm of the egg, these diseases are **always inherited from the mother**.

Multiple-Choice Questions

1. Which is TRUE about a testcross?

   (A) It is a mating between two hybrid individuals.
   (B) It is a mating between a hybrid individual and a homozygous recessive individual.
   (C) It is a mating between a homozygous dominant individual and a homozygous recessive individual.
   (D) It is a mating to determine which individual is homozygous recessive.
   (E) It is a mating between an individual of unknown genotype and a homozygous recessive individual.

2. All are true of crossover EXCEPT

   (A) it normally occurs between sister chromatids
   (B) it is the site of a chiasma
   (C) it is responsible for linked genes being inherited separately
   (D) it cannot occur between sex chromosomes in human males
   (E) it is a common event

3. A round watermelon is crossed with a long watermelon and all the offspring are oval. If two oval watermelons are crossed, what is the percent of watermelons that will be round?

   (A) 0
   (B) 25%
   (C) 50%
   (D) 75%
   (E) 100%

4. A diploid cell has three pairs of homologous chromosomes: *AaBaCc*. How many different gametes can this cell produce?

   (A) 4
   (B) 8
   (C) 16
   (D) 32
   (E) 64

5. In peas, the trait for tall plants is dominant (*T*) and the trait for short plants is recessive (*t*). The trait for yellow seeds is dominant (*Y*) and the trait for green seeds is recessive (*y*). A cross between two plants results in 292 tall yellow plants and 103 short green plants. Which of the following are most likely to be the genotypes of the parents?

   (A) *TtYY* × *Ttyy*
   (B) *TTYy* × *TTYy*
   (C) *TTyy* × *TTYy*
   (D) *TtYy* × *TtYy*
   (E) *TtYy* × *TTYy*

6. A child is born with blood type O. All of the following could be the blood type of the parents EXCEPT

   (A) A and B
   (B) A and A
   (C) O and O
   (D) AB and O
   (E) A and O

7. *ABCDEF* → *ABEDCF*

   A rearrangement in the linear sequence of genes as shown in the diagram above is known as a/an

   (A) translocation
   (B) deletion
   (C) addition
   (D) polyploidy
   (E) inversion

8. A diploid organism has 36 chromosomes per cell. How many linkage groups does it have?

   (A) 9
   (B) 18
   (C) 36
   (D) 72
   (E) It cannot be determined without crossing some of the organisms.

9. An organism has three independently assorting traits: *AaBbCc*. What fraction of its gametes will contain the dominant genes *ABC*?

   (A) 1
   (B) $\frac{1}{8}$
   (C) $\frac{1}{4}$
   (D) $\frac{1}{2}$
   (E) $\frac{3}{8}$

10. All of the following are true of a person who has Klinefelter's syndrome EXCEPT

    (A) the cheek cells of such a person contain two Barr bodies
    (B) the condition resulted from nondisjunction
    (C) the person is a male
    (D) the person contains somatic cells with 47 chromosomes
    (E) the person is most likely sterile

11. Which of the following is NOT caused by an autosomal recessive genetic mutation?

    (A) PKU
    (B) Angelman syndrome
    (C) cystic fibrosis
    (D) Tay-Sachs disease
    (E) sickle cell anemia

12. The expression of both alleles for a trait in a hybrid individual is

    (A) pleiotropy
    (B) epistasis
    (C) codominance
    (D) complementation
    (E) incomplete dominance

13. How many autosomes does the human male normally have?

    (A) 1
    (B) 2
    (C) 22
    (D) 23
    (E) 44

14. A couple has 6 children, all girls. If the mother gives birth to a seventh child, what is the probability that the seventh child will be a girl?

    (A) $^6/_7$
    (B) $^1/_{128}$
    (C) $^1/_2$
    (D) 1
    (E) Not enough information is given because the sperm determines the sex of the child.

15. Assume that two genes, *A* and *B*, are not linked. If the probability of allele *A* being in a gamete is $\frac{1}{2}$ and the probability of allele *B* being in a gamete is $\frac{1}{2}$, then the probability of BOTH *A* and *B* being in the same gamete is

    (A) $\frac{1}{2}$
    (B) $\frac{1}{4}$
    (C) 1
    (D) $\frac{1}{8}$
    (E) 0.5

16. Gene *R* controls the formation of feathers on a bird. In addition, it seems to be responsible for traits in several other body systems. What is the best explanation for this type of inheritance?

    (A) blending inheritance
    (B) codominance
    (C) pleiotropy
    (D) epistasis
    (E) mutation

17. In one strain of mice, fur color ranges from white to darkest brown with every shade of brown in between. This pattern of inheritance for fur color is probably controlled by

    (A) multiple genes
    (B) a single gene with many alleles
    (C) pleiotropy
    (D) one gene with hundreds of incidences of crossover
    (E) incomplete dominance

18. How is Huntington's disease inherited?

    (A) It is caused by a virus inherited from either parent.
    (B) It is sex-linked recessive.
    (C) It is autosomal recessive.
    (D) It is sex-linked dominant.
    (E) It is autosomal dominant.

19. Two traits, *A* and *B*, are linked, but they are not always inherited together. The most likely reason is

    (A) they are not on the same chromosome
    (B) they are not sex-linked
    (C) they are on the same chromosome but are far apart
    (D) they are close together on the same chromosome
    (E) *A* is dominant over *B*

20. A cross was made between two fruit flies, a white-eyed female and a wild male (red eyed). One hundred F_1 offspring were produced. All the males were white eyed and all the females were wild. When these F_1 flies were allowed to mate, the F_2 flies were observed and the following data was collected.

	Females		**Males**
P:	White eyed	×	Wild (red eyed)
F_1:	59 wild		51 white eyed
F_2:	24 wild		23 wild
	26 white eyed		27 white eyed

What is the most likely pattern of inheritance for the white-eyed trait?

(A) autosomal dominant
(B) autosomal recessive
(C) sex-linked dominant
(D) sex-linked recessive
(E) holandric

21. A man who has a sex-linked allele will pass it on to

(A) all his daughters
(B) all his sons
(C) ½ of his daughters
(D) ½ of his sons
(E) all of his children

22. The figure below shows a pedigree for a family that carries the gene for Huntington's disease. Individuals who express a particular trait are shown shaded in.

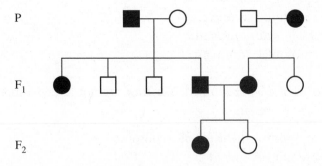

What is the genotype of the daughter in the F_2 generation who does not have the disease?

(A) *H/H*
(B) *H/h*
(C) *h/h*
(D) *X–X*
(E) *X–X–*

23. This figure shows a pedigree of the blood types for a family. What is the genotype for person number 14?

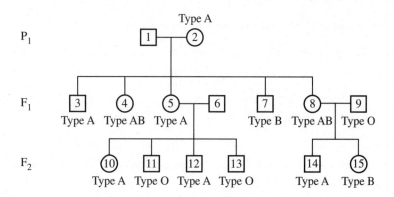

(A) *A/A*
(B) *A/i*
(C) *i/i*
(D) *A/B*
(E) *AB/i*

Answers to Multiple-Choice Questions

1. **(E)** A testcross is an actual mating between a pure recessive animal and an animal that shows the dominant phenotype but whose genotype is unknown.

2. **(A)** Crossover occurs between homologous chromosomes, not sister chromatids. Sister chromatids are genetically identical. If crossover did occur between sister chromosomes, it would be undetectable.

3. **(B)** Here is the cross.

	R	*L*
R	*RR*	*RL*
L	*RL*	*LL*

RR is round, *RL* is oval and *LL* is long.
The inheritance is codominant.

4. **(B)** You can count on your fingers: *ABC, ABC, ABc,* and so on. Alternatively, you can use the equation 2^n where n equals the number of pairs. In this case, $2^3 = 8$.

5. **(D)** Since there are four genes (*T, t, Y,* and *y*) but only two phenotypes in the offspring, the traits for height and seed color must be linked, that is, on the same chromosome. So solve the problem this way. Consider the traits separately at first. The phenotype ratio in the offspring for height is $3:1$, tall to short. Therefore, the parents must be *Tt* and *Tt*. The phenotype ratio in the offspring for seed color is $3:1$, yellow to green. Therefore, the parents must be *Yy* and *Yy*. Now, put both genotypes together. The parents must be *TtYy* and *TtYy*.

6. **(D)** Blood type O is homozygous recessive, *ii* or *I°I°*. The child must receive an *O* allele from each parent. Blood type AB has no *O* allele. Therefore, choice D is excluded as a parent.

7. **(E)** The section of the strand shows genes *CDE* inverted.

8. **(C)** Linked genes are located on one chromosome. Since there are 36 chromosomes, there are 36 linkage groups.

9. **(B)** Since there are 8 possible different gametes and only 1 that is *ABC*, the answer is $\frac{1}{8}$. See question #4.

10. **(A)** One Barr body forms when there are two X chromosomes in a cell. Since a person with Klinefelter's has the genotype XXY, there will be only one Barr body in each body cell. The other choices are all true of a person with Klinefelter's syndrome. He is male with 47 chromosomes and is most likely sterile. Klinefelter's is caused by nondisjunction.

11. **(B)** Angelman syndrome is caused by two things: a chromosome deletion and genomic imprinting. The other diseases result from mutations in the genes.

12. **(C)** In codominance, both traits show. An example is blood type in humans, where a person who has a gene for blood antigen A and another for blood antigen B has the AB blood type.

13. **(E)** Autosomes are the chromosomes other than sex chromosomes (X and Y).

14. **(C)** No matter how many children a couple has, the chance that the child will be a boy *or* a girl is always $\frac{1}{2}$. Although it is true that whether the sperm carries an X or a Y sex chromosome determines the sex of the child, that it is irrelevant to the question here.

15. **(B)** Since *A* and *B* are not linked, they assort independently. To find the probability of two independent events happening, multiply the chance of one happening by the chance of the other happening.

16. **(C)** When one gene seems to control the expression of several traits, the pattern of inheritance is pleiotropy. A classic example of pleiotropy in humans is Marfan's syndrome.

17. **(A)** Examples of polygenic inheritance in humans are genes for skin color and height.

18. **(E)** Although the trait is autosomal dominant, symptoms do not usually appear until later in life, long after the person with the condition has passed the trait on to children and, perhaps, grandchildren.

19. **(C)** If genes are on the same chromosome but far apart, they will often be inherited separately because they will often be separated by crossover.

20. **(D)** Here is the first cross.

	X	Y
X–	X–X	X–Y
X–	X–X	X–Y

All the female offspring are carriers (X–X), and all the male offspring have white eyes (X–Y).

Here is the second cross.

	X–	Y
X–	X–X–	X–Y
X	X–X	XY

There is a 50 percent chance that a male will be white eyed and a 50 percent chance he will be red eyed. There is a 50 percent chance a female will be white eyed and a 50 percent chance she will be red eyed (a carrier).

21. **(A)** A man gives his sons the Y chromosome and passes sex-linked traits to all his daughters.

22. **(C)** Females are represented as circles; males as squares unless otherwise stated. You should know that Huntington's disease is inherited as autosomal dominant. The F_2 daughter who does not have the condition must have inherited one healthy gene from each parent. She must be *h/h*, normal.

23. **(B)** Person 14 has blood type A. It can be *A/A* or *A/i*. Since his father has type O blood, person 14 must have inherited the A from his mother and the O from his father. His genotype therefore is *A/i*.

Free-Response Questions

Directions: Answer all questions. You must answer the question in essay—**not** outline—form. You may use labeled diagrams to supplement your essay, but diagrams alone are *not* sufficient. Before you start to write, read each question carefully so that you understand what the question is asking.

1. A person with Turner's syndrome has a genotype of XO, while a person with Klinefelter's syndrome has the genotype XXY. Explain how these two mutations come about.

NOTE

The explanations of these terms can be found in this review chapter

2. Scientists understand that comparatively few genes are inherited in a simple Mendelian fashion. The expression of most genes is altered by many things, including other genes or the environment. Choose five of the terms below and explain what they mean.
   a. Pleiotropy
   b. Penetrance and expressivity
   c. Epistasis
   d. Gene collaboration
   e. Polygenic inheritance
   f. Complementary genes
   g. The effect of the environment on genes

Typical Free-Response Answers

Note: See the free-response question on page 157 for an essay related to Mendel's laws of segregation and independent assortment.

1. Both of these conditions, Turner's and Klinefelter's syndrome, are mutations that arise as a result of nondisjunction. Normally, during anaphase I of meiosis, homologous pairs separate (disjoin), with one homologue going into each of two daughter cells. Occasionally, one homologous pair does not separate as it should during anaphase I. As a result, both homologues go into one of the daughter cells, giving it an extra chromosome, while leaving the other daughter cell missing one chromosome. Klinefelter's syndrome results when an egg with two X chromosomes fuses with a sperm carrying a Y chromosome. The resulting zygote has the genotype XXY. Turner syndrome results when a gamete without any sex chromosome fuses with a normal gamete with one X chromosome. The resulting zygote has the genotype XO, where O means missing chromosome. A person with Turner's syndrome has 45 chromosomes, while a person with Klinefelter's syndrome has 47 chromosomes.

The Molecular Basis of Inheritance

CHAPTER 9

- The search for inheritable material
- Structure of nucleic acids
- DNA replication in eukaryotes
- DNA makes RNA makes protein
- Gene mutation
- The genetics of viruses and bacteria
- Prions

- Transposons
- The human genome
- Recombinant DNA
- Cloning genes
- Tools and techniques of recombinant DNA
- Ethical considerations

INTRODUCTION

Today, everyone knows that DNA is the molecule of heredity. We know that DNA makes up chromosomes and that genes are located on the chromosomes. Today, we can even see the location of particular genes by tagging them with fluorescent dye.

However, until the 1940s, many scientists believed that proteins, not DNA, were the molecules that make up genes and constitute inherited material. Several factors contributed to that belief. First, proteins are a major component of all cells. Second, they are complex macromolecules that exist in seemingly limitless variety and have great specificity of function. Moreover, a great deal was known about the structure of proteins and very little was known about DNA. The work of many brilliant scientists has transformed our knowledge of the structure and function of the DNA molecule and led to the acceptance of DNA as the molecule responsible for heredity.

This chapter includes the history of the search for the heritable material, the structure of nucleic acids and how DNA makes proteins. It also includes an extensive review of genetic engineering and recombinant DNA techniques. Because this is such a complex topic and because it is 9 percent of the AP Exam, many multiple-choice questions are included at the end of this chapter.

THE SEARCH FOR INHERITABLE MATERIAL

Griffith (1927) performed experiments with several different strains of the bacterium *Diplococcus pneumoniae*. Some strains are virulent and cause pneumonia in humans and mice, and some strains are harmless. Griffith discovered that *bacteria have the ability to transform harmless cells into virulent ones by transferring some genetic factor from one bacteria cell to another*. This phenomenon is known as **bacterial transformation**, and the experiment is known as the **transformation experiment**. See the information about bacterial transformation later in this chapter.

Avery, MacLeod, and McCarty (1944) published their classic findings that Griffith's **transformation factor** is, in fact, DNA. This research proved that DNA was the agent that carried the genetic characteristics from the virulent dead bacteria to the living nonvirulent bacteria. *This provided direct experimental evidence that DNA, not protein, was the genetic material.*

Hershey and Chase (1952) carried out experiments that lent strong support to *the theory that DNA is the genetic material*. They tagged bacteriophages with the radioactive isotopes ^{32}P and ^{35}S. Since proteins contain sulfur but not phosphorous and DNA contains phosphorous but not sulfur, the radioactive ^{32}P labeled the DNA of the phage viruses while ^{35}S labeled the protein coat of the phage viruses. Hershey and Chase found that when bacteria were infected with phage viruses, the radioactive phosphorous in the phage that always entered the bacterium while the radioactive sulfur remained outside the cells. This proved that *DNA from the viral nucleus, not protein from the viral coat, was infecting bacteria and producing thousands of progeny.*

Rosalind Franklin (1950–53), while working in the lab of Maurice Wilkins, carried out the X-ray crystallography analysis of DNA that showed DNA to be a helix. Her work was critical to Watson and Crick. Without it, they would not have been able to develop their now-famous model of DNA. Although Maurice Wilkins shared the Nobel prize with Watson and Crick, Rosalind Franklin did not. She had died by the time the prize was awarded, and the prize is not awarded posthumously.

Watson and Crick (1953), while working at Cambridge University, *proposed the double helix structure of DNA* in a one-page paper in the British journal *Nature*. Throughout the 1940s, until 1953, many scientists worked to understand the structure of DNA. All the data that Watson and Crick used to build their model of DNA derived from other scientists who published earlier. Two major pieces of information they used were the biochemical analysis of DNA (from Erwin Chargaff) and the X-ray diffraction analysis of DNA (from Rosalind Franklin). The fact that much of the components of DNA were known before Watson and Crick began their model building does not detract from the brilliance of their achievement. Understanding the structure of DNA gives a foundation to understand how DNA could replicate itself. Watson and Crick received the Nobel prize in 1962 for correctly describing the structure of DNA.

Meselsohn and Stahl (1958) *proved that DNA replicates in a semiconservative fashion*, as Francis Crick predicted. They cultured bacteria in a medium containing heavy nitrogen (^{15}N), allowing the bacteria to incorporate this heavy nitrogen into their DNA as they replicated and divided. These bacteria were then transferred to a medium containing light nitrogen (^{14}N) and allowed to replicate and divide only once. The bacteria that resulted from this final replication were spun in a centrifuge and found to be midway in density between the bacteria grown in heavy nitrogen and those grown in light nitrogen. This demonstrated that the new bacteria contained DNA consisting of one heavy strand and one light strand. See Figure 9.1 of semiconservative replication.

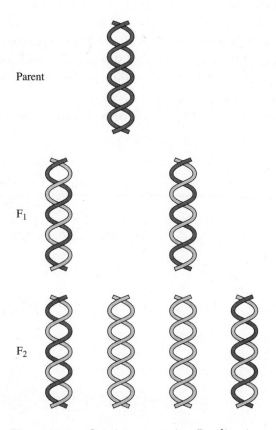

Figure 9.1   Semiconservative Replication

STRUCTURE OF NUCLEIC ACIDS

X-Ray Crystallography

X-ray crystallography is a technique used to determine the three-dimensional structure of a molecule. To carry out X-ray diffraction analysis, purified samples of DNA are crystallized and bombarded X-rays. These rays are scattered by the DNA molecule, and the diffraction pattern is captured on photographic film and analyzed. This process was used by Linus Pauling in his analysis of the alpha-helix molecules. Rosalind Franklin analyzed DNA by X-ray diffraction and proved it to be some sort of helix.

Deoxyribonucleic Acid (DNA)

The DNA molecule is a **double helix**, shaped like a twisted ladder, consisting of two strands running in opposite directions (antiparallel), see Figure 9.2. One runs **5′ to 3′** (right side up), the other **3′ to 5′** (upside down). DNA is a polymer of repeating units of **nucleotides**. In DNA, these consist of a **5-carbon sugar** (**deoxyribose**), a **phosphate**, and a **nitrogen base**. The carbon atoms in deoxyribose are numbered 1 to 5. There are four nitrogenous bases in DNA: **adenine (A)**, **thymine (T)**, **cytosine (C)**, and **guanine (G)**. Of the four nitrogenous bases, adenine and guanine are **purines**, and thymine and cytosine are **pyrimidines**. The nitrogenous bases of opposite chains are paired to one another by **hydrogen bonds**. The adenine nucleotide bonds by a **double hydrogen bond** to the thymine nucleotide, and the cytosine nucleotide bonds by a **triple hydrogen bond** to the guanine nucleotide.

> **REMEMBER**
>
> The two strands of DNA run in opposite directions.

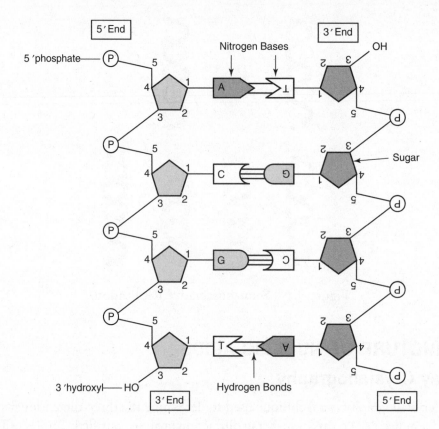

Figure 9.2   DNA

Ribonucleic Acid (RNA)

RNA is a single-stranded helix consisting of repeating nucleotides: adenine, cytosine, guanine, and **uracil (U)**, which replaces thymine. The 5-carbon sugar in RNA is **ribose**.

Figure 9.3 shows structural formulas for the purines—adenine and guanine—and for the pyrimidines—cytosine, thymine, and uracil.

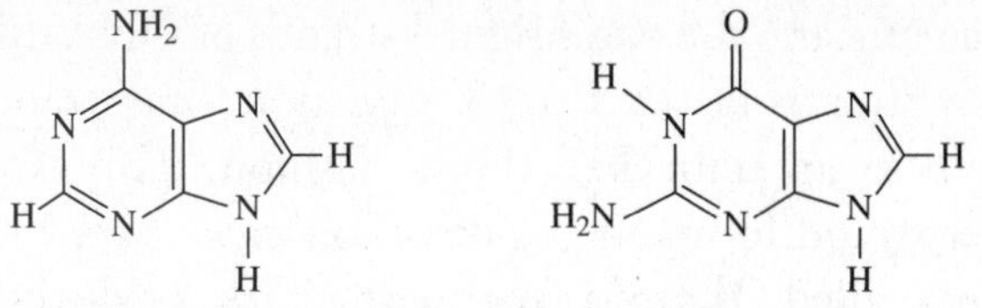

adenine (A)   guanine (G)

Purine Nitrogen Bases

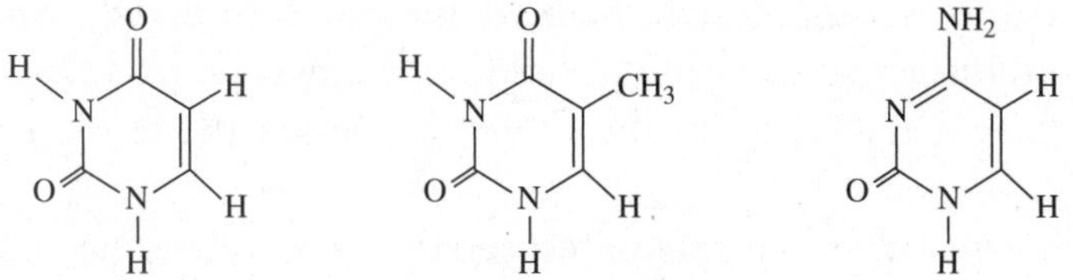

uracil (U)   thymine (T)   cytosine (C)

Pyrimidine Nitrogen Bases

Figure 9.3

DNA REPLICATION IN EUKARYOTES

DNA replication, the making of an exact replica of the DNA molecule by semiconservative replication, was predicted by Watson and Crick and proven by Meselsohn and Stahl. The DNA double helix unzips, and each strand serves as a **template** for the formation of a new strand composed of complementary nucleotides: A with T and C with G. The two new molecules each consist of one old strand and one new strand. The following describes the important steps in DNA replication in eukaryotes, see also Figure 9.4.

- Replication begins at special sites called **origins of replication,** where the two strands of DNA separate to form **replication bubbles**. Thousands of these bubbles are located along the DNA molecule. They speed up the process of replication along the giant DNA molecule that consists of **3 billion base pairs**.
- A *replication bubble expands as replication proceeds in opposite directions*. At each end of the replication bubble is a **replication fork**, a Y-shaped region where the new strands of DNA are elongating. Eventually, all the replication bubbles fuse.
- The enzyme **DNA polymerase** catalyzes the elongation of the new DNA strands. It builds a new strand from **5′ to 3′**, moving along the template strand and pushing the replication fork ahead of it. In humans, the rate of **elongation** is about 50 nucleotides per second.
- DNA polymerase cannot initiate synthesis; it can only add *nucleotides to the 3′ end of a preexisting chain*. This preexisting chain actually consists of RNA and is called an **RNA primer**. An enzyme called **primase** joins RNA nucleotides to make the primer.

- DNA polymerase replicates the two original strands of DNA differently. Although it builds both new strands in the 5′ to 3′ direction, one strand is formed *toward the replication fork* in an unbroken, linear fashion. This is called the **leading strand**. The other strand forms in the direction *away from the replication fork* in a series of segments called **Okazaki fragments**. This is called the **lagging strand**. Okazaki fragments are about 100–200 nucleotides long and will be joined into one continuous strand by the enzyme **DNA ligase**.

- Other proteins and enzymes assist in replication of the DNA. **Helicase** is an enzyme that untwists the double helix at the replication fork. **Single-stranded binding proteins** act as scaffolding, holding the two DNA strands apart. DNA **topoisomerases** make "cuts" in the DNA that lessen the tension on the tightly wound helix.

- DNA polymerase carries out **mismatch repair**, or proofreading, and corrects any errors.

- Each time the DNA replicates, some nucleotides from the ends of the chromosomes are lost. To protect against the possible loss of genes at the ends of the chromosomes, eukaryotes have special nonsense nucleotide sequences (TTAGGG) at the ends of the chromosomes that repeat thousands of times. These protective ends are called **telomeres**. Telomeres are created and maintained by the enzyme **telomerase**. Normal body cells contain little telomerase, so every time the DNA replicates, the telomeres get shorter. This may serve as a clock that counts cell divisions and causes the cell to stop dividing as the cell ages.

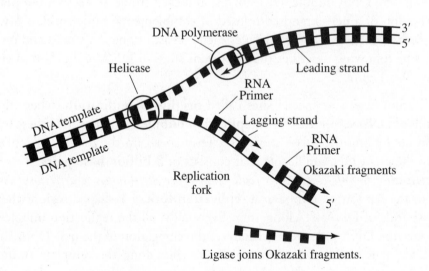

Figure 9.4   DNA Replication

DNA MAKES RNA MAKES PROTEIN

The process whereby DNA makes proteins has been worked out in great detail. To summarize, the **triplet code** in DNA is **transcribed** into a **codon sequence** in messenger-RNA (mRNA) inside the nucleus. Next, this newly formed strand of **RNA**, known as pre-RNA, **is processed** in the nucleus. Then the codon sequence is **translated** into an amino acid sequence (a polypeptide) in the cytoplasm at the ribosome.

If the strand of DNA triplets to be transcribed is   5′-AAA TAA CCG GAC-3′

Then the strand of mRNA codons that forms is   3′-UUU AUU GGC CUG-5′

The transfer RNA (tRNA) anticodon strand complementary to the mRNA strand is   AAA UAA CCG GAC

Figure 9.5 shows an overview of transcription, RNA processing, and translation.

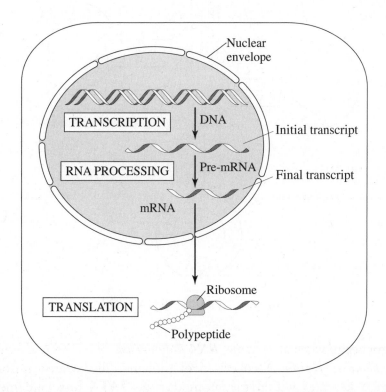

Figure 9.5 Transcription and Translation in a Eukaryotic Cell

Transcription

Transcription is the process by which DNA makes RNA. There are three types of RNA: **mRNA, tRNA** and **rRNA. Messenger RNA** (mRNA) carries messages directly from DNA to the cytoplasm and varies in length, depending on the length of the message. **Transfer RNA** (tRNA) is shaped like a cloverleaf and carries amino acids to the mRNA at the ribosome; see Figure 9.6. **Ribosomal RNA** (rRNA) is structural. Along with proteins, rRNA makes up the ribosome, which is formed in the **nucleolus**.

Transcription consists of three stages: **initiation, elongation**, and **termination**.

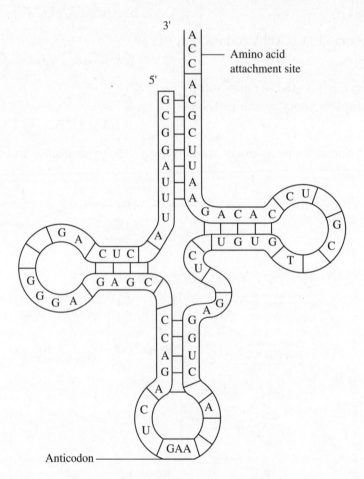

Figure 9.6   Transfer-RNA (tRNA)

- **Initiation** begins when an enzyme, *RNA polymerase, recognizes and binds to DNA at the promoter region*. A collection of proteins called **transcription factors** recognize a key area within the promoter, the **TATA box**, and mediate the binding of RNA polymerase to the DNA. The completed assembly of transcription factors and RNA polymerase bound to the promoter is called a **transcription initiation complex**. Once RNA polymerase is attached to the promoter, DNA transcription of the DNA **template** begins.
- **Elongation** of the strand continues as *RNA polymerase adds nucleotides to the 3' end of a growing chain*. RNA polymerase pries the two strands of DNA apart and attaches RNA nucleotides according to the base pairing rules: C with G and A with U. The stretch of DNA that is transcribed into an mRNA molecule is called

a **transcription unit**. Each unit consists of triplets of bases called **codons** (for example, AAU, CGA) that code for specific amino acids. A single gene can be transcribed into mRNA simultaneously by several molecules of RNA polymerase following each other in a caravan fashion.

- **Termination** is the final stage in transcription. Elongation continues for a short distance after the RNA polymerase transcribes the **termination sequence** (AAUAAA). At this point, mRNA is cut free from the DNA template.

RNA Processing

Before the newly formed pre-RNA strand is shipped out to the ribosome in the cytoplasm, it is altered or **processed** by a series of enzymes.

- A **5′ cap** consisting of a modified guanine nucleotide is added to the 5′ end. This cap helps the RNA strand bind to the ribosome in the cytoplasm.
- A **poly (A) tail**, consisting of a string of adenine nucleotides, is added to the 3′ end. This tail protects the RNA strand from degradation by hydrolytic enzymes, helps the ribosome attach to the RNA, and facilitates the release of the RNA into the cytoplasm.
- Noncoding regions of the mRNA called **introns** or **intervening sequences** are removed by **snRNPs** (pronounced "snurps"), small nuclear ribonucleoproteins, and **splicesomes**. This removal allows only **exons**, which are **expressed sequences**, to leave the nucleus. As a result of this processing, the mRNA that leaves the nucleus is a great deal shorter than the original transcription unit. See Figure 9.5 on page 189.

Translation of mRNA—Synthesis of a Polypeptide

Translation *is the process by which the codons of an mRNA sequence are changed into an amino acid sequence,* see Figure 9.7. Amino acids present in the cytoplasm are carried by tRNA molecules to the codons of the mRNA strand at the ribosome according to the base pairing rules (A with U and C with G). One end of the tRNA molecule bears a specific amino acid, and the other end bears a nucleotide triplet called an **anticodon**. Unlike mRNA which is broken down immediately after it is used, tRNA is used repeatedly. The energy for this process is provided by **GTP (guanosine triphosphate)**, a molecule closely related to ATP. Each amino acid is joined to the correct tRNA by a specific enzyme called **aminoacyl-tRNA synthetase**. There are only 20 different aminoacyl-tRNA synthetases, one for each amino acid. There are 64 codons, 61 of them code for amino acids. One codon, **AUG**, has two functions; it codes for methionine and is also a **start codon**. Three codons, **UAA**, **UGA**, and **UAG**, are **stop codons**, and terminate all sequences. Some tRNA molecules have anticodons that can recognize two or more different codons. This occurs because the pairing rules for the third base of a codon are not as strict as they are for the first two bases. This relaxation of base pairing rules is known as **wobble**. For example, the codons UCU, UCC, UCA, and UCG all code for the amino acid **serine**. The process of translation consists of three stages: **initiation**, **elongation** and **termination**. See Figure 9.6 for a sketch of tRNA and Figure 9.7 for a sketch of translation.

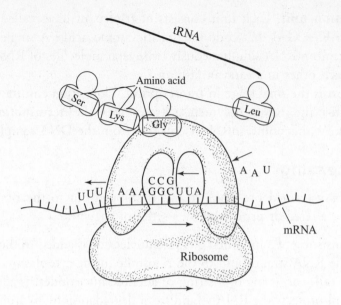

Figure 9.7   Translation

- **Initiation** begins when mRNA becomes attached to a subunit of the ribosome. This first codon is always **AUG**. It codes for **methionine** and must be positioned correctly in order for transcription of an amino acid sequence to begin.
- **Elongation** continues as tRNA brings amino acids to the ribosome and a polypeptide chain is formed.
- **Termination** of an mRNA strand is complete when a ribosome reaches one of three **termination** or **stop codons**. A **release factor** breaks the bond between the tRNA and the last amino acid of the polypeptide chain. The polypeptide is freed from the ribosome, and mRNA is broken down.

GENE MUTATION

Mutations are changes in genetic material. They occur *spontaneously* and at *random*. They can be caused by **mutagenic agents** including toxic chemicals and radiation. If a mutation has an adverse effect in **somatic** or body cells of an organism, it is referred to as a **genetic disorder** or **hereditary disease**. Mutations that occur in gametes can be transmitted to offspring and immediately change the **gene pool** of a population. *Mutations are very important because they are the raw material for natural selection.*

Point Mutation

The simplest mutation is a **point mutation**. This is a **base-pair substitution**, a chemical change in just one base pair in a single gene. Here is an example of a change in an English sentence analogous to a point mutation in DNA:

Point Mutation

THE FAT CAT SAW THE **D**OG → THE FAT CAT SAW THE **H**OG.

The inherited genetic disorder **sickle cell anemia** results from a single point mutation in a single base pair in the gene that codes for hemoglobin. This point mutation is responsible for the production of abnormal hemoglobin that can cause red blood cells to sickle when oxygen tension is low. When red blood cells sickle, a variety of tissues may be deprived of oxygen and suffer severe, permanent damage. The possibility exists, however, that a point mutation could result in a beneficial change for an organism or, because of **wobble** in the genetic code, result in no change in the proteins produced. (Wobble is the relaxation of the base-pairing rules for the third base in a codon.) Here is an example of wobble:

DNA	mRNA	Amino Acid Produced
AAA	UUU	Phenylalanine
AAG	UUC	Phenylalanine
↑ mutation		No change occurs in the amino acid.

Insertion or Deletion

A second type of gene mutation results from a single nucleotide **insertion** or **deletion**. To continue the three-letter word analogy, a deletion is the loss of one letter, and an insertion is the addition of a letter into the DNA sentence. Both mutations result in a **frameshift**, because the entire reading frame is altered.

Deletion of the Letter E
↓
THE FAT CAT SAW THE DOG → THF ATC ATS AWT HED OG

Insertion of the Letter T
↓
THE FAT CAT SAW THE DOG → THE FTA TCA TSA WTH EDO G

As a result of the frameshift, one of two things can happen. Either a mutated polypeptide is formed or no polypeptide is formed.

Missense Mutations

When point mutations or frameshifts change a codon *within a gene* into a stop codon, translation will be altered into a **missense** or **nonsense mutation**.

THE GENETICS OF VIRUSES AND BACTERIA

Since the early part of the twentieth century when Griffith discovered the transformation factor, knowledge of genetics has been based on work with the simplest biological systems—viruses and bacteria. Scientists' understanding of replication, transcription, and translation of DNA was worked out using bacteria as a model. Their understanding of how viruses and bacteria infect cells is the basis for how diseases are treated and how vaccines are developed. A worldwide industry of genetic engineering and recombinant DNA relies on bacteria like *Escherichia coli* and viruses like the phage viruses for research and therapeutic endeavors. Whereas Gregor

Mendel depended on the garden pea and Thomas Hunt Morgan on the fruit fly, researchers now depend on bacteria and viruses.

The Genetics of Viruses

A virus is a parasite that can live only inside another cell. It commandeers the host cell machinery to transcribe and translate all the proteins it needs to fashion new viruses. In the process, thousands of new viruses are formed and the host cell is often destroyed. A virus consists of DNA or RNA enclosed in a protein coat called a **capsid**. Some viruses also have a viral **envelope** that is derived from membranes of host cells, cloaks the capsid, and aids the virus in infecting the host. Each type of virus can infect only one specific cell type because it gains entrance into a cell by binding to *specific receptors* on the cell surface. For example, the virus that causes colds in humans infects only the membranes of the respiratory system, and the virus that causes AIDS infects only one type of white blood cell. In addition, one virus can usually only infect one species. The range of organisms that a virus can attack is referred to as the **host range** of the virus. A sudden emergence of a new viral disease that affects humans, such as AIDS or hantavirus, may result from a mutation in the virus that expands its host range.

- **Bacteriophages**—The most complex and best understood virus is the one that infects bacteria, the **bacteriophage**, or **phage** virus. The bacteriophage can reproduce in different ways.

  1. In the **lytic cycle**, the phage enters a host cell, takes control of the cell machinery, replicates itself, and then causes the cell to burst, releasing a new generation of infectious phage viruses. These new viruses infect thousands of cells in the same manner.
  2. In the **lysogenic cycle**, viruses replicate without destroying the host cell. The phage virus becomes incorporated into a specific site in the host's DNA. It remains dormant within the host genome and is called a **prophage**. As the host cell divides, the phage is replicated along with it and a single infected cell gives rise to a population of infected cells. At some point, an environmental trigger causes the prophage to switch to the **lytic phase**. Viruses capable of both modes of reproducing, lytic and lysogenic, within a bacterium are called **temperate viruses**.

- **Retroviruses** are viruses that contain RNA instead of DNA and replicate in an unusual way. Following infection of the host cell, their RNA serves as a template for the synthesis of complementary DNA (cDNA) because it is complementary to the RNA from which it was copied. *Thus, these retroviruses reverse the usual flow of information from DNA to RNA.* This reverse transcription occurs under the direction of an enzyme called **reverse transcriptase**. A retrovirus usually inserts itself into the host genome, becomes a permanent resident, called a prophage, and is capable of making multiple copies of the viral genome for years. Examples of retroviruses are the polio virus and the HIV (human immunodeficiency virus), which causes AIDS.

- **Transduction**—Phage viruses acquire bits of bacterial DNA as they infect one cell after another. This process, which leads to genetic recombination is called

transduction. Two types of transduction occur, **generalized** and **restricted (specialized)**. Generalized transduction moves random pieces of bacterial DNA as the phage lyses one cell and infects another during the lytic cycle. **Restricted transduction** involves the transfer of specific pieces of DNA. During the lysogenic cycle, a phage integrates into the host cell at a specific site. At a later time, when the phage ruptures out of the host DNA, it sometimes carries a piece of adjacent host DNA with it and inserts this host DNA into the next host it infects.

The Genetics of Bacteria

The bacterial chromosome is a circular, double-stranded DNA molecule, tightly condensed into a structure called a **nucleoid**, which has no nuclear membrane. Bacteria replicate their DNA in **both directions** from a **single point of origin** by a process known as **theta replication**, owing to the similarity in appearance to the Greek letter theta, θ.

Although bacteria can reproduce by a primitive sexual method called **conjugation**, the main mode of reproduction is asexual, by **binary fission**. Binary fission results in a population with all identical genes, but mutations do occur spontaneously. Although mutations are rare, bacteria reproduce by the millions, and even one mutation in every 1,000 replications can amount to significant variation in the population as a whole.

- **Bacterial transformation** was discovered by **Frederick Griffith** in 1927 when he performed experiments with several different strains of the bacterium *Diplococcus pneumoniae*.

  Transformation is either a natural or an artificial process that provides a mechanism for the recombination of genetic information in some bacteria. Small pieces of extracellular DNA are taken up by a living bacterium, ultimately leading to a stable genetic change in the recipient cell. Bacterial transformation is very easy to carry out today, in fact, it is one of the recommended labs from the College Board. See Lab #6.

- A **plasmid** is a foreign, small, circular, self-replicating DNA molecule that inhabits a bacterium. A bacterium can harbor many plasmids and will express the genes carried by the plasmid. The first plasmid discovered was the **F plasmid**. F stands for fertility. Bacteria that contain the F plasmid are called F^+; those that do not carry the plasmid are called F^-. The F plasmid contains genes for the production of **pili**, cytoplasmic bridges that connect to an adjacent cell and that allow DNA to move from one cell to another in a form of primitive sexual reproduction called **conjugation**. Another plasmid, the **R plasmid** makes the cell in which it is carried resistant to specific antibiotics, such as ampicillin or tetracycline. In addition, the R plasmid can be transferred to other bacteria by conjugation. Bacteria that carry the R plasmid have a distinct evolutionary advantage over bacteria that are not resistant to antibiotics. Resistant bacteria will be selected for (survive) and their populations will increase while nonresistant bacteria die out. This is exactly what is happening today as an increasing number of populations of pathogenic bacteria, such as the one that causes tuberculosis, are becoming resistant to antibiotics. This is cause for serious concern in the health community.

The Operon

The **operon** was discovered in the bacterium, *E. coli* and is an important model of **gene regulation**. An operon is essentially a set of genes and the switches that control the expression of those genes. **Jacob and Monod**, who discovered the operon in the 1940s, described two different types: the **inducible (Lac)** operon and the **repressible (tryptophan)** operon. The Lac operon is switched off until it is induced to turn on. The tryptophan operon, in contrast, is always in the on position until it is not needed and becomes repressed or switched off. Figure 9.8 shows two sketches of the two systems with an explanation beneath.

STUDY TIP

If an essay question on the AP Exam is about regulation, the operon is a perfect example.

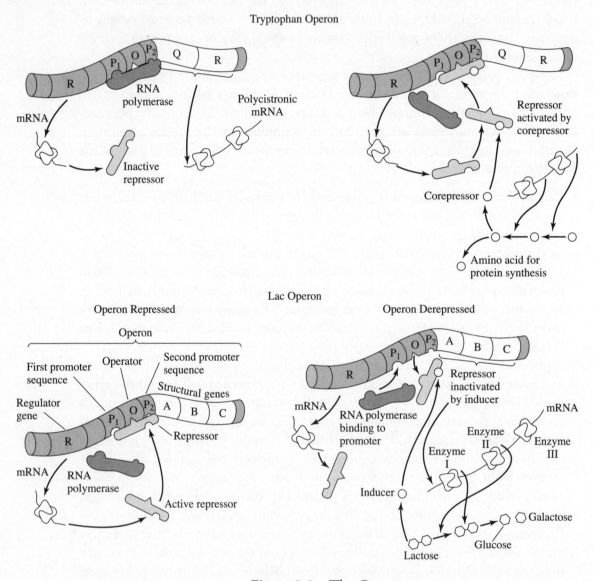

Figure 9.8   The Operon

- **The Lac operon**—Normally, lactose is not available to bacteria as an energy source and the genes necessary to utilize lactose are not transcribed. In order for *E. coli* to utilize lactose, three structural genes must be transcribed to produce the enzymes necessary for the breakdown of lactose into glucose and galactose, sugars in a form the cell can readily use. The three enzymes, b-galactosidase, permease, and transacetylase, are coded for by three structural genes in the Lac operon (A, B, and C shown in Figure 9.8). In order for transcription to occur, **RNA polymerase** must bind to DNA at the **promoter**. If a **repressor** binds to the **operator**, RNA polymerase is prevented from binding to the promoter and transcription of the structural genes is blocked or repressed. The relationship between RNA polymerase and the repressor is an example of **noncompetitive inhibition** because both substances are competing for two active sites, one of which coincidentally blocks the other. If **allolactose**, which is similar to lactose, is present in the environment, it acts as an **inducer** or **allosteric effector**. It binds to the repressor, causing the repressor to change its shape or **conformation**. With the shape altered, the repressor cannot bind to the operator and RNA polymerase is free to bind to the promoter. When RNA polymerase binds to the promoter, the structural genes transcribe and lactose can be utilized.

- The **tryptophan operon** is a repressible operon; it is continuously switched *on* unless turned *off* by a **corepressor**. It consists of five structural genes that code for the enzymes necessary to synthesize the amino acid tryptophan (Q and R shown in Figure 9.8). The **repressor** molecule encoded by the **regulator gene** (located at some distance from the structural genes) is initially inactive. Therefore RNA polymerase is free to bind to the **promoter** and transcribe the structural genes, resulting in the production of tryptophan. When the inactive **repressor** combines with a specific corepressor molecule (**tryptophan**), it changes its conformation and binds to the operator, preventing RNA polymerase from binding to the promoter and, thus, blocking transcription of the structural genes. If tryptophan levels surrounding the bacterium are high, no more is needed, so no more is synthesized. Tryptophan, like lactose in the lac operon, is an **allosteric effector**.

PRIONS

Prions are not cells and are not viruses. They are misfolded versions of a protein normally found in the brain. If prions get into a normal brain, they cause all the normal versions of the protein to misfold in the same way. Prions are infectious and cause several brain diseases: **scrapie** in sheep, **mad cow disease** in cattle, and **Creutzfeldt-Jakob disease** in humans. All known prion diseases are fatal.

TRANSPOSONS

Transposons are transposable genetic elements and are sometimes called **jumping genes**. They were discovered by **Barbara McClintock**, who was studying the genetics of corn in the 1940s and 1950s. Some transposons jump in a cut-and-paste fashion from one part of the **genome** to another. Others make copies of themselves that move to another region of the genome, leaving the original behind. There are two classes of jumping genes, **insertion sequences** and **complex transposons**.

- **Insertion sequences** consist of only one gene, which codes for **transposase**, the enzyme responsible for moving the sequence from one place to another. These can cause a **mutation** if they happen to land within a DNA region that regulates gene expression.
- **Complex transposons** are longer than insertion sequences and include extra genes, such as a gene for antibiotic resistance or for seed color. McClintock hypothesized the existence of transposons when she saw patterns in corn color that made sense only if some genes were mobile. At the time, very few scientists believed that genes could move at all, and her work was ignored. In 1983, however, her work was recognized and she was awarded the Nobel prize.

THE HUMAN GENOME

The human genome consists of 3 billion base pairs of DNA and about 30,000 genes. Surprisingly, **97 percent** of human DNA does NOT code for protein product and has often been called **junk**. Of the **noncoding DNA**, some are **regulatory sequences** that control gene expression, some are **introns** that interrupt genes, and most are **repetitive sequences** that never get transcribed. A number of genetic disorders, including Huntington's disease, are caused by abnormally long stretches of **tandem repeats** (back-to-back repetitive sequences) *within* affected genes. Many of these tandem repeats make up the **telomeres**. Scientists have also identified certain non-coding regions of DNA, **polymorphic regions**, that are highly variable from one region to the next.

RECOMBINANT DNA

Recombinant DNA means taking DNA from two sources and combining them into one molecule. This occurs in nature during **viral transduction**, **bacterial transformation**, and **conjugation** and when **transposons** jump around the genome. Scientists can also manipulate and engineer genes in vitro (in the laboratory). The branch of science that uses **recombinant DNA techniques** for practical purposes is called **biotechnology** or **genetic engineering**.

Many tools and techniques have been developed to manipulate and engineer genes. The following sections contains a discussion of the uses for genetic engineering, an explanation of some techniques that are used, and a brief discussion of some ethical issues within the field.

CLONING GENES

The potential uses of **recombinant DNA** or **gene cloning** are many:

- To produce a **protein product**, such as human insulin, in large quantities as an inexpensive pharmaceutical.
- To **replace** a **nonfunctioning gene** in a person's cells with a functioning gene by **gene therapy**. Scientists are currently conducting clinical trials in this area with disappointing results. Sometimes the human subjects become ill from the viral **vector** used to carry the gene. Other times, the gene is inserted successfully and begins to produce the necessary protein but stops working in a short time. If scientists can master this technique, many lives will be changed.

- To prepare **multiple copies of a gene** itself **for analysis**. Since most genes exist in only one copy on a chromosome, the ability to make multiple copies is of great value as a research tool.
- To **engineer bacteria** to clean up the environment. Scientists have engineered many bacteria; one can even eat **toxic waste**.

The Technique of Gene Cloning

- Isolate a gene of interest, for example, the gene for human insulin.
- Insert the gene into a **plasmid**.
- Insert the plasmid into a **vector**, a cell that will carry the plasmid, such as a bacterium. To accomplish this, a bacterium must be made **competent**, which means be able to take up a plasmid.
- **Clone** the gene. As the bacteria reproduce themselves by **fission**, the plasmid and the selected gene are also being **cloned**. Millions of copies of the gene are produced.
- Identify the bacteria that contain the selected gene and harvest it from the culture.

TOOLS AND TECHNIQUES OF RECOMBINANT DNA

Restriction Enzymes

Restriction enzymes were discovered in the late 1960s and are a basic biotechnology tool. They are extracted from bacteria, which use them to fend off attacks by invading bacteriophages. Restriction enzymes cut DNA at specific **recognition sequences** or **sites**, such as GAATTC. Often these cuts are staggered, leaving single-stranded **sticky ends** to form a temporary union with other sticky ends. The fragments that result from the cuts made by restriction enzymes are called **restriction fragments**.

Scientists have now isolated hundreds of different restriction enzymes. They are named for the bacteria in which they were found. Common examples are *Eco*RI (which was discovered in *E. coli*), *Bam*HI and *Hind*III. Restriction enzymes have many uses, including **gene cloning**.

> **STUDY TIP**
>
> Restriction enzymes and gel electrophoresis are important topics. Study them well.

Gel Electrophoresis

Gel electrophoresis separates large molecules of DNA on the basis of their rate of movement through an **agarose gel** in an electric field. The smaller the molecule, the faster it runs. DNA, which is negative (due to the presence of phosphate groups, PO_4^{3-}), flows from **cathode** (−) to **anode** (+). The concentration of the gel can be altered to provide a greater impediment to the DNA, allowing for finer separation of smaller pieces.

Electrophoresis is also commonly used to separate proteins and amino acids. If DNA is going to be run through a gel, it must first be cut up by **restriction enzymes** into pieces small enough to migrate through the gel. Once separated on a gel, the DNA can be analyzed in many ways. The DNA strands can be **sequenced** to determine the sequence of bases A, C, T, and G. The gel can also be used in a comparison with other DNA samples. A **DNA probe** can also identify the location of a specific sequence within the DNA.

Figure 9.9 shows gel electrophoresis of DNA that was cut with restriction enzymes. Lane 1 has four bands of DNA, three larger pieces and one short piece. Lane 2 contains two pieces of DNA, one large and one tiny one. Lane 3 contains one very large, uncut piece of DNA. Lane 4 contains two pieces of DNA. *The smaller the piece of DNA, the farther it has traveled from the well.*

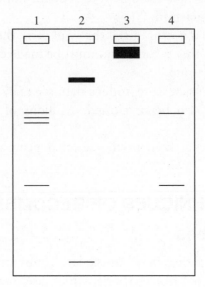

Figure 9.9   Gel Electrophoresis of DNA Fragments

DNA Probe

A **DNA probe** is a **radioactively labeled single strand** of nucleic acid molecule used to tag a specific sequence in a DNA sample. The probe bonds to the complementary sequence wherever it occurs, and the radioactivity enables scientists to detect its location. The DNA probe is used to identify a person who carries an inherited genetic defect, such as sickle cell anemia, Tay-Sachs disease, Huntington's disease, and hundreds of others.

Polymerase Chain Reaction (PCR)

Devised in 1985, **polymerase chain reaction** is a cell-free, automated technique by which a piece of DNA can be rapidly copied or amplified. Billions of copies of a fragment of DNA can be produced in a few hours. The DNA piece that is to be amplified is placed into a test tube with *Taq* **polymerase** (a heat-stable form of DNA polymerase extracted from extremophile bacteria), along with a supply of **nucleotides** (**A**, **C**, **T**, and **G**) and **primers** necessary for DNA synthesis. Once the DNA is amplified, these copies can be studied or used in a comparison with other DNA samples.

The PCR technique has limitations:

• Some information about the nucleotide sequence of the target DNA must be known in advance in order to make the necessary primers.

- The size of the piece that can be amplified must be very short.
- Contamination is a major problem. A few skin cells from the technician working with the sample could make obtaining accurate results difficult or impossible. This could have dire consequences with a crime scene sample.

Restriction Fragment Length Polymorphisms (RFLPs)

A **restriction fragment** is a segment of DNA that results when DNA is treated with restriction enzymes. When scientists compared **noncoding regions** (junk DNA) of human DNA across a population, they discovered that the **restriction fragment pattern** is different in every individual. These differences have been named **restriction fragment length polymorphisms** or **RFLPs**, pronounced "riflips." A RFLP analysis of someone's DNA gives a human **DNA fingerprint** that looks like a bar code.

Each person's RFLPs are unique, except in identical twins, and are inherited in a Mendelian fashion. Because they are inherited in this way, they can be used very accurately in **paternity suits** to determine, with absolute certainty, if a particular man is the father of a particular child. In addition, RFLPs are routinely used to identify the perpetrator in rape and murder cases. DNA from the crime scene and the victim are compared against DNA from the suspect. Because of the accuracy of RFLP analysis, these cases can be solved with a high degree of certainty, and some suspects have been convicted on DNA evidence alone. By contrast, several incidents have occurred where men who have been jailed for many years for violent crimes were proven innocent by DNA evidence and released.

Complementary DNA (cDNA)

When scientists try to clone a human gene in a bacterium, the introns (long intervening, noncoding sequences) present a problem. Bacteria lack introns and have no way to edit them out after transcription. Therefore, in order to clone a human gene in a bacterium, scientists must insert a gene with no introns. To do this, scientists extract fully processed mRNA from cells and then use the enzyme **reverse transcriptase** (obtained from **retroviruses**) to make DNA transcripts of this RNA. The resulting DNA molecule carries the complete coding sequence of interest but without introns. The DNA produced by retroviruses in this way is called **complementary DNA** or **cDNA**.

ETHICAL CONSIDERATIONS

Many people are worried about potential problems arising from genetic engineering. This section discusses some of those concerns.

Safety

Much of the milk available comes from cows that have been given a genetically engineered bovine growth hormone (BGH) to increase the quantity of milk they produce. Many people are concerned that this hormone will find its way into the milk and cause problems for the people who drink it.

Vegetable seeds have been genetically engineered to produce special characteristics in the vegetables people eat. Again, individuals are concerned that the genes that have been inserted into the vegetables may be dangerous to those who eat them.

Privacy

DNA probes are being coupled with the technology of the semiconductor industry to produce **DNA chips** that are about 1/2 inch square and can hold personal information about someone's genetic makeup. The chips scan a person for mutations in over 7,000 genes, including mutations in the immune system or the breast cancer genes (BRCA I and II) and for a predisposition to other cancers or heart attacks.

Although many people might like to know if they carry these mutations, the possibility that the personal information on a DNA chip might not remain private has caused much controversy. For example, if your health insurance company learned that you carry a harmful gene, it might not insure you or might charge you a higher premium. Similarly, if a company where you have applied for employment learns about a defect in your personal genetic makeup, based on the possibility that you might be disabled with a serious illness in the future, it might refuse to hire you.

Multiple-Choice Questions

Questions 1–4
Questions 1–4 refer to the following five choices.

(A) Translation
(B) Replication
(C) Transcription
(D) Transformation
(E) Termination

1. The process in which DNA makes messenger RNA

2. The process in which DNA is synthesized from a template strand

3. The process in which foreign DNA is taken up by a bacterial cell

4. The process in which a polypeptide strand is synthesized using mRNA as a template

5. All of the following are true about electrophoresis EXCEPT

    (A) it can be used to analyze only DNA
    (B) the heavier the fragment, the slower it moves
    (C) the fragments of DNA are negatively charged and migrate to the positive pole
    (D) a buffer must cover the gel to allow a current to pass through the system
    (E) restriction enzymes cut DNA in only certain sites on the strand

6. *Eco*RI is

    (A) a bacterium
    (B) a bacteriophage
    (C) a type of DNA used extensively in research
    (D) a protein that mimics DNA for use in research
    (E) a restriction enzyme

7. In DNA replication, the role of DNA polymerase is to

    (A) bring two separate strands back together after new ones are formed
    (B) join the RNA nucleotides together to make the primer
    (C) build a new strand of DNA from 5′ to 3′
    (D) unwind the tightly wound helix
    (E) join the Okazaki fragments

8. Which is NOT used in the normal replication of DNA?

    (A) RNA primer
    (B) ligase
    (C) restriction enzymes
    (D) polymerase
    (E) single-stranded binding protein

9. DNA replication can best be described as

    (A) semiconservative
    (B) conservative
    (C) degenerate
    (D) comparative
    (E) the same in eukaryotes and prokaryotes

Questions 10–13

Questions 10–13 refer to these scientists famous for their work with DNA.

(A) Hershey and Chase
(B) Watson and Crick
(C) Meselsohn and Stahl
(D) Rosalind Franklin
(E) Griffith

10. Discovered transformation in bacteria

11. Proved that DNA replicates by semiconservative replication

12. Proved that the nuclear material in a bacteriophage, not the protein coat, infects a bacterium

13. The first to analyze DNA by X-ray crystallography

---

14. If a segment of DNA is 5'-TGA AGA CCG-3', the RNA that results from the transcription of this segment will be

(A) 5'-TGA AGA CCG-3'
(B) 3'-GCC AGA AGT-5'
(C) 3'-ACT TCT GGC-5'
(D) 3'-CGG UCU UCA-5'
(E) 3'-ACU UCU GGC-5'

15. Once transcribed, eukaryotic RNA normally undergoes substantial alteration that results primarily from

(A) removal of exons
(B) removal of introns
(C) addition of introns
(D) combining of RNA strands by a ligase
(E) editing by the Golgi apparatus

16. Which of the following contain a pyrimidine and a purine?

(A) adenine and guanine
(B) uracil and thymine
(C) cytosine and uracil
(D) adenine and cytosine
(E) cytosine and thymine

17. What happens when T7 bacteriophages are grown in radioactive phosphorous?

(A) They can no longer infect bacteria.
(B) They die.
(C) Their DNA becomes radioactive.
(D) Their protein coat becomes radioactive.
(E) Both the DNA and the protein coat become radioactive.

18. Which of the following acts as a primer that initiates the synthesis of a new strand of DNA?

    (A) single-strand binding protein
    (B) RNA
    (C) DNA
    (D) topoisomerases
    (E) ligase

19. If guanine makes up 28% of the nucleotides in a sample of DNA from an organism, then thymine would make up _____ % of the nucleotides.

    (A) 28
    (B) 56
    (C) 22
    (D) 44
    (E) 0

20. If AUU is the codon, what is the anticodon?

    (A) AUU
    (B) TAA
    (C) UUA
    (D) UAA
    (E) TUU

21. Which of the following is an example of wobble?

    (A) amino acids carried to the ribosome to form a polypeptide chain
    (B) the excision of introns from mRNA
    (C) the binding of a primer to DNA
    (D) four codons can all code for the same amino acid
    (E) the attachment of mRNA to the ribosome

22. Which of the following is TRUE about sickle cell anemia?

    (A) It is caused by a chromosome mutation that resulted from nondisjunction.
    (B) It is common in people from the Middle East.
    (C) It is never found in Caucasians.
    (D) A person with sickle cell anemia is resistant to many other genetic disorders.
    (E) It is caused by a point mutation.

23. A particular triplet code on DNA is AAA. What is the anticodon for it?

    (A) AAA
    (B) TTT
    (C) UUU
    (D) CCC
    (E) GGG

24. What are the regions of DNA called that code for proteins?

    (A) introns
    (B) codons
    (C) anticodons
    (D) exons
    (E) transposons

25. Prions are

    (A) bacteriophages that cause disease
    (B) infectious proteins
    (C) a bacterium that infects viruses
    (D) the cause of sickle cell anemia
    (E) introns that have been excised from a strand of mRNA

26. Which word would **best** describe the operon?

    (A) respiration
    (B) transport
    (C) regulation
    (D) nutrition
    (E) photosynthesis

Questions 27–30
Questions 27–30 refer to the operon.

    (A) Lactose
    (B) Repressor
    (C) Regulator
    (D) Promoter
    (E) RNA polymerase

27. Acts as an inducer in the *Lac* operon

28. Binding site for RNA polymerase

29. Codes for the repressor

30. Binds at the operator

---

31. Which is TRUE of biotechnology techniques?

    (A) PCR is used to cut DNA molecules.
    (B) A DNA probe consists of a radioactive single strand of DNA.
    (C) Restriction enzymes were first discovered in bacteriophage viruses.
    (D) *Eco*RI is a name for a DNA probe.
    (E) All humans contain the same RFLPs.

32. Gel electrophoresis is used to

    (A) amplify small pieces of DNA
    (B) make bacterial cells competent
    (C) cut DNA into small pieces
    (D) cause DNA to twist back into a helix after amplification
    (E) separate DNA that has already been cut up by restriction enzymes

33. Mad cow disease is caused by a

    (A) virus
    (B) prion
    (C) bacterium
    (D) genetic mutation
    (E) plant toxin

34. Which enzyme permanently seals together DNA fragments that have complementary sticky ends?

    (A) DNA polymerase
    (B) single-stranded binding protein
    (C) reverse transcriptase
    (D) DNA ligase
    (E) RNA polymerase

Answers to Multiple-Choice Questions

1. **(C)** Transcription is the process by which DNA makes RNA. There are three types of RNA; mRNA, tRNA and rRNA. Transcription occurs in three stages: initiation, elongation, and termination.

2. **(B)** DNA makes an exact copy of itself during replication. This process occurs in a semiconservative fashion, as proven by Meselsohn and Stahl.

3. **(D)** Griffith was the first to recognize the phenomenon of transformation while working with pneumococcus bacteria.

4. **(A)** Translation is the process by which the codons of mRNA sequence are changed into an amino acid sequence. Amino acids present in the cytoplasm are carried by tRNA molecules to the codons of the mRNA strand at the ribosome according to the base-pairing rules (A with U and C with G).

5. **(A)** Electrophoresis is commonly used in the separation of proteins as well as in the separation of DNA.

6. **(E)** *Eco*RI stands for *E. coli* restriction enzyme #1. It was the first restriction enzyme discovered.

7. **(C)** DNA polymerase builds a new strand of DNA from the 5′ end to the 3′ end (of the new strand). DNA polymerase can only add nucleotides to an existing strand of DNA.

8. **(C)** Restriction enzymes are a laboratory tool for cutting pieces of DNA at specific restriction sites.

9. **(A)** Replication of DNA is semiconservative. This means that when one double helix makes a copy of itself, the two new DNA molecules each consist of one new strand and one old strand. This was hypothesized by Watson and Crick and confirmed experimentally by Meselsohn and Stahl.

10. **(E)** Griffith discovered bacterial transformation in 1927.

11. **(C)** Semiconservative replication was hypothesized by Watson and Crick and confirmed experimentally by Meselsohn and Stahl.

12. **(A)** Hershey and Chase carried out experiments where they tagged bacteriophages with ^{32}P and ^{35}S. They proved that the DNA from the viral (phage) nucleus, not protein from the viral coat, was infecting bacteria and producing thousands of progeny.

13. **(D)** Rosalind Franklin, while working in the lab of Maurice Wilkins, carried out the X-ray crystallography analysis of DNA that showed DNA to be a helix.

14. **(E)** You are given a strand of DNA, which makes a strand of mRNA. Follow the base-pairing rules: T with A, C with G, C with G, and A with U. Remember, RNA contains uracil instead of thymine. If the DNA segment is 5'-TGA AGA CCG-3', then the mRNA strand complementary to that is 3'-ACU UCU GGC-5'.

15. **(B)** Once transcription has occurred, the new RNA molecule undergoes RNA processing. During this process, introns (intervening sequences) are removed with the help of snRNPs and a 5' cap and poly(A) tail are added.

16. **(D)** Pyrimidines often have the letter *y* in them. They are thymine, cytosine, and uracil, which replaces thymine in RNA. Adenine is the purine; thymine is the pyrimidine.

17. **(C)** A bacteriophage virus consists of nuclear material surrounded by a protein coat. Proteins contain sulfur, and DNA contains phosphorous (in the phosphates). When a phage virus is grown in radioactive phosphorous, the phosphorous gets incorporated into the DNA, not into the protein coat.

18. **(B)** DNA polymerase can only add nucleotides to an existing strand of nucleotides. RNA primer binds to the DNA, and DNA polymerase attaches nucleotides to the RNA primer.

19. **(C)** If guanine makes up 28% of the DNA, then there must be an equal amount of cytosine (28%), for a total of 56%. That leaves 44% for adenine and thymine. Divide 44 by 2 = 22%, which is the percentage of thymine in the DNA.

20. **(D)** The codon is the nucleotide triplet associated with mRNA; the anticodon is the nucleotide sequence associated with tRNA. Codons and anticodons are complementary to each other.

21. **(D)** The pairing rules are not as strict for the third codon in mRNA as they are for the first two. One example is that UUU and UUA both code for phenylalanine.

22. **(E)** Sickle cell anemia is caused by a gene mutation in the gene that codes for hemoglobin. Sickle cell disease is common where malaria is endemic, in West Africa and southeast Asia. People who are carriers for the sickle cell trait are resistant to malaria. Sickle cell disease does occur in the Middle East but is not common there. Sickle cell does occur in Caucasians, but rarely.

23. **(A)** DNA (triplet code) makes RNA (codon) makes protein (anticodon). If the triplet in DNA is AAA, then the codon on mRNA is UUU, and the anticodon on tRNA is AAA.

24. **(D)** The regions of DNA that code for proteins are called exons or expressed sequences.

25. **(B)** A prion is a misfolded version of a protein normally found in the brain. Prions are infectious. If they get into the brain, they will convert the normal proteins to abnormal ones. They have been identified as the infectious agent in several diseases, including scrapie in sheep, Creutzfeldt-Jakob in humans, and mad cow disease.

26. **(C)** The operon is the means by which prokaryotes regulate gene expression. An operon consists of a cluster of related genes and the DNA that controls them, such as, a promoter and operator.

27. **(A)**

28. **(D)**

29. **(C)**

30. **(B)**

31. **(B)** A DNA probe is a single radioactive strand of DNA used to tag and identify a specific sequence in a strand of DNA. PCR is a cell-free system that amplifies small pieces of DNA rapidly. Restriction enzymes are found in bacteria. *Eco*RI was the first restriction enzyme discovered. Every person has a unique set of RFLPs.

32. **(E)** Restriction enzymes cut DNA at specific recognition sites. Gel electrophoresis separates DNA that has already been cut up by restriction enzymes. Single-stranded binding proteins, helicases, and topoisomerases help DNA twist during replication.

33. **(B)** A prion is a misfolded and infectious protein. See question 25.

34. **(D)** DNA ligase seals together DNA fragments that have complementary sticky ends.

Free-Response Questions

Directions: Answer all questions. You must answer the question in essay—**not** outline—form. You may use labeled diagrams to supplement your essay, but diagrams alone are *not* sufficient. Before you start to write, read each question carefully so that you understand what the question is asking.

1. Explain the process by which DNA makes proteins.

2. By using techniques of genetic engineering, scientists are able to learn more about the human genome and to use these techniques for the betterment of humankind. Describe these techniques or procedures below, and explain how each contributes to our understanding of the human genome or for what practical purpose it can be used.
    a. Polymerase chain reaction
    b. Restriction fragment length polymorphism (RFLP) analysis
    c. Gene cloning

3. All humans are almost genetically identical. However, every person has a unique DNA fingerprint. Explain this contradiction.

Typical Free-Response Answers

Note: This is an essay that has appeared on the AP Exam before and could appear again in some form. You must understand molecular biology thoroughly because it is the basis of so much of modern biological theory. Once again, key words are in boldface to remind you that you must include correct scientific terminology and clear definitions to get full credit.

1. The process whereby DNA makes proteins has been worked out in great detail. To summarize, the **triplet code** in DNA is transcribed into a **codon sequence** in **messenger RNA (mRNA)** inside the nucleus. Next, this newly formed strand of **RNA is processed** in the nucleus. Then the codon sequence is **translated into an amino acid sequence** in the cytoplasm at the ribosome by **tRNA**.

    The first stage of the process is **transcription,** where DNA makes mRNA that carries the message directly to the ribosome in the cytoplasm. Transcription consists of three stages: **initiation, elongation**, and **termination**. During initiation, the enzyme **RNA polymerase** recognizes and binds to DNA at the **promoter**. Once RNA polymerase is attached to the promoter, DNA transcription of the DNA template begins. The next step, elongation, continues as RNA polymerase adds nucleotides to the 3′ end of a growing chain. RNA polymerase pries the two strands of DNA apart and attaches RNA nucleotides according to the base-pairing rules: **C (cytosine)** with **G (guanine)** and **A (adenine)** with **U (uracil)**. The stretch of DNA that is transcribed into an mRNA molecule is called a **transcription unit** and consists of triplets of bases called **codons** that code for specific amino acids. When RNA transcribes a **termination sequence**, the process stops.

Before the newly formed **transcription unit** is shipped out to the ribosome from the nucleus, it is altered or processed by a series of enzymes. A **5′ cap** is added to the 5′ end, which helps protect the RNA strand from degradation by hydrolytic enzymes and which also helps the RNA strand bind to the ribosome in the cytoplasm. In addition, a **poly(A) tail**, is added to the 3′ end to protect the RNA strand from degradation by hydrolytic enzymes and to facilitate the release of the RNA into the cytoplasm. A major part of the processing is the removal of **introns** (noncoding regions) from the transcription unit by **SnRNPs** (small nuclear ribonucleoproteins) and spliceosomes. With processing complete, only **exons**, expressed sequences, move out to the ribosome for translation.

Translation is the process by which the codons of an mRNA sequence are changed into an amino acid sequence. Amino acids present in the cytoplasm are carried by **tRNA** molecules to the codons of the mRNA strand at the ribosome according to the base-pairing rules (**A** with **U** and **C** with **G**). One end of the tRNA molecule bears a specific amino acid, and the other end bears a nucleotide triplet called an anticodon. Some tRNA molecules have **anticodons** that can recognize two or more different codons. This is because the pairing rules for the third base of a codon are not as strict as they are for the first two bases. This relaxation of base-pairing rules is known as **wobble**. For example, codons UCU, UCC, UCA, and UCG all code for the amino acid serine. The process of translation consists of three stages: initiation, elongation, and termination. Translation of an mRNA strand is complete when a ribosome reaches one of the **termination or stop codons**. The mRNA is broken down, the tRNA is reused, and the polypeptide is freed from the **ribosome**.

Note: There have been two free-response questions on the AP Biology Exam about genetic engineering in the last few years. This reflects the importance of this topic today. You can assume it will appear again. Here is a similar sample question. The answers to Part a are taken directly from the text above.

2a. **Polymerase chain reaction**

Devised in 1985, polymerase chain reaction is a cell-free, automated technique by which a piece of DNA can be rapidly copied or amplified. Billions of copies of a fragment of DNA can be produced in a few hours. The DNA piece that is to be amplified is placed into a test tube with *Taq* **polymerase** (a heat-stable form of DNA polymerase extracted from bacteria that live in very hot places, such as hot springs), along with a supply of nucleotides (A, C, T, and G) and primers necessary for DNA synthesis. Once the DNA is amplified, these copies can be studied or used in a comparison with other DNA samples.

The PCR technique has limitations. First, some information about the nucleotide sequence of the target DNA must be known in advance in order to make the necessary primers. Second, the size of the piece that can be amplified must be very short. Contamination is a major problem. Third, a few skin cells from the technician working with the sample could make obtaining accurate results difficult or impossible. This could have dire consequences with a crime scene sample.

2b. **Restriction length polymorphisms (RFLP) analysis**

A restriction fragment is a segment of DNA resulting from treatment of DNA with restriction enzymes. When scientists compared noncoding regions of human DNA across a population, they discovered that the restriction fragment pattern is different in every individual. These differences have been named restriction fragment length polymorphisms or RFLPs, pronounced "riflips." A RFLP analysis of someone's DNA gives a human DNA fingerprint that looks like a bar code.

Each person's RFLPs are unique, except in identical twins, and are inherited in a Mendelian fashion. Because they are inherited in this way, they can be used very accurately in paternity suits to determine, with absolute certainty, if a particular man is the father of a particular child. In addition, RFLPs are routinely used to identify the perpetrator in rape and murder cases. DNA from the crime scene and the victim are compared against DNA from the suspect. Because of the accuracy of RFLP analysis, these cases can be solved with a high degree of certainty, and some suspects have been convicted on DNA evidence alone. By contrast, several incidents have occurred where men who have been jailed for many years for violent crimes were proven innocent by DNA evidence and released.

2c. **Gene cloning**

The potential uses of recombinant DNA or gene cloning are many. Here are several.

- To produce a protein product such as insulin or human growth hormone, in large quantities as an inexpensive pharmaceutical.
- To replace a nonfunctioning gene in a person's cells with a functioning gene by gene therapy. Scientists are currently conducting clinical trials in this area with disappointing results. If scientists can master this technique, many lives will be changed.
- To prepare multiple copies of a gene itself for analysis. Since most genes exist in only one copy in a cell, the ability to make multiple copies is of great value.
- To engineer bacteria to clean up the environment. Scientists have engineered many bacteria; one can even eat toxic waste.

In order to clone a gene, you must first isolate a gene of interest, for example, the gene for human insulin. Next insert the gene into a **plasmid.** Then insert the plasmid into a **vector**, a cell that will carry the plasmid, such as a bacterium like *E. coli*. To accomplish this, a bacterium must be made **competent**, be able to take up a plasmid. As the bacteria reproduce themselves by fission, the plasmid and the selected gene are also being cloned.

3. Genetically, all humans are almost identical. We can say this because only about 3 percent of all DNA contains genes that code for proteins. Furthermore, all the genes for a particular trait are identical. For example, everyone's gene for brown eyes is identical, and all genes for normal hemoglobin are identical.

    If all genes for a particular trait are identical, then how can each person have a unique DNA fingerprint? The answer lies not in the DNA that codes for genes but in the DNA that does not code for proteins—the **introns**, junk and **repetitive sequences**. Scientists have identified certain noncoding regions of DNA that are highly variable from one individual to the next. These regions are referred to as **polymorphic regions**. By analyzing many polymorphic regions, scientists have identified and agreed upon certain standard regions to analyze. This standardized analysis produces the DNA fingerprint that looks like a bar code and that is unique for every individual.

Classification

CHAPTER 10

- The three-domain classification system
- Evolutionary trends in animals
- Nine common animal phyla
- Characteristics of mammals
- Characteristics of primates

INTRODUCTION

Taxonomy or **classification** is the naming and classification of species. It began in the eighteenth century when Linnaeus developed the system used today, the **system of binomial nomenclature**. This system has two main characteristics: a two-part name for every organism (for example, human is *Homo sapiens* and lion is *Panthera leo*); and a hierarchical classification of species into broader groups of organisms. These broader groups or **taxa**, in order from the general to the specific are **kingdom**, **phylum**, **class**, **order**, **family**, **genus**, and **species**.

THE THREE-DOMAIN CLASSIFICATION SYSTEM

In the twentieth century, our system of classification went through many changes. Prior to the 1950s and 1960s, all organisms were placed into only three kingdoms. From the 1960s to around 1990, scientists classified all organisms into five kingdoms: **Monera**, **Protista**, **Fungi**, **Plantae**, and **Animalia**. In 1990, some scientists added a sixth kingdom, the **Archaebacteria**, which included **extremophiles**, microorganisms that seemed so different from bacteria that they had to be placed into a separate kingdom.

Today, however, most scientists use another system, based on DNA analysis, which more accurately reflects evolutionary history and the relationships among organisms. This system is called the **three-domain system**. In it, all life is organized into three domains: **Bacteria**, **Archaea**, and **Eukarya**.

Figure 10.1 shows the organization of the current three-domain system of classification. (*The kingdom **Monera** is no longer used in this system. Instead, prokaryotes are spread across two different domains, Archaea and Bacteria.*)

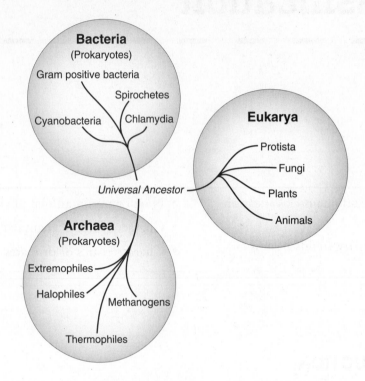

Figure 10.1   The Three-Domain Classification System

Table 10.1 shows a comparison of Bacteria, Archaea, and Eukaryotes. Notice, for some characteristics, the archaea resemble eukaryotes more than they resemble prokaryotes.

TABLE 10.1

Bacteria, Archaea, and Eukaryotes

Feature	Bacteria	Archaea	Eukaryotes
Membrane-enclosed organelles	Absent	Absent	Present
Peptidoglycan in cell wall	Present	Absent	Absent
RNA Polymerase	One type	Several kinds	Several kinds
Introns (noncoding regions of genes)	Absent	Present in some genes	Present
Antibiotic sensitivity to streptomycin, chloramphenicol	Inhibited	Not inhibited	Not inhibited

Domain Bacteria

- All are single-celled **prokaryotes** with no internal membranes (no nucleus, mitochondria, or chloroplasts).
- Some are anaerobes; some are aerobes.
- Bacteria play a vital role in the ecosystem as **decomposers** that recycle dead organic matter.
- Many are **pathogens**, disease causing.
- Bacteria play a vital role in **genetic engineering**. The bacteria from the human intestine, *Escherichia coli*, are used to manufacture human insulin.
- Some bacteria carry out **conjugation**, a primitive form of sexual reproduction where individuals exchange genetic material.
- Have a thick, rigid cell wall containing a substance known as **peptidoglycan**.
- Some carry out photosynthesis, but others do not.
- No introns (noncoding regions within the DNA).
- Corresponds roughly to the old grouping Eubacteria and includes blue-green algae, bacteria like *E. coli* that live in the human intestine, those that cause disease like *Clostridium botulinum* and *Streptococcus*, and those necessary in the nitrogen cycle, like nitrogen-fixing bacteria and nitrifying bacteria.
- Viruses are placed here because we do not know where else to place them.

Domain Archaea

- Unicellular
- Prokaryotic—no internal membranes such as a nucleus
- Includes **extremophiles**, organisms that live in extreme environments, like

  1. **Methanogens**—obtain energy in a unique way by producing methane from hydrogen
  2. **Halophiles**—thrive in environments with high salt concentrations like Utah's Great Salt Lake
  3. **Thermophiles**—thrive in very high temperatures, like in the hot springs in Yellowstone Park or in deep-sea hydrothermal vents

- Introns present in some genes
- No peptidoglycan

Domain Eukarya

Eukarya is a superkingdom that includes four of the original kingdoms, protista, fungi, plants, and animals.

- All organisms have a nucleus and internal organelles
- No peptidoglycan in cells
- Includes the four remaining kingdoms: Protista, Fungi, Plantae, and Animalia (see Table 10.2)

Today, taxonomy as a separate discipline is being replaced by the study of **systematics**, which includes taxonomy but considers biological diversity in an evolutionary context. Systematics focuses on tracing the ancestry of organisms. This is particularly important in light of current advances in DNA techniques that allow scientists to compare two species at the molecular level.

TABLE 10.2

The Four Kingdoms of Eukarya

Kingdom	Characteristics
Protista	Includes the widest variety of organisms, but all are eukaryotes Includes organisms that do not fit into the fungi or plant kingdoms, such as seaweeds and slime molds Consists of single and primitive multicelled organisms Includes **heterotrophs** and **autotrophs** Amoeba and paramecium are heterotrophs Euglena are primarily autotrophic with red **eyespot** and chlorophyll to carry out photosynthesis Protozoans like amoeba and paramecium are classified by how they move Mobility by varied methods: amoeba—pseudopods; paramecium—cilia; euglena—flagella Some carry out **conjugation**, a primitive form of sexual reproduction Some cause serious diseases like amoebic dysentery and malaria
Fungi	All are heterotrophs and eukaryotes Secrete hydrolytic enzymes outside the body where **extracellular** digestion occurs, then the building blocks of the nutrients are absorbed into the body of the fungus by diffusion Are important in the ecosystem as **decomposers** Cell walls are composed of **chitin**, not cellulose Examples: yeast, mold, mushrooms, the fungus that causes athlete's foot
Plantae	All are autotrophic eukaryotes Some plants have vascular tissue (Tracheophytes), some do not have any vascular tissue (Bryophytes) Examples: mosses, ferns, cone-bearing and flowering plants See separate sections on plants for specifics
Animalia	All are heterotrophic, multicellular eukaryotes Are grouped in 35 phyla; but this book discusses 9 main phyla: **Porifera, Cnidaria, Platyhelminthes, Nematoda, Annelida, Mollusca, Arthropoda, Echinodermata,** and **Chordata** Most animals reproduce sexually with a dominant **diploid** stage In most species, a small, flagellated sperm fertilizes a larger, nonmotile egg Is **monophyletic**, meaning all animal lineages can be traced back to one common ancestor Traditionally classified based primarily on **anatomical features** (**homologous structures**) and embryonic **development**

For the AP Exam, you should have a general knowledge of the characteristics of each of the four kingdoms as well as sample organisms belonging to each. These are in Table 10.2. For more details about plants, see the separate chapter on plants.

EVOLUTIONARY TRENDS IN ANIMALS

Organisms began as tiny, primitive, single-celled organisms that lived in the oceans. The first **multicellular** eukaryotic **organisms** evolved about 1.5 billion years ago. The appearance of each phylum of animal represents the evolution of a new and successful body plan. These important trends include: specialization of tissues, germ layers, body symmetry, cephalization, and body cavity formation. Specifics of these trends are summarized below. Table 10.3 summarizes this information.

TABLE 10.3

Trends in Animal Development from the Primitive to the Complex

From the Primitive	To the Complex
No symmetry or radial symmetry with little or no sensory apparatus	Bilateral symmetry with a head end and complex sensory apparatus
No cephalization	Cephalization
Two cell layers: ectoderm and endoderm (diploblastic)	Three cell layers: ectoderm, mesoderm, and endoderm (triploblastic)
No coelom	Pseudocoelom to coelom
No true tissues	True tissues, organs, and organ systems
Life in water	Life on land and all the modification it requires
Sessile	Motile
Few organs, but no organ systems	Many organ systems and much specialization

Specialized Cells, Tissues, and Organs

We need to begin with some definitions.

- The **cell** is the basic unit of all forms of life. A neuron is a cell.
- A **tissue** is a group of similar cells that perform a particular function. The sciatic nerve is a tissue.
- An **organ** is a group of tissues that work together to perform related functions. The brain is an organ.

Sponges (Porifera) consist of a loose federation of cells, which are not considered tissue because the cells are relatively unspecialized. They possess cells that can sense and react to the environment but do not have real nerve or muscular tissue.

Cnidarians like the hydra and jellyfish possess only the most primitive and simplest forms of tissue.

As larger and more complex animals evolved, specialized cells joined to form real tissues, organs, and organ systems. Flatworms have organs, but no organ systems.

More complex animals, like annelids (segmented worms) and arthropods, have organ systems.

Germ Layers

Germ layers are the main layers that form various tissues and organs of the body. They are formed early in embryonic development as a result of gastrulation. Complex animals are **triploblastic**. They consist of the ectoderm, endoderm, and mesoderm.

- The **ectoderm**, or outermost layer, becomes the skin and nervous system, including the nerve cord and brain.
- The **endoderm**, the innermost layer, becomes the viscera (guts) or the digestive system.
- The **mesoderm**, middle layer, becomes the blood and bones.

Primitive animals, like the **Porifera** and **Cnidarians**, have only two cell layers and are called **diploblastic**. Their bodies consist of ectoderm, endoderm, and **mesoglea** (middle glue), which connects the two layers together.

Bilateral Symmetry

Whereas primitive animals exhibit radial symmetry (see Figure 10.2), most sophisticated animals exhibit **bilateral symmetry** (see Figure 10.3). The echinoderms seem to be an exception because they exhibit bilateral symmetry only as larvae and revert to radial symmetry as adults. In bilateral symmetry, the body is organized along a **longitudinal axis** with right and left sides that mirror each other.

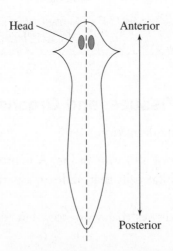

Figure 10.2   Bilateral Symmetry in Flatworm-Planaria

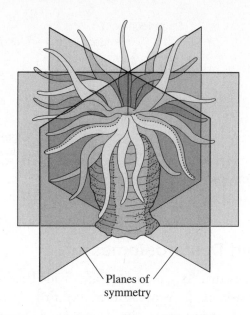

Planes of
symmetry

Figure 10.3   Radial Symmetry

Cephalization—Development of a Head End

Along with bilateral symmetry comes a front end, **anterior**, and a rear end, **posterior**. Sensory apparatus and a brain are clustered at the anterior, while digestive, excretory, and reproductive are located posterior. This enables animals to move faster to flee or to capture prey more effectively. Bilateral animals are all **triploblastic**, with an **ectoderm**, **mesoderm**, and **endoderm**.

The Coelom

The **coelom** is a fluid-filled body cavity (see Figure 15.6). A true coelom arises from within the **mesoderm** and is completely surrounded by mesoderm tissue. It is a significant advance in the course of animal evolution because it provides space for elaborate body systems like a transport or respiratory system. Consider how much space lungs need to expand in the chest cavity or where 30 feet (9 m) of intestines would coil without space in the abdomen. These major organs could not have evolved without the coelom.

Primitive animals (**Porifera, Cnidaria**, and **Platyhelminthes**) have no coelom at all and are called the **acoelomates**. Their three germ layers are packed together with no body cavity except the digestive cavity. **Nematodes** or roundworms are **pseudocoelomates** with a fluid-filled tube between the **endoderm** and the **mesoderm**. The pseudocoelom acts as a **hydrostatic skeleton**, increasing the effectiveness of the animal's muscular contractions in movement. See Figure 10.4.

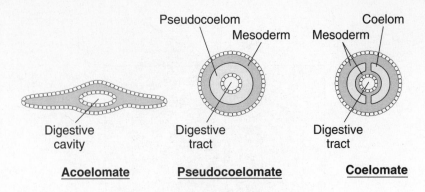

Figure 10.4

Protostomes and Deuterostomes

The coelomates, animals that have a coelom, are divided into two categories: protostomes and deuterostomes. In **protostomes**, the first opening, the blastopore, becomes the *mouth*. In **deuterostomes**, the second opening becomes the mouth and the first opening, the blastopore, becomes the *anus*. The protostome animals include the annelids, mollusks, and arthropods. The deuterostomes include the echinoderms (sea stars) and chordates.

NINE COMMON ANIMAL PHYLA

Zoologists recognize about 35 phyla of animals. You must be familiar with **nine common phyla** with representative animals in each. Since much of animal classification is based on embryonic development, in order to understand the classification of animals, you must also understand the basics of embryonic development. See Figure 10.5. As you study the individual animal phyla, think in terms of strategies animals have evolved to adapt to a particular environment. Also, notice the trends in development in animals from the primitive to the complex, Table 10.3. For more information, see the review chapter entitled "Animal Reproduction and Development."

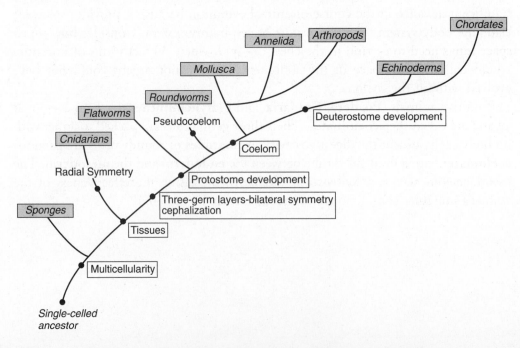

Figure 10.5   Trends in Nine Common Animal Phyla

Porifera—The Sponges—Invertebrates

- No symmetry
- Have no nerves or muscles; are **sessile**, meaning they do not move
- Filter nutrients from water drawn into a central cavity called a **spongocoel**
- Consist of two cell layers only: **ectoderm** and **endoderm** connected by noncellular **mesoglea**
- Have **no true tissues or organs** although they have different types of cells

  ✔ **Choanocytes**, collar cells, line the body cavity and have flagella that circulate water
  ✔ **Spicules** for support—sponges are classified by the material that makes up the spicules
  ✔ **Amoebocytes** are cells that move on their own and perform numerous functions: reproduction, carrying food particles to nonfeeding cells, and secretion of material that forms the **spicules**

- Evolved from colonial organisms; if a sponge is squeezed through fine cheesecloth, it will separate into individual cells that will spontaneously reaggregate into a sponge
- Reproduce asexually by **fragmentation** as well as sexually; are **hermaphrodites**

Cnidarians—Hydra and Jellyfish

- Invertebrate
- Radial symmetry
- Body plan is the **polyp** (vase shaped) or the **medusa** (upside down bowl shaped)
- Life cycle—some go through a **planula larva** (free-swimming) stage then go through two reproductive stages: asexually reproducing (polyp) and sexually reproducing (medusa)
- Two cell layers: **ectoderm** and **endoderm** connected by noncellular **mesoglea**
- Have a **gastrovascular cavity** where **extracellular digestion** occurs
- Also carry out **intracellular digestion** inside body cells, carried out in lysosomes
- Have no transport system because every cell is in contact with environment.
- All members have **stinging cells: cnidocytes**

The Hydra

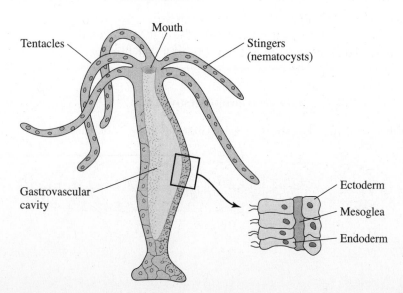

Tentacles · Mouth · Stingers (nematocysts) · Gastrovascular cavity · Ectoderm · Mesoglea · Endoderm

Platyhelminthes—Flatworms Including Tapeworms

- Invertebrate
- Simplest animals with bilateral symmetry, an anterior end, three distinct cell layers, and cephalization
- Have true tissues and organs
- Digestive cavity has only one opening for both ingestion and egestion so food cannot be processed continuously
- Flatworms are **acoelomate**, they have no coelom; they have a solid body with no room for true digestive or respiratory systems to circulate food molecules or oxygen; have solved this problem in two ways

    ✔ The body is very flat, which keeps the body cells in direct contact with oxygen in the environment
    ✔ The digestive cavity is branched so that food can be spread to all regions of the body

Nematoda—Roundworms

- Invertebrate
- Unsegmented worms with bilateral symmetry but little sensory apparatus
- Protostome pseudocoelomate
- **Pseudocoelom** transports nutrients, but there is inadequate room for a circulatory system
- Many are parasitic, *Trichinella* causes **trichinosis** acquired from uncooked pork
- One species, *Caenorhabditis elegans,* is widely used as a model in studying the link between genes and development

Annelida—Segmented Worms: Earthworms, Leeches

- Invertebrate
- **Protostome coelomates** with bilateral symmetry but little sensory apparatus
- Digestive tract is a **tube within a tube** consisting of **crop**, **gizzard**, and **intestine**
- **Nephridia** for excretion of nitrogenous waste, urea
- **Closed circulatory system**—heart consists of five pairs of **aortic arches**
- Blood contains **hemoglobin** and carries oxygen
- Diffusion of oxygen and carbon dioxide through moist skin
- Are **hermaphroditic**, but the animal does not self-fertilize

Earthworm-Digestive Tract

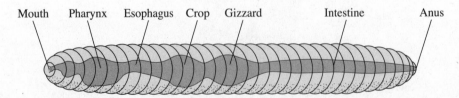

Mouth   Pharynx   Esophagus   Crop   Gizzard   Intestine   Anus

Mollusca—Squids, Octopuses, Slugs, Clams, and Snails

- Invertebrate
- **Protostome coelomates**
- Have a **soft body** often protected by a hard **calcium-containing shell**
- Have bilateral symmetry with three distinct body zones

  ✔ **Head-foot**, which contains both sensory and motor organs
  ✔ **Visceral mass**, which contains the organs of digestion, excretion, and reproduction
  ✔ **Mantle**, a specialized tissue that surrounds the visceral mass and secretes the shell

- **Radula**, a movable, tooth-bearing structure, acts like a tongue
- **Open circulatory system** with blood-filled spaces called **hemocoels**; lack capillaries
- Most have **gills and nephridia**

Arthropoda—Insects (Grasshoppers), Crustaceans (Shrimp, Crabs), and Arachnids (Spiders)

- Invertebrate
- **Protostome coelomates**
- Jointed appendages
- Segmented: head, thorax, abdomen
- Having more sensory apparatus than the annelids gives them more speed and freedom of movement
- **Chitinous exoskeleton** protects the animal and aids in movement.
- **Open circulatory system** with a **tubular heart** and **hemocoels**; lack capillaries
- **Malpighian tubules** for removal of nitrogenous wastes, uric acid
- Air ducts called **trachea** bring air from the environment into **hemocoels**
- Some have **book lungs** or book gills

Echinodermata—Sea Stars (Starfish) and Sea Urchins

- Invertebrate
- **Deuterostome coelomates**
- Most are sessile or slow moving
- Bilateral symmetry as an embryo but reverts to the primitive radial symmetry as an adult; the radial anatomy is an adaptation to a sedentary lifestyle
- **Water vascular system**, which is a modified coelom, creates hydrostatic support for **tube feet**, the locomotive structures
- Reproduces by sexual reproduction where fertilization is external
- Also reproduces by fragmentation and **regeneration**; any piece of a sea star that contains part of the central canal will form an entirely new organism
- Sea stars have an **endoskeleton** consisting of calcium plates, which grow with the body

Chordata—Fishes, Amphibians, Reptiles, Birds, and Mammals

- Vertebrate
- **Deuterostome coelomates**
- Have a **notochord**—a rod that extends the length of the body and serves as a flexible axis
- Dorsal, hollow nerve cord
- **Tail** aids in movement and balance—the coccyx bone in humans is a vestige of a tail
- Birds and mammals are endotherms and homeotherms—maintain a consistent body temperature; all others are ectotherms, although some reptiles can raise their body temperature to a limited extent

Table 10.4 shows the general characteristics of the nine common animal phyla.

TABLE 10.4

General Characteristics of Nine Common Animal Phyla

Phylum	Sample Animal	Deutero-stome	Proto-stome	Acoelo-mate	Pseudo-coelomate	Coelo-mate	Symmetry of Embryo	Number of Germ Cell Layers
Porifera	Sponge		X	X			None	2
Cnidaria	Hydra		X	X			Radial	2
Platyhelminthes	Planaria		X	X			Bilateral	3
Nematoda	Roundworm, pinworm		X		X		Bilateral	3
Annelida	Earthworm		X			X	Bilateral	3
Mollusca	Slug, clam		X			X	Bilateral	3
Arthropoda	Insect, spider		X			X	Bilateral	3
Echinodermata	Starfish, sea urchin	X				X	Bilateral	3
Chordata	Dog, human	X				X	Bilateral	3

CHARACTERISTICS OF MAMMALS

- Mothers nourish their babies with milk
- Have hair or fur, both made of **keratin**
- Are **homeotherms**
- **Placental mammals** (eutherians) are born and the embryo develops internally in a uterus connected to the mother by a **placenta**, where nutrients diffuse from mother to embryo

- Some, the **marsupials**, including kangaroos, are born very early in embryonic development and the joey completes its development while nursing in the mother's pouch
- **Monotremes**, egg-laying mammals, like the duck-billed platypus and the spiny anteater, derive nutrients from a shelled egg

Table 10.5 shows a classification of three mammals using the current system of taxonomy. The domain Eukarya is the new addition.

TABLE 10.5

Taxonomic Classification of Three Mammals

Taxon	Human	Lion	Dog
Domain	Eukarya	Eukarya	Eukarya
Kingdom	Animalia	Animalia	Animalia
Phylum	Chordata	Chordata	Chordata
Class	Mammalia	Mammalia	Mammalia
Order	Primates	Carnivora	Carnivora
Family	Hominidae	Felidae	Canidae
Genus	*Homo*	*Panthera*	*Canus*
Species	*sapiens*	*leo*	*familiaris*

CHARACTERISTICS OF PRIMATES

Primates descended from insectivores, probably from small, tree-dwelling mammals. Primates have **dexterous hands** and **opposable thumbs**, which allow them to do fine motor tasks. Claws have been replaced by **nails**, and hands and fingers contain many nerve endings and are sensitive. The **eyes** of a primate are **front facing** and set close together. Front-facing eyes enhances face-to-face communication, depth perception, and hand-eye coordination. Although mammals devote much energy to **parenting** young, primates engage in the most intense parenting of any mammal. Primates usually have single births and **nurture their young for a long time**. The primates include humans, gorillas, chimpanzees, orangutans, gibbons, and the old-world and new-world monkeys.

Multiple-Choice Questions

1. Which of the following is the least inclusive?

   (A) kingdom
   (B) species
   (C) family
   (D) domain
   (E) genus

2. Which is not true of Protista?

   (A) Some are autotrophs.
   (B) Some are heterotrophs.
   (C) An example is euglena.
   (D) They do not have internal membranes.
   (E) Some have flagella; some move by pseudopods.

3. Which of the following contains prokaryote organisms capable of surviving extreme conditions of heat and salt concentration?

   (A) archaea
   (B) viruses
   (C) protists
   (D) fungi
   (E) plants

4. Which of the following is best characterized as being eukaryotic, heterotrophic, and having cell walls made of chitin?

   (A) plants
   (B) animals
   (C) archaea
   (D) fungi
   (E) viruses

5. In which of the following pairs are the organisms most closely related?

   (A) fruit fly—lobster
   (B) sea stars—jellyfish
   (C) earthworm—tape worm
   (D) clam—earthworm
   (E) sponge—hydra

Questions 6–11

Questions 6–11 refer to this list of animals below:

    (A) Platyhelminthes
    (B) Nematoda
    (C) Chordata
    (D) Echinodermata
    (E) Cnidaria

6. Bilateral symmetry, protostome, acoelomate, triploblastic

7. Bilateral symmetry, deuterostome. An example is a frog.

8. Bilateral symmetry, pseudocoelomate, triploblastic

9. Radial symmetry, diploblastic, acoelomate

10. Bilateral symmetry in the larval stage, radial symmetry as an adult, deuterostome, endoskeleton

11. All organisms in the kingdom contain stinging cells

---

12. A randomly selected group of organisms from a family would show more genetic variation than a randomly selected group from a

    (A) genus
    (B) kingdom
    (C) class
    (D) domain
    (E) phylum

13. The only taxon that actually exists as a natural unit is

    (A) class
    (B) phylum
    (C) order
    (D) species
    (E) kingdom

14. Which of the following is NOT generally a characteristic of complex, advanced organisms?

    (A) mesoderm
    (B) bilateral symmetry
    (C) true tissues and organs
    (D) a true coelom
    (E) sessile

Answers to Multiple-Choice Questions

1. **(B)** The least inclusive grouping includes the one where the organisms are the most similar, the species. The order of taxa from the most general to the most specific is domain, kingdom, phylum, class, order, family, genus, and species.

2. **(D)** Protista are all eukaryotes and have cells with internal membranes. They include paramecium and amoeba which are heterotrophs, and euglena, which is an autotroph. The prokaryotes are organisms without internal membranes.

3. **(A)** Archaea is the group that includes the extremophiles. Most biologists believe that the prokaryotes and the archaea diverged from each other in very ancient times. One basic way in which the two differ is in their nucleic acids.

4. **(D)** Fungi are all heterotrophs. They secrete digestive enzymes and digest food outside the organism. This is called extracellular digestion. Once the food is digested and broken down into building blocks, it is absorbed into the body of the organism by diffusion. The cell walls of mushrooms consist of chitin.

5. **(A)** Both fruit flies and lobsters are in the phylum Arthropoda—animals with jointed appendages. Arthropoda includes arachnids, crustaceans, and insects. Sea stars are Echinodermata, jellyfish are Cnidaria, earthworms are Annelida, tapeworms are Platyhelminthes, and clams are Mollusca.

6. **(A)**

7. **(C)**

8. **(B)**

9. **(E)**

10. **(D)**

11. **(E)**

12. **(A)** The genus is a narrower grouping than the family, so it contains animals with fewer differences.

13. **(D)** Whether you call it a species or something else; it still only contains one type of organism. The other taxa contain many different types of organisms.

14. **(E)** It is generally true that bilateral symmetry and motility are characteristics of advanced organisms. Echinoderms are an exception. Motility, the ability to move from place to place, is an important characteristic of an advanced organism because it enables an organism to search for food, a mate, or safety.

Free-Response Questions

Directions: Answer all questions. You must answer the question in essay—**not** outline—form. You may use labeled diagrams to supplement your essay, but diagrams alone are *not* sufficient. Before you start to write, read each question carefully so that you understand what the question is asking.

1. Describe the differences between the terms in each of the following pairs:

    a. Acoelomate—Coelomate
    b. Radial Symmetry—Bilateral Symmetry

2. Explain how each of the features listed in question 1 can be used to construct an evolutionary history of these common animal phyla:
    Porifera
    Cnidaria
    Platyhelminthes
    Nematoda
    Annelida
    Mollusca
    Arthropoda
    Echinodermata
    Chordata
    (See Table 10.4 on page 226 for answers.)

Typical Free-Response Answers

1a. A coelom is a fluid-filled body cavity and arises from within **mesoderm** tissue early in embryonic development. It is a significant advance in the course of animal evolution because it provides space for elaborate body systems like a transport system. More advanced phyla, such as echinodermata and chordata have a coelom. An animal that does not have a coelom (acoelomate) lacks internal cavities and complex organs. Primitive acoelomate phyla are Cnidaria, Porifera, and Platyhelminthes. Some animals have a pseudocoelom, an internal cavity that is only partly lined with mesoderm tissue. This fluid-filled pseudocoelom functions as a **hydrostatic skeleton**, providing support and making movement easier. An animal phylum with a pseudocoelom is the Nematoda.

1b. If symmetry is **radial**, several planes can pass through the long axis and divide the animal into similar parts. An example of an animal with radial symmetry is the hydra. If only one plane can bisect the animal into left and right halves, the symmetry is **bilateral**. Primitive organisms show radial symmetry, advanced organisms show bilateral symmetry. Chordates all have bilateral symmetry. The embryo of the echinoderm demonstrates bilateral symmetry but reverts to the primitive radial symmetry as an adult.

2. There is great diversity among all the animals, and they can be grouped and distinguished by several characteristics. These characteristics include the number of cell layers, whether they have a true coelom, symmetry at early cleavage, body plan, and whether they are protostomes or deuterostomes. The phyla listed above begin with the most primitive and oldest on top and end with the most advanced and most recently evolved at the bottom.

The most primitive animals, the **Porifera**, or sponges, and **Cnidaria** are diploblastic; they lack a mesoderm. Instead, they have a mesoglea, a noncellular, gluey layer between the ectoderm and endoderm that helps keep them together. All of the other phyla are triploblastic with three germ layers. The ectoderm will become the skin and nervous system. The endoderm will become the internal organs. The mesoderm will become the blood, bones, and muscle. The most primitive phylum, the Porifera, includes the sponges. They have no symmetry. The slightly more advanced Cnidaria have radial symmetry. All triploblastic animals have bilateral symmetry. Animals with no true coelom, for example, Porifera, Cnidaria, and **Platyhelminthes**, are primitive. The **Nematoda** lack a true coelom, but they have a pseudocoelom. All the others have a true coelom. The two most advanced phyla are the **Echinodermata** and **Chordata**; they are both deuterostomes. All the remaining phyla are more primitive and are protostomes.

Evolution

CHAPTER 11

- Evidence for evolution
- Historical context for evolutionary theory
- Darwin's theory of natural selection
- Types of selection
- Sources of variation in a population
- Causes of evolution of a population
- Hardy-Weinberg equilibrium—characteristics of stable populations
- Speciation and reproductive isolation
- Patterns of evolution
- Modern theory of evolution
- The origin of life

INTRODUCTION

Evolution is the change in allelic frequencies in a population. **Microevolution** refers to the changes in a single gene pool. **Macroevolution** refers to the appearance of a major evolutionary development or a new species. Speciation can occur in two ways. **Anagenesis** or **phyletic evolution** occurs when one species replaces another. **Cladogenesis** or **branching evolution** occurs when a new species branches out from a parent species; see Figure 11.1.

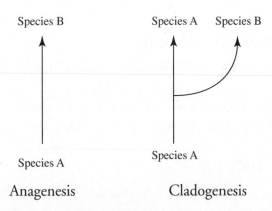

Figure 11.1

EVIDENCE FOR EVOLUTION

Six areas of scientific study provide evidence for evolution

1. Fossil Record

By studying the fossil record, paleontology reveals the existence of species that have become extinct or have evolved into other species. Studies using **radioactive dating** and **half-life** indicate that the earth is about **4.6 billion years old**. Prokaryotes were the first organisms to develop on earth, and they are the oldest fossils. Paleontologists have discovered many transitional forms that link older fossils to modern species; such as the transition from *Eohippus* to the modern horse, *Equus*. *Archaeopteryx* is a fossil that links reptiles and birds.

2. Comparative Anatomy

The study of different structures contributes to scientists' understanding of the evolution of anatomical structures and of evolutionary relationships.

- The wing of a bat, the lateral fin of a whale, and the human arm all have the same internal bone structure, although the function of each varies. These structures, known as **homologous structures**, have a common origin and reflect a common ancestry.
- **Analogous structures**, such as a bat's wing and a fly's wing, have the same function. However, the similarity is superficial and reflects an adaptation to similar environments, not descent from a recent common ancestor.
- **Vestigial structures**, such as the appendix, are evidence that structures have evolved. The appendix is a vestige of a structure needed when human ancestors ate a very different diet.

3. Comparative Biochemistry

Organisms that have a common ancestor will have common biochemical pathways. The more closely related the organisms are to each other, the more similar their biochemistry is. Humans and mice are both mammals. This close relationship is the reason that medical researchers can test new medicines on mice and extrapolate the results to humans.

4. Comparative Embryology

Closely related organisms go through similar stages in their embryonic development. For example, all vertebrate embryos go through a stage in which they have gill pouches on the sides of their throats. In fish, the gill pouches develop into gills. In mammals, they develop into eustachian tubes in the ears.

5. Molecular Biology

Since all aerobic organisms contain cells that carry out aerobic cell respiration, they all contain the polypeptide **cytochrome c**. A comparison of the amino acid sequence of cytochrome c among different organisms shows which organisms are most closely related. The cytochrome c in human cells is almost identical to that of our closest relatives, the chimpanzee and gorilla, but differs from that of a pig.

6. Biogeography

The theory of **continental drift** states that about 200 million years ago, the continents were locked together in a single supercontinent called Pangea, which slowly separated into seven continents over the course of 150 million years. Study of the location of marsupial fossils and the **geographic distribution** of living marsupials, which is limited almost exclusively to Australia, confirms this theory.

HISTORICAL CONTEXT FOR EVOLUTIONARY THEORY

Aristotle spoke for the ancient world with his theory of *Scala Natura.* According to this theory, all life-forms can be arranged on a ladder of increasing complexity, each with its own allotted rung. The species are permanent and do not evolve. Humans are at the pinnacle of this ladder of increasing complexity.

Carolus Linnaeus or **Carl von Linné** (1707–1778) specialized in **taxonomy**, the branch of biology concerned with naming and classifying the diverse forms of life. He believed that scientists should study life and that a classification system would reveal a divine plan. He developed the naming system used today: **binomial nomenclature**. In this system, every organism has a unique name consisting of two parts: a **genus** name and a **species** name. For example, the scientific name of humans is *Homo sapiens*.

Cuvier, who died in 1832 before Darwin published his thesis, studied fossils and realized that each stratum of earth is characterized by different fossils. He believed that a series of **catastrophes** was responsible for the changes in the organisms on earth and was a strong opponent of evolution. Cuvier's detailed study of fossils, however, was very important in the development of Darwin's theory.

James Hutton, one of the most influential geologists of his day, published his theory of **gradualism** in 1795. He stated that the earth had been molded, not by sudden, violent events, but by slow, gradual change. The effects of wind, weather, and the flow of water that he saw in his lifetime were the same forces that formed the various geologic features on earth, such as mountain ranges and canyons. His theories were important because they were based on the idea that the earth had a very long history and that change is the normal course of events.

Lyell was a leading geologist of Darwin's era. He stated that geological change results from slow, continuous actions. He believed that the earth was much older than the 6,000 years thought by early theologians. His text, *Principles of Geology*, was a great influence on Darwin.

Lamarck was a contemporary of Darwin who also developed a theory of evolution. He published his theory in 1809, the year Darwin was born. His theory relies on the ideas of **inheritance of acquired characteristics** and **use and disuse**. He stated that individual organisms change in response to their environment. According to Lamarck, the giraffe developed a long neck because it ate leaves of the tall acacia tree for nourishment and had to stretch to reach them. The animals stretched their necks and passed the acquired trait of an elongated neck onto their offspring. Although this theory may seem funny today, it was widely accepted in the early nineteenth century.

Wallace, a naturalist and author, published an essay discussing the process of natural selection identical to Darwin's, which had not yet been published. Many people credit Wallace, along with Darwin, for the theory of natural selection.

Darwin was a naturalist and author who, when he was 22, left England aboard the HMS Beagle to visit the Galapagos Islands, South America, Africa, and Australia. By the early 1840s, Darwin had worked out his **theory of natural selection** or **descent with modification** as the mechanism for how populations evolve, but he did not publish them. Perhaps he was afraid of the furor his theories would cause. He finally published "**On the Origin of the Species**" in 1859 because he was spurred on to publish by the appearance of a similar treatise by Wallace.

DARWIN'S THEORY OF NATURAL SELECTION

Here are the tenets of **Darwin's theory of natural selection**:

- **Populations tend to** grow exponentially, **overpopulate**, and exceed their resources. Darwin developed this idea after reading Malthus's work, a treatise on population growth, disease, and famine published in 1798.
- **Overpopulation results in competition and a struggle for existence**.
- **In any population, there is variation and an unequal ability of individuals** to survive and reproduce. Darwin, however, could not explain the origin of variation in a population. (Mendel's theory of genetics, published in 1865, would have given Darwin a basis for variation in a population. However, the ramifications of Mendel's theories were not understood until many years after he presented them.)
- **Only the best-fit individuals survive and get to pass on their traits to off-spring**.
- **Evolution occurs as advantageous traits accumulate in a population**. No individual organism changes in response to pressure from the environment. Rather, the frequency of an allele within a population changes.

How the Giraffe Got Its Long Neck

According to **Darwin's theory**, ancestral giraffes were short-necked animals although neck length varied from individual to individual. As the population of animals competing for the limited food supply increased, the taller individuals had a better chance of surviving than those with shorter necks. Over time, the proportion of giraffes in the population with longer necks increased until only long-necked giraffes existed.

How Insects "Become" Resistant to Pesticides

Insects do not actually become resistant to pesticides. Instead, some insects are resistant to a particular chemical insecticide. When the environment is sprayed with that insecticide, the resistant insects have the selective advantage. All the insects not resistant to the insecticide die, and the remaining resistant ones breed quickly with no competition. The entirely new population is resistant to the insecticide. Insecticide resistance is just an example of **directional selection**.

TYPES OF SELECTION

Natural selection can alter the frequency of inherited traits in a population in five different ways, depending on which phenotypes in a population are favored. Five types of selection are **stabilizing**, **diversifying**, **directional**, **sexual**, and **artificial**; see Figure 11.2.

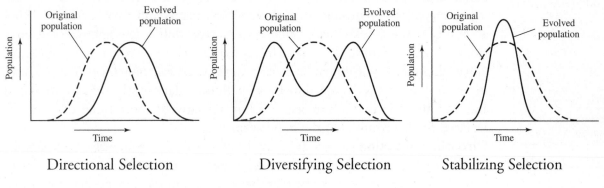

Figure 11.2

1. Stabilizing Selection

Stabilizing selection eliminates the extremes and favors the more common intermediate forms. Many mutant forms are weeded out in this way.

- In humans, stabilizing selection keeps the majority of birth weights in the 6–8 pound (2.7–3.6 kg) range. For babies much smaller and much larger, infant mortality is greater.

- In Swiss starlings, genotypes that lead to a clutch size (the number of eggs a bird lays) of up to five will have more surviving young than birds of the same species that lay a larger or smaller number of eggs.

2. Disruptive or Diversifying Selection

Disruptive selection increases the extreme types in a population at the expense of intermediate forms. What may result is called **balanced polymorphism**, one population divided into two distinct types. Over great lengths of time, disruptive selection may result in the formation of two new species.

Imagine that an environment with very light rocks and very dark soil is colonized by light, intermediate-colored, and dark mice. The frequency of very light and very dark mice, which are camouflaged, would increase, while the intermediate-colored mice would die out because of predation. Pressure from the environment selects for two extreme characteristics.

3. Directional Selection

Changing environmental conditions give rise to **directional selection**, where one phenotype replaces another in the gene pool. Here are two examples of directional selection.

- One example of **directional selection** is **industrial melanism** in **peppered moths**, *Biston betularia*. Until 1845 in England, most peppered moths were light; a few individuals were found to be dark. With increasing industrialization, smoke and soot polluted the environment, making all the plants and rocks dark. By 1900, all moths in the industrialized regions were dark; only a few light-colored individuals could be found. Before the industrial revolution, white moths were camouflaged in their environment and dark moths were easy prey for predators. After the environment was darkened by heavy pollution, dark moths were camouflaged and had the selective advantage. Within a relatively short time, dark moths replaced the light moths in the population.
- **Directional selection** can produce rapid shifts in allelic frequencies. For example, soon after the discovery of **antibiotics**, bacteria appeared that were resistant to these drugs. Scientists now know that the genes for antibiotic resistance are carried on **plasmids**, small DNA molecules, which can be transferred from one bacterial cell to another and which can spread the mutation for **antibiotic resistance** very rapidly within the bacterial population. The appearance of antibiotics themselves does not induce mutations for resistance; it merely selects against susceptible bacteria by killing them. Since only resistant individuals survive to reproduce, the next generation will all be resistant. Joshua Lederberg carried out the experiment that proved that some bacteria are resistant to antibiotics prior to any exposure.

4. Sexual Selection

Sexual selection is selection based on variation in secondary sexual characteristics related to competing for and attracting mates. In males, the evolution of horns, antlers, large stature, and great strength are the result of sexual selection. Male elephant seals fight for supremacy of a harem that may consist of as many as fifty females. In baboons, long canines are important for male-male competition. Differences in appearance between males and females are known as **sexual dimorphism**. In many species of birds, the females are colored in a way to blend in with their surroundings, thus protecting them and their young. The males, on the other hand, have bright, conspicuous plumage because they must compete for the attention of the females.

5. Artificial Selection

Humans breed plants and animals by seeking individuals with desired traits as breeding stock. This is known as **artificial selection**. Racehorses are bred for speed, and laying hens are bred to produce more and larger eggs. Humans have bred cabbage, brussels sprouts, kale, kohlrabi, cauliflower, and broccoli all from the wild mustard plant by selecting for different traits.

SOURCES OF VARIATION IN A POPULATION

Variation in a population is necessary in order for a population to evolve as the environment changes. Although Darwin could not explain the origin of variation, he knew that variation exists in every population. A good example of this is the existence of hundreds of breeds of dogs. All dogs belong to one species, *Canus familiaris.*

While it would seem that natural selection would tend to reduce genetic variation by removing unfavorable genotypes from a population, nature has many mechanisms to preserve it. Here are eight mechanisms that *preserve* diversity or variation in a gene pool or population: balanced polymorphism, geographic variation, sexual reproduction, outbreeding, diploidy, heterozygote superiority, frequency-dependent selection, and evolutionary neutral traits.

1. Balanced Polymorphism

Balanced polymorphism is the presence of two or more phenotypically distinct forms of a trait in a single population of a species. The shells of one genus of land snail exhibit a wide range of colors and banding patterns. Banded snails living on dark, mottled ground are less visible than unbanded ones and therefore are preyed upon less frequently. In areas where the background is fairly uniform, unbanded snails have the selective advantage. Each **morph** is better adapted in a different area, but both varieties continue to exist.

2. Geographic Variation

Two different varieties of rabbit continue to exist in two different regions in North America. Rabbits in the cold, snowy northern regions are camouflaged with white fur and have short ears to conserve body heat. Rabbits living in warmer, southern regions have mottled fur to blend in with surrounding woodsy areas and long ears to radiate off excess body heat. Such a graded variation in the phenotype of an organism is known as a **cline**. Because the variation in rabbit appearance is due to differences in northern and southern environments, this is an example of a **north-south cline**.

3. Sexual Reproduction

UNDERSTAND THIS

The more variation in a population, the greater the capacity for change or evolution.

Sexual reproduction provides variation due to the shuffling and recombination of alleles during meiosis and fertilization.

- **Independent assortment of chromosomes**, during metaphase I, results in the recombination of unlinked genes.
- **Crossing-over** is the exchange of genetic material of homologous chromosomes and occurs during meiosis I. It produces individual chromosomes that combine genes inherited from two parents. In humans, two or three crossover events occur per homologous pair.
- The **random fertilization** of one ovum by one sperm out of millions results in enormous variety among the offspring.

4. Outbreeding

Outbreeding is the mating of organisms within one species that are not closely related. It is the opposite of inbreeding, the mating of closely related individuals. Outbreeding maintains both variation within a species and a strong gene pool. Inbreeding weakens the gene pool because if organisms that are closely related interbreed, detrimental recessive traits tend to appear in homozygous recessive individuals. Many mechanisms have evolved to promote outbreeding.

In lions, the dominant male of a pride chases away the young maturing males before they become sexually mature. This ensures that these young males will not inbreed with their female siblings. These young males roam the land, often over great distances, looking for another pride to join. If one of these young male lions can successfully overthrow the king of another pride, he will inseminate all the females of that new pride and develop his own lineage.

5. Diploidy

Diploidy, the $2n$ condition, maintains and shelters a hidden pool of alleles that may not be suitable for present conditions but that could be advantageous when conditions change in the future.

6. Heterozygote Advantage

Heterozygote advantage preserves multiple alleles in a population. It is a phenomenon in which the hybrid individual is selected for because it has greater reproductive success. The hybrids are sometimes better adapted than the homozygotes.

In the case of **sickle cell anemia** in West Africa, people who are hybrid (*Ss*) for the sickle cell trait have the selective advantage over other individuals. Those who are hybrid have normal hemoglobin and do not suffer from sickle cell disease. However, they are resistant to malaria, which is endemic in West Africa. Those individuals who are homozygous for the sickle cell trait (*ss*) are at a great disadvantage because they have abnormal hemoglobin and suffer from and may die of sickle cell disease. People who are homozygous for normal hemoglobin (*SS*) do not have sickle cell disease but are susceptible to and may die of **malaria**. Thus, the mutation for sickling is retained in the gene pool.

7. Frequency-Dependent Selection

Another mechanism that preserves variety in a population is known as **frequency-dependent selection** or the **minority advantage**. This acts to decrease the frequency of the more common phenotypes and increase the frequency of the less common ones. In predator-prey relationships, predators develop a **search image**, or standard representation of prey, that enables them to hunt a particular kind of prey effectively. If the prey individuals differ, the most common type will be preyed upon disproportionately while the less common individuals will be preyed upon to a lesser extent. Since these rare individuals have the selective advantage, they will become more common for a time, will lose their selective advantage, and will eventually be selected against.

8. Evolutionary Neutral Traits

Evolutionary neutral traits are traits that seem to have no selective value. Examples are **blood type** and **fingerprint variation** in humans. Scientists do not understand where they evolved from or why they have remained (been conserved) in the human population. Perhaps they actually influence survival and reproductive success in ways that are difficult to perceive or measure.

CAUSES OF EVOLUTION OF A POPULATION

The agents of change for a population, that is, those things that cause **evolution** of a population are **genetic drift**, **gene flow**, **mutations**, **nonrandom mating**, and **natural selection**.

Genetic Drift

Genetic drift is change in the gene pool due to chance. It is a fluctuation in frequency of alleles from one generation to another and is unpredictable. It tends to limit diversity. There are two examples: the **bottleneck effect** and the **founder effect**. Here are two examples.

- **Bottleneck effect**: Natural disasters such as fire, earthquake, and flood reduce the size of a population unselectively, resulting in a loss of genetic variation. The resulting population is much smaller and not representative of the original one. Certain alleles may be under or overrepresented compared with the original population. This is known as the **bottleneck effect**.

  The high rate of Tay-Sachs disease among Eastern European Jews is attributed to a population bottleneck experienced by Jews in the Middle Ages. During that period, many Jews were persecuted and killed, and the population was reduced to a small fraction of its original size. Of the individuals who remained alive, there happen to have been a disproportionate percentage of people who carried the Tay-Sachs gene. Since Jews in Europe remained isolated and did not intermarry with other Europeans to any great extent, the incidence of the trait remained unusually high in that population.

  From the 1820s to the 1880s along the California coast, the northern elephant seal was hunted almost to extinction. Since 1884, when the seal was placed under government protection, the population has increased to about 35,000; all are descendants from that original group and have little genetic variation.

- **The founder effect**: When a small population breaks away from a larger one to colonize a new area, it is most likely not genetically representative of the original larger population. Rare alleles may be overrepresented. This is known as the **founder effect** and occurred in the Old Order of Amish of Lancaster, Pennsylvania. All of the colonists descended from a small group of settlers who came to the United States from Germany in the 1770s. Apparently one or more of the settlers carried the rare but do minant gene for **polydactyly**, having extra fingers and toes. Due to the extreme isolation and intermarriage of the close community, this population now has a high incidence of polydactyly.

Notice that whether a trait is dominant or recessive merely determines if it is expressed or remains hidden. It does not determine how common the trait is in a population. The change in allelic frequency of a trait results from genetic drift or from the trait being advantageous or disadvantageous.

Gene Flow

Gene flow is the movement of alleles into or out of a population. It can occur as a result of the migration of fertile individuals or gametes between populations. For example, pollen from one valley can be carried by the wind across a mountain to another valley. It tends to increase diversity.

Mutations

Mutations are changes in genetic material and are the raw material for evolutionary change. They increase diversity. A single point mutation can introduce a new allele into a population. *Although mutations at one locus are rare, the cumulative effect of mutations at all loci in a population can be significant.*

Nonrandom Mating

Individuals choose their mates for a specific reason. The selection of a mate *serves to eliminate the less-fit individuals.* Snow geese exist in two phenotypically distinct forms, white and blue. Blue snow geese tend to mate with blue geese, and white geese tend to mate with white geese. If, for some reason, the blue geese became more attractive and both blue and white geese began mating with only blue geese, the population would evolve quickly, favoring the blue geese. The white goose might disappear.

Natural Selection

Natural selection is the major mechanism of evolution in any population. Those individuals who are better adapted in a particular environment exhibit *better reproductive success.* They have more offspring that survive and pass their genes on to more offspring.

HARDY-WEINBERG EQUILIBRIUM—CHARACTERISTICS OF STABLE POPULATIONS

Hardy and **Weinberg**, two scientists, described a **stable**, **nonevolving population**, that is, one in which allelic frequencies do not change. For example, if the frequency of an allele for a particular trait is 0.5 and the population is not evolving, in 1,000 years the frequency of that allele will still be 0.5.

According to Hardy-Weinberg, if the population is stable, the following must be true:

1. **The population must be very large**. In a small population, the smallest change in the gene pool will have a major effect in allelic frequencies. In a large population, a small change in the gene pool will be diluted by the sheer number of individuals and no change in the frequency of alleles will occur.

2. **The population must be isolated from other populations**. There must be no migration of organisms into or out of the gene pool because that could alter allelic frequencies.

3. **There must be no mutations in the population**. A mutation in the gene pool could cause a change in allelic frequency by introducing a new allele.

4. **Mating must be random**. If individuals select mates, then those individuals that are better adapted will have a reproductive advantage and the population will evolve.

5. **No natural selection**. Natural selection causes changes in relative frequencies of alleles in a gene pool.

The Hardy-Weinberg Equation

The Hardy-Weinberg equation enables us *to calculate frequencies of alleles in a population*. Although it can be applied to complex situations of inheritance, for the purpose of explanation here, we will discuss a simple case—a gene locus with only two alleles. Scientists use the letter p to stand for the **dominant allele** and the letter q to stand for the **recessive allele**.

The Hardy-Weinberg equation is

$$p^2 + 2pq + q^2 = 1 \qquad \text{or} \qquad p + q = 1$$

The monohybrid cross is the basis for this equation.

	A	a
A	AA	Aa
a	Aa	aa

$$p^2 = AA$$
$$2pq = 2(Aa)$$
$$q^2 = aa$$

Here are three sample problems.

PROBLEM 1

If 9% of the population has blue eyes, what percent of the population is hybrid for brown eyes? Homozygous for brown eyes?

To solve this problem, follow these steps:

1. The trait for blue eyes is homozygous recessive, *b/b*, and is represented by q^2.

   $q^2 = 9\%$ Converting to a decimal: $q^2 = .09$

2. To solve for q, the square root of .09 = 0.3

3. Since $p + q = 1$, if $q = 0.3$, then $p = 0.7$

4. The hybrid brown condition is represented by $2pq$. So—

5. To solve for the percent of the population that is hybrid, substitute values for $2(p)(q)$.

    The percentage of the population that is hybrid brown is $2(.7)(.3) = 42\%$.

6. Homozygous dominant is represented by p^2.

    The percentage of the population that is homozygous brown is $p^2 = (.7)^2 = 49\%$

PROBLEM 2

Determine the percent of the population that is homozygous dominant if the percent of the population that is homozygous recessive is 16%.

1. Homozygous recessive $= q^2 = 16$. Therefore, $q^2 = .16$ and $q = 0.4$

2. If $q = 0.4$, then $p = 0.6$

3. Therefore, the percentage of the population that is homozygous dominant $= p^2 = .36 = 36\%$.

PROBLEM 3

Determine the percent of the population that is hybrid if the allelic frequency of the recessive trait is 0.5.

1. In this example, you are given the value of q (not q^2). You only need to subtract from 1 to get the value of p.

2. If $p + q = 1$ and $q = 0.5$; then $p + 0.5 = 1$

3. Since both $p = 0.5$ and $p = 0.5$. The percentage of the population that is hybrid is $2pq = 0.5 \times 0.5 \times 2 = 50\%$.

SPECIATION AND REPRODUCTIVE ISOLATION

The definition of a **species** is a population whose members have the potential to interbreed in nature and produce viable, fertile offspring. Lions and tigers can be induced to interbreed in captivity but would not do so naturally. Therefore, they are considered separate species. Horses and donkeys can interbreed in nature and produce a mule that is not fertile. Therefore, the horse and donkey belong to different species.

A **species** is defined in terms of **reproductive isolation**, meaning that one group of genes becomes isolated from another to begin a separate evolutionary history. Once separated, the two isolated populations may begin to diverge genetically under the pressure of different selective forces in different environments. If enough time elapses and differing selective forces are sufficiently great, the two populations may become so different that, even if they were brought back together, interbreeding would not naturally occur. At that point, **speciation** is said to have taken place. *Anything that fragments a population and isolates small groups of individuals may cause speciation.* The following describes different modes of speciation due to different modes of isolation, and Figure 11.3 shows diagrams of allopatric and sympatric speciation.

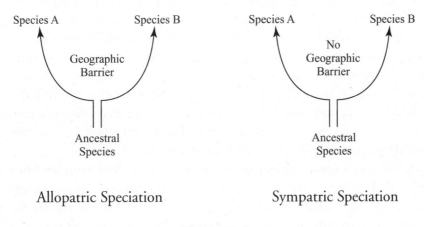

Figure 11.3

Allopatric Speciation

Allopatric speciation is caused by **geographic isolation**: separation by mountain ranges, canyons, rivers, lakes, glaciers, altitude, or longitude.

Sympatric Speciation

Under certain circumstances, speciation may occur without geographic isolation, in which case the cause of the speciation is **sympatric**. Examples of **sympatric speciation** are **polyploidy**, **habitat isolation**, **behavioral isolation**, **temporal isolation**, and **reproductive isolation**.

- **Polyploidy** is the condition where a cell has more than two complete sets of chromosomes ($4n$, $8n$, etc.). It is common in plants and can occur naturally or through breeding. Nondisjunction during meiosis can result in gametes with the $2n$ chromosome number that, when fertilized by another normal ($2n$) gamete, results in a daughter cell with $4n$ chromosomes. Plants that are polyploid cannot breed with others of the same species that are not polyploidy and are functionally isolated from them. See Figure 11.4.

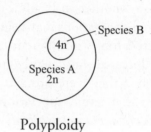

Polyploidy

Figure 11.4

- **Habitat isolation**: Two organisms live in the same area but encounter each other rarely. Two species of one genus of snake can be found in the same geographic area, but one inhabits the water while the other is mainly terrestrial.
- **Behavioral isolation**: Sticklebacks, small saltwater fish that have been studied extensively, have elaborate mating behavior. At breeding time, in response to increased sunlight, the males change in color and develop a red underbelly. The male builds a nest and courts the female with a dance that triggers a complex set of movements between the partners. If either partner fails in any step of the mating dance, no mating occurs and no young are produced.
  Male fireflies of various species signal to females of their kind by blinking the lights on their tails in a particular pattern. Females respond only to characteristics of their own species, flashing back to attract males. If, for any reason, the female does not respond with the correct blinking pattern, no mating occurs. The two animals become isolated from each other.
- **Temporal isolation**: Temporal refers to time. A flowering plant colonizes a region with areas that are warm and sunny and areas that are cool and shady. Flowers in the regions that are warmer become sexually mature sooner than flowers in the cooler areas. This separates flowers in the two different environments into two separate populations.
- **Reproductive isolation**: Closely related species may be unable to mate because of a variety of reasons. Differences in the structure of genitalia may prevent insemination. Difference in flower shape may prevent pollination. Things that prevent mating are called **prezygotic barriers**. For example, a small male dog and a large female dog cannot mate because of the enormous size differences between the two animals. Things that prevent the production of fertile offspring, once mating has occurred, are called **postzygotic barriers**. One example might be that a particular zygote is not viable. Both prezygotic and postzygotic barriers result in **reproductive isolation**.

PATTERNS OF EVOLUTION

The evolution of different species is classified into five patterns: divergent, convergent, parallel, coevolution, and adaptation radiation, see Figure 11.5.

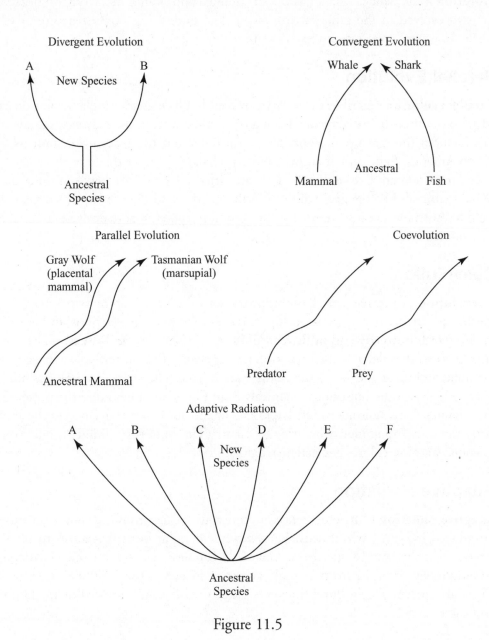

Figure 11.5

Divergent Evolution

Divergent evolution occurs when a population becomes isolated (for any reason) from the rest of the species, becomes exposed to new selective pressures, and evolves into a new species. All the examples of allopatric and sympatric speciation on the previous pages are examples of divergent evolution.

Convergent Evolution

When unrelated species occupy the same environment, they are subjected to similar selective pressures and show similar adaptations. The classic example of **convergent evolution** is the **whale**, which has the streamlined appearance of a large fish because the two evolved in the same environment. The underlying bone structure of the whale, however, reveals an ancestry common to mammals, not to fish.

Parallel Evolution

Parallel evolution describes two related species that have made similar evolutionary adaptations after their divergence from a common ancestor. The classic example of this includes the marsupial mammals of Australia and the placental mammals of North America. (The only mammals in Australia are the ones that have been introduced from abroad, like rabbits.) There are striking similarities between some placental mammals like the gray wolf of North America and the marsupial Tasmanian wolf of Australia because they share a common ancestor and evolved in similar environments.

Coevolution

Coevolution is the reciprocal evolutionary set of adaptations of two interacting species. All predator-prey relationships are examples and the relationship between the **monarch butterfly** and **milkweed plant** is another. The **milkweed plant** contains poisons that deter herbivores from eating them. The butterfly lays its eggs in the milkweed plant and when the larvae (caterpillars) hatch, they feed on the milkweed and absorb the poisonous chemicals from the plant. They store the poison in their tissues. This poison, which is present in the adult butterfly, makes the butterfly toxic to any animal who tries to eat it. (The butterfly exhibits bright conspicuous warning colors that deter predators.)

Adaptive Radiation

Adaptive radiation is the emergence of numerous species from a common ancestor introduced into an environment. Each newly emerging form specializes to fill an ecological niche. All 14 species of **Darwin's finches** that live on the **Galapagos Islands** today diverged from a single ancestral species perhaps 10,000 years ago. There are currently six ground finches, six tree finches, one warbler finch, and one bud eater.

MODERN THEORY OF EVOLUTION

Gradualism

Gradualism is the theory that organisms descend from a common ancestor gradually, over a long period of time, in a linear or branching fashion. Big changes occur by an accumulation of many small ones. According to this theory, fossils should exist as evidence of every stage in the evolution of every species with no missing links. However, the fossil record is at odds with this theory because scientists rarely find **transitional forms** or **missing links**.

Punctuated Equilibrium

The favored theory of evolution today is called **punctuated equilibrium** and was developed by **Stephen J. Gould** and **Niles Eldridge** after they observed that the gradualism theory was not supported by fossil record. The theory proposes that new species appear suddenly after long periods of stasis. A new species changes most as it buds from a parent species and then changes little for the rest of its existence. The sudden appearance of the new species can be explained by the **allopatric model** of speciation. A new species arises in a different place and expands its range, outcompeting and replacing the ancestral species. See Figure 11.6, which has sketches showing gradualism and punctuated equilibrium.

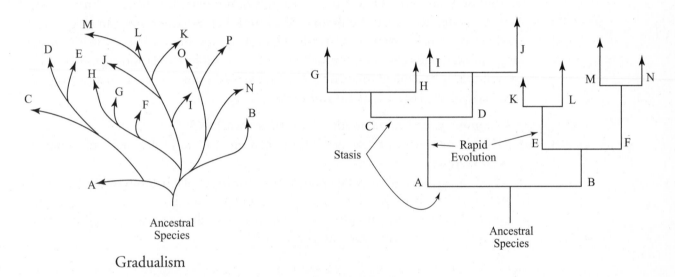

Gradualism

Figure 11.6   Punctuated Equilibrium

THE ORIGIN OF LIFE

The age of the earth is estimated to be 4.6 billion years old. The ancient atmosphere consisted of CH_4, NH_3, CO, CO_2, N_2 and H_2O, but lacked free O_2. There was probably intense lightning and ultraviolet (U.V.) radiation that penetrated the primitive atmosphere, providing energy for chemical reactions. Scientists have tried to mimic this early atmosphere to determine how the first organic molecules and earliest life developed. Here is a synopsis of those experiments.

A. I. Oparin and **J. B. S. Haldane**, in the 1920s, hypothesized separately that under the conditions of early earth, organic molecules could form. Without corrosively reactive molecular oxygen present to react with and degrade them, organic molecules could form and remain.

Stanley Miller and **Harold Urey**, in the 1950s, tested the Oparin-Haldane hypothesis and proved that almost any energy source would have converted the molecules in the early atmosphere into a variety of organic molecules, including **amino acids**. They used electricity to mimic lightning and U.V. light that must have been present in great amounts in the early atmosphere.

Sidney Fox, in more recent years, carried out similar experiments. However, he began with organic molecules (not the original inorganic ones) and was able to produce membrane-bound, cell-like structures he called proteinoid microspheres that would last for several hours.

The Heterotroph Hypothesis and the Theory of Endosymbiosis

The first cells on earth were **anaerobic heterotrophic prokaryotes**. They simply absorbed organic molecules from the surrounding primordial soup to use as a nutrient source. They probably began to evolve about 3.5 billion years ago. Eukaryotes did not evolve until another 2 billion years after the evolution of prokaryotes (about 1.5 billion years ago). They arose as a result of **endosymbiosis** according to Lynn Margulis, who developed the **theory**. She states that mitochondria and chloroplasts (and perhaps nuclei) were once free-living prokaryotes that took up residence inside larger prokaryotic cells. The mutually beneficial symbiotic relationship worked out so well that it became permanent. There are several points to prove that mitochondria and chloroplasts are **endosymbionts**.

- Chloroplasts and mitochondria have their own DNA.
- DNA is more like prokaryotic DNA than eukaryotic DNA. It is not wrapped with **histones**.
- These organelles have double membranes. The inner one belongs to the symbiont; the outer one belongs to the host plasma membrane. The theory states that the chloroplast and mitochondria were taken up by the host cell by some sort of endosymbiosis process, such as phagocytosis.

Multiple-Choice Questions

Questions 1–5

Questions 1–5 refer to the fields of study listed below. Choose the one that has provided each of the following pieces of evidence that biological evolution has occurred.

    (A) Comparative biochemistry
    (B) Comparative anatomy
    (C) Comparative embryology
    (D) Geographic distribution
    (E) Paleontology

1. Giraffes have the same number of vertebrae in the neck as do humans.

2. Kangaroos are found in only Australia.

3. Human embryos have tails.

4. Humans and sea stars both have radial cleavage in early embryonic development.

5. Humans can be made temporarily immune to various human diseases by receiving antibodies against those diseases from horses.

---

6. The condition in which there are barriers to successful interbreeding between individuals of different species in the same community is referred to as

    (A) sexual dependency
    (B) reproductive isolation
    (C) geographic isolation
    (D) adaptive radiation
    (E) balanced polymorphism

7. The wing of the bat and a human's arm have different functions and appear very different. Yet, the underlying anatomy is basically the same. Therefore, these structures are examples of

    (A) geographic isolation
    (B) analogous structures
    (C) homology
    (D) reproductive isolation
    (E) balanced polymorphism

8. In a population that is in Hardy-Weinberg equilibrium, the frequency of a particular recessive allele a is 0.4. What is the percentage of the population heterozygous for this allele?

    (A) 4%
    (B) 16%
    (C) 32%
    (D) 48%
    (E) 64%

9. According to the Hardy-Weinberg equation, the dominant trait is represented by

    (A) p
    (B) q
    (C) q^2
    (D) p^2
    (E) $2pq$

Questions 10–14
Matching Column

    (A) Stabilizing selection
    (B) Disruptive selection
    (C) Directional selection
    (D) Sexual selection
    (E) Artificial selection

10. The population of peppered moths in England changed from white to black in fifty years.

11. Human newborns usually weigh between 6–8 pounds (2.7–3.6 kg).

12. Humans have bred dairy cows to produce 100 pints of milk per day.

13. In one region of New Jersey there exist two distinct types of one species of snake.

14. Large horns and giant antlers are characteristic of the male.

---

15. Miller's classic experiment demonstrated that a discharge of sparks through a mixture of gases could result in the formation of a large variety of organic compounds. Miller used all of the following gases in his experiment EXCEPT

    (A) methane
    (B) ammonia
    (C) water
    (D) oxygen
    (E) hydrogen

16. Which is an example of a cline?

    (A) Males of a species have long antlers to fight other males of that species.
    (B) In many species of birds, males have bright plumage to attract the female.
    (C) In one species of rabbit, the ones that evolved in the cold, snowy north are white, while the ones that evolved in the south are brown.
    (D) The hybrid tomato plant is stronger and produces better fruit than the pure genotype.
    (E) There are two distinct varieties in one population of snail that inhabits an island in the Pacific Ocean.

17. Who synthesized proteinoid microspheres in the laboratory using an apparatus that mimicked the early earth?

    (A) Haldane
    (B) Fox
    (C) Urey
    (D) Miller
    (E) Oparin

18. The differences in sparrow songs among sympatric species of sparrows are examples of

    (A) geographic isolation
    (B) convergent evolution
    (C) parallel evolution
    (D) physiological isolation
    (E) behavioral isolation

19. Which part of the theory of evolution did Darwin develop after reading Thomas Malthus?

    (A) Evolution occurs as advantageous traits accumulate in a population.
    (B) In any population, there is variation and an unequal ability of individuals to survive and reproduce.
    (C) Only the best-fit individuals survive and get to pass on their traits to offspring.
    (D) Populations tend to grow exponentially, overpopulate, and exceed their resources.
    (E) Overpopulation results in competition and a struggle for existence.

20. In a population of 1,000 people, 90 have blue eyes. What percent of the population has hybrid brown eyes?

    (A) 3%
    (B) 9%
    (C) 21%
    (D) 42%
    (E) 49%

21. The average length of a rabbit's ears decreases the farther north the rabbits live. This variation is an example of a

    (A) genetic drift
    (B) a cline
    (C) geographic isolation
    (D) founder effect
    (E) bottleneck effect

Questions 22–25
Matching Column

    (A) Founder effect
    (B) Competitive exclusion
    (C) Adaptive radiation
    (D) Convergent evolution
    (E) Parallel evolution

22. Darwin's finches

23. The establishment of a genetically unique population through genetic drift

24. The independent development of similarities between unrelated groups resulting from adaptation to similar environments

25. The Tasmanian wolf in Australia is a marsupial but looks very similar to the gray wolf, a placental mammal of North America

Answers to Multiple-Choice Questions

1. **(B)**

2. **(D)**

3. **(C)**

4. **(C)**

5. **(A)**

6. **(B)** Any barrier that isolates organisms fosters evolution. Barriers to interbreeding are caused by reproductive isolation. Geographic isolation refers to organisms being isolated by geography, such as mountains or rivers. Balanced polymorphism refers to two different versions of the same species living in one area, such as a speckled snail and a plain snail. Both are camouflaged in different environments. Sexual dependency is not related to the topic in any way.

7. **(C)** Homologous structures demonstrate a common ancestry. They may not look alike, but they have an underlying common structure. Analogous structures may have the same function and look alike, but they do not have a common structure, nor do they have a common ancestry.

8. **(D)** The question provides you with the frequency of the allele. It provides you with q. Since the frequency of the recessive allele is 0.4, the frequency of the dominant allele is 0.6. The formula for the hybrid = $2pq$. Therefore, substituting, $2 \times 0.4 \times 0.6 = 48\%$.

9. **(A)** According to Hardy-Weinberg equilibrium, p is the dominant allele and q is the recessive allele.

10. **(C)** The black peppered moths replaced the white peppered moths. Since one characteristic replaced another, this is directional selection.

11. **(A)** Stabilizing selection tends to eliminate the extremes in a population.

12. **(E)** Humans controlled the breeding that produced a cow that gives a certain amount of milk. This is an example of artificial selection; it would not have occurred without human intervention.

13. **(B)** Disruptive selection tends to select for the extremes. Originally, there was probably a range of coloration of snakes in the area in question. Over time, pressure from the environment selected against different colorations until only two remained.

14. **(D)** Sexual selection has to do with the selection for traits that attract a mate.

15. **(D)** Free oxygen was not available in the early earth's atmosphere. Scientists believe that since oxygen is very reactive, had it been present in the ancient atmosphere, it would have reacted with and degraded all the other chemicals in the atmosphere. The consequence would be that evolution of the early earth would not have occurred as it did.

16. **(C)** A cline is a change in some trait along some geographic axis, such as a north-south cline. In this example, the animal is camouflaged by its colorings.

17. **(B)** Sidney Fox was able to produce these cell-like structures, which he called proteinoid microspheres, when he began with amino acids in his experiment.

18. **(E)** A bird's song is a behavior; therefore, this is an example of behavioral isolation.

19. **(D)** Malthus was a mathematician studying populations. He stated that populations tend to overpopulation and exceed their resources. This leads to starvation, disease, and death.

20. **(D)** Of the total population, 9% have blue eyes (90 out of 1000), so $q^2 =$ 0.09 and $q = 0.3$. Therefore, $p = 0.7$ and the frequency of hybrid brown = $2pq = 42\%$.

21. **(B)** The rabbit's ears get shorter and grow closer to the head to retain heat. This is an example of a north-south cline.

22. **(C)** Adaptive radiation is the emergence of numerous species from one common ancestor introduced into a new environment. Today, 13 different species of finches are on the Galapagos Islands where originally there was only 1 species. Each species fills a different niche.

23. **(A)** Genetic drift is evolution through chance. The founder effect is one example of genetic drift. Another is the bottleneck effect.

24. **(D)** The classic example of this can be seen in the whale and the shark. The two animals are unrelated; the whale is a mammal and the shark is a fish. However, they look alike because they experience the same environmental pressures. They both have a streamlined appearance with fins because that design is best for living in the ocean, not because they are related or have a recent common ancestor.

25. **(E)** Eutherians (placental mammals) and marsupials are closely related although they diverged several million years ago. Although they live thousands of miles apart, these two animals live in similar environments and are under the same selective pressures from their respective environments. As a result, they have evolved along similar parallel lines.

Free-Response Questions

Directions: Answer all questions. You must answer the question in essay—**not** outline—form. You may use labeled diagrams to supplement your essay, but diagrams alone are *not* sufficient. Before you start to write, read each question carefully so that you understand what the question is asking.

1. Explain Charles Darwin's theory of evolution by natural selection.

2. Each of the following refers to one aspect of evolution. Explain each in terms of natural selection.
   a. Convergent evolution and the similarities among species in a particular biome.
   b. Insecticide resistance
   c. Speciation and isolation
   d. Heterozygote advantage

Typical Free-Response Answers

Note: This essay has two main sections (a and b), and section b is divided into four parts. Since the question demands much more from part b than part a, assume b is worth more and answer it accordingly.

If you run into trouble because you find that you understand the concept but cannot remember the name of the organism you are writing about, that is not very important. Call it organism "X." Showing that you understand the concept is the important thing.

1. Darwin's theory of evolution is known as the theory of natural selection. According to Darwin, populations tend to grow exponentially, to overpopulate, and to exceed the carrying capacity of their environment. This overpopulation results in a struggle for existence, where only the best-adapted organisms survive and gain a greater share of limited resources. Those who survive long enough

pass their traits to the next generation. The less-fit individuals will not survive and will not reproduce. Therefore, the genes that pass to the next generation will be best adapted for that particular environment.

2a. Convergent evolution

The existence of convergent evolution demonstrates the power of the environment to select the direction of evolution. In convergent evolution, unrelated species often come to resemble one another because they are subject to the same environmental pressures. A perfect example can be seen with two unrelated families of plants, the cactus and the euphorbs, which are found in deserts in different parts of the world, the southwest of North America, and Central Asia. Both families of plants developed fleshy photosynthetic stems adapted for water storage, protective spines, and greatly reduced leaves. It is likely that there was once an array of different kinds of plants in each region, but the climate changed and became very dry. The only plants that could survive were the ones that bore adaptations that enabled them to survive in the dry environment. Traits that were well adapted for the environment were selected for, while the ones that were disadvantageous died out.

2b. Insecticide resistance

Insecticide resistance is an example of directional selection. Some insects are resistant to a particular chemical insecticide. The origin of this resistance is unknown, but it probably derives from mutation. When the environment is sprayed with that insecticide, the resistant insects have the selective advantage. All the nonresistant insects die and the remaining ones breed quickly, without any competition for resources. The new population is entirely resistant to the insecticide.

2c. Speciation and isolation

Anytime a population becomes isolated from another, the two isolated populations may begin to diverge genetically under the pressure of different selective forces in different environments. If enough time elapses and differing selective forces are sufficiently great, the two populations may become so different that, even if they were brought back together, interbreeding would not occur. The two populations would have become two different species.

There are two types of isolation, **allopatric** and **sympatric**. Allopatric isolation is caused by geographic separation, such as mountain ranges, rivers, glaciers, or canyons. One population of wildebeest might become separated from the larger group during a migration and remain isolated forever. Sympatric isolation occurs without geographic isolation. Examples are polyploidy, **habitat isolation**, **temporal** or **behavioral isolation**, and **reproductive isolation**.

An example of reproductive isolation can be seen in the stickleback fish, which has an elaborate premating behavior. If either partner fails to respond correctly in any step of the behavior, no mating occurs and the two individuals are effectively isolated from one another. Another cause of isolation is polyploidy, which is common in plants. It results from **nondisjunction** during gametogenesis and can produce gametes that are diploid. When these diploid gametes fuse with normal gametes, the resulting plants are triploid and unable to mate with any diploid individuals. This is an explanation of how a small population has become isolated.

2d. Heterozygote advantage

There are many instances where the hybrid individual is more fit than the homozygous condition. A perfect example can be seen in West Africa where sickle cell anemia is endemic and so is malaria. People who are homozygous for the sickle cell trait (*ss*) have abnormal hemoglobin and suffer with sickle cell disease. People who are homozygous for normal hemoglobin (*SS*) are susceptible to malaria. However, people who are hybrid for the sickle cell trait do not have sickle cell disease and are resistant to malaria. Without serious medical intervention, the population homozygous for the sickle cell trait may die of sickle cell disease and the population homozygous for normal hemoglobin may die of malaria. The hybrid condition (*Ss*) has the selective advantage.

Plants

12

- Classification of plants
- Bryophytes
- Tracheophytes
- Strategies that enabled plants to move to land
- Plant tissue
- Roots
- Stems
- The leaf
- Transport in plants
- Plant reproduction
- Alternation of generations
- Plant responses to stimuli

INTRODUCTION

Plants are defined as multicelled, eukaryotic, photosynthetic **autotrophs**. Their cell walls are made of cellulose, and their surplus carbohydrate is stored as starch. The life cycle of plants is characterized by **alternation of generations**. One generation is the gametophyte generation, where all the cells of the plant body are haploid (n). The other, alternate generation is the sporophyte generation, where the cells of the plant body are diploid ($2n$). See Figure 12.10.

Plants evolved from aquatic green algae about 500 million years ago. Along with fungi and animals, they colonized the land during the Paleozoic era. This gradual move from the ancestral aquatic environment to the land occurred as organisms evolved adaptations to a dry environment.

Today, most plants live on land. They have diversified into almost 300,000 different species inhabiting all but the harshest environment. Plants stabilize the soil they live in and provide a home for billions of insects and larger animals. They release oxygen into the atmosphere and absorb carbon dioxide. Most of the world depends on the following plants for survival: rice, beans, soy, corn, and wheat.

Plants are organized into two groups. Those plants with no transport vessels are called **bryophytes**. Those with transport vessels are called **tracheophytes**.

CLASSIFICATION OF PLANTS

1. **Bryophytes**—Non-vascular plants

    Ex: mosses, liverworts, hornworts

2. **Tracheophytes**—Vascular plants

    A. **Seedless plants**, like ferns, that reproduce by **spores**

    B. **Seed plants**

    1. **Gymnosperms**—cone-bearing

        Ex: cedars, sequoias, redwoods, pines, yews, and junipers

    2. **Angiosperms** or **Anthophyta**—the **Flowering plants**

        Ex: roses, daisies, apples, and lemons

        -   **Monocotyledon** (monocots)

            Ex: grasses such as corn, wheat, rye, and oats

        -   **Dicotyledon** (dicots)

            Ex: peanuts

BRYOPHYTES

Bryophytes are primitive plants that lack transport vessels (xylem and phloem) and must therefore absorb water by diffusion from the air. In addition, their flagellated sperm must swim through water to fertilize an egg. They also lack any **lignin**-fortified tissue that is necessary to support a tall plant. As a result, bryophytes are restricted to moist habitats and are tiny. Bryophytes play a significant role in diverse terrestrial ecosystems. They grow on rocks, soil, and trees. Sphagnum or peat moss is used as fuel in much of the world.

Like all plants they exhibit alternation of generations. (see page 273.)

TRACHEOPHYTES

Tracheophytes are plants with vascular tissue. Prior to the evolution of tracheophytes, bryophytes were the dominant form of plant life on Earth. However, having an efficient transport system enabled the tracheophytes to outcompete bryophytes. The characteristics of tracheophytes include:

-   Xylem and phloem for transport
-   Lignified transport vessels to support the plant
-   Roots to absorb water while also anchoring and supporting the plant
-   Leaves that increase the photosynthetic surface
-   Life cycle with a dominant sporophyte generation

Tracheophytes are divided into two groups, those with seeds and those without. Those with seeds, the seed plants, are more advanced and far more numerous than the seedless plants. Examples of seedless plants are ferns. The seed plants can be further divided into **gymnosperms**—those bearing cones, and **angiosperms**—those bearing flowers and fruits.

Ferns—Seedless Plants

The ferns are the most widespread **seedless tracheophytes**. They are primitive plants and reproduce by spores instead of by seeds. They are **homosporous**, which means that they produce only one type of spore which then develops into a bisexual gametophyte. Although they have transport tissues and can grow several feet tall, ferns are still restricted to moist habitats. This is true because their sperm are flagellated and must swim from the antheridium to the archegonium to fertilize the egg.

Seed Plants

In contrast to seedless tracheophytes like ferns, seed plants are **heterosporous**. That is, they produce two kinds of spores, megaspores and microspores. **Megaspores** develop into female gametophytes. **Microspores** develop into male gametophytes. In addition, the sperm of seed plants have no flagella and therefore do not require a watery environment in order for fertilization to occur. There are two types of seed plants, gymnosperms and angiosperms.

GYMNOSPERMS: THE CONIFERS

Gymnosperms were the first seed plants to appear on Earth. The seeds of gymnosperm are said to be *naked* because they are not enclosed inside a fruit as are seeds in angiosperms. Instead, they are exposed on modified leaves that form cones, which are better adapted for a dry environment. These modifications for a dry environment include needle-shaped leaves, which have a thick, protective cuticle and a relatively small surface area. In addition, gymnosperms depend on wind for pollination. Examples of gymnosperms are pines, firs, redwoods, junipers, and sequoia.

ANGIOSPERMS: THE FLOWERING PLANTS

Angiosperms are seed plants whose reproductive structures are flowers and fruits. Today, these are the most diverse plant species, including about 90 percent of all plants. The color and scent of a flower attracts animals that will carry pollen from one plant to another over great distances. After pollination and fertilization, the **ovary** becomes the **fruit** and the **ovule** becomes the **seed**. Fruit protects dormant seeds and aids in their dispersal. Maple trees have seeds with wings that enable them to be dispersed great distances by the wind. Some plants have burrs on their fruits that cling to an animal's fur or a person's clothing. Many plants have fruit that are brightly colored and very sweet. An animal eats and digests the fleshy part of the fruit while the tough seed passes through the animal's digestive tract and is deposited with its feces as a package of fertilizer. There are two groups of angiosperms: monocots and dicots.

Table 12.1

The Principal Differences Between Monocots and Dicots		
Characteristic	Monocots	Dicots
Cotyledons (seed leaves)	One	Two
Vascular bundles in stem	Scattered	In a ring
Leaf venation	Parallel	Netlike
Floral parts	Usually in 3s	Usually in 4s or 5s
Roots	Fibrous Roots	Taproots

STRATEGIES THAT ENABLED PLANTS TO MOVE TO LAND

Plants began life in the seas and moved to land as competition for resources increased. The biggest problems a plant on land faces are supporting the plant body and absorbing and conserving water. Here are some modifications that evolved that enable plants to live on land.

- **Cell walls** made of cellulose lend support to the plant whose cells, unsupported by a watery environment, must maintain their own shape.
- **Roots** and **root hairs** absorb water and nutrients from the soil.
- **Stomates** open to exchange photosynthetic gases and close to minimize excessive water loss.
- The waxy coating on the leaves, **cutin**, helps prevents excess water loss from the leaves.
- In some plants, gametes and zygotes form within a protective jacket of cells called **gametangia** that prevents drying out.
- **Sporopollenin**, a tough polymer, is resistant to almost all kinds of environmental damage and protects plants in a harsh terrestrial environment. It is found in the walls of spores and pollen.
- **Seeds** and **pollen** have a protective coat that prevents desiccation. They are also a means of dispersing offspring.
- The gametophyte generation has been reduced.
- Xylem and phloem vessels enable plants to grow tall.
- Lignin embedded in xylem and other plant cells provides support.

Primary and Secondary Growth

Plants continue to grow as long as they live because they contain tissue called **meristem** that continually divides and generates new cells.

    Primary growth is the *elongation of the plant down into the soil and up into the air.* **Apical meristem** at the tips of the roots and in the buds of shoots is the source of primary growth. See Figure 12.4.

    Lateral meristem provides **secondary growth** which is increase in girth. In **herbaceous** (nonwoody) plants, there is only primary growth. In **woody plants**, secondary growth is responsible for the gradual thickening of the roots and shoots formed from earlier primary growth.

PLANT TISSUE

A plant consists of three types of tissue, each with different functions: **dermal tissue**, **vascular tissue**, and **ground tissue**.

Dermal Tissue

Dermal tissue covers and protects the plant. It includes **epidermis** and modified cells like guard cells, root hairs, and cells that produce a waxy cuticle.

Vascular Tissue

Vascular tissue consists of **xylem** and **phloem**. These transport water and nutrients around the plant; see Figure 12.1.

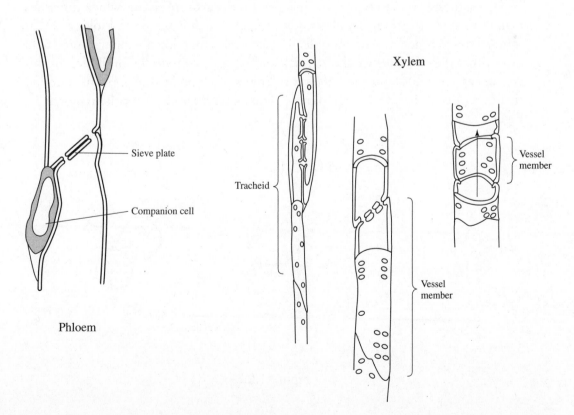

Figure 12.1

XYLEM

Xylem, the water- and mineral-conducting tissue, consists of two types of elongated cells: **tracheids** and **vessel elements**. (See Figure 12.1.) Both tracheids and vessels are dead at functional maturity. **Tracheids** are long, thin cells that overlap and are tapered at the ends. Water passes from one cell to another through **pits**, areas with no secondary cell wall. Because their secondary cell walls are hardened with **lignin**, tracheids function to support the plant as well as to transport nutrients and water. **Vessel elements** are generally wider, shorter, thinner walled, and less tapered than tracheids. Vessel elements are aligned end to end and differ from tracheids in that the ends are perforated to allow free flow through the vessel tubes. Seedless vascular plants and most gymnosperms have only tracheids; most angiosperms have both tracheids and vessel members. Xylem is what makes up wood.

PHLOEM

Phloem carries sugars from the photosynthetic leaves to the rest of the plant by active transport. The phloem vessels consist of chains of **sieve tube members** or **elements** whose end walls contain **sieve plates** that facilitate the flow of fluid from one cell to the next. In contrast to the xylem elements, these cells are alive at maturity, although they lack nuclei, ribosomes, and vacuoles. Connected to each sieve tube member is at least one **companion cell** that does contain a full complement of cell organelles and nurtures the sieve tube elements. (See Figure 12.1.)

Ground Tissue

By far the most common tissue type in a plant is the **ground tissue**, which functions mainly in support, storage, and photosynthesis. A major theme in biology is that form relates to function. For example, if a plant cell's function is support, you would expect that the cell walls would be thick and perhaps contain lignin. Think of this concept as you read about the different cell types of ground tissue in plants. Ground tissue consists of three cell types: **parenchyma**, **sclerenchma**, and **collenchyma**; see Figure 12.2.

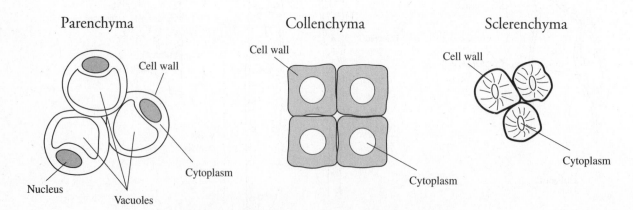

Parenchyma

Cell wall

Nucleus

Vacuoles

Cytoplasm

Collenchyma

Cell wall

Cytoplasm

Sclerenchyma

Cell wall

Cytoplasm

Figure 12.2

PARENCHYMAL CELLS

Parenchymal cells look like classic plant cells. They have primary cell walls that are thin and flexible. They lack secondary cell walls. The protoplasm contains one large vacuole, and the cell carries out most metabolic functions. Some, like mesophyll cells in the leaf, contain chloroplasts and carry out photosynthesis. In contrast, parenchymal cells in roots contain plastids and store starch. When turgid with water, they give support and shape to the plant. Most parenchymal cells retain the ability to divide and differentiate into other cell types after a plant has been injured in some way. In a laboratory, an entire plant can be regenerated or cloned from one parenchymal cell.

> **STUDY TIP**
>
> The most common cell type in a plant is a parenchymal cell.

COLLENCHYMAL CELLS

Collenchymal cells have unevenly thickened primary cell walls but lack secondary cell walls. Mature collenchymal cells are alive and their function is to support the growing stem. The "strings" of a stalk of celery, for example, consists of collenchymal cells.

SCLERENCHYMAL CELLS

Sclerenchymal cells have very thick primary and secondary cell walls fortified with **lignin**. Their function is to support the plant. There are two forms of these cells: fibers and sclereids. Fibers are long, thin, and fibrous like, and they usually occur in bundles. They are used commercially to make rope and flax fibers, which are used to make linen. Sclereids are short and irregular in shape. They make up tough seed coats and pits, and they give the pear its gritty texture.

ROOTS

Function and Structure

The three **functions** of roots are to **absorb nutrients** from the soil, **anchor** the plant, and **store food**. The root consists of specialized tissues and structures organized to carry out these various functions of the roots. See the sketch of monocot and dicot roots in cross section in Figure 12.3.

- The **epidermis** covers the entire surface of the root and is modified for **absorption**. Slender cytoplasmic projections from the epidermal cells called **root hairs** extend out from each cell and greatly increase the absorptive surface area.
- The **cortex** consists of **parenchymal cells** that contain many **plastids** for the storage of starch and other organic substances.
- **Stele**: The **vascular cylinder** or **stele** of the root consists of **vascular tissues** (xylem and phloem) surrounded by one or more layers of tissue called the **pericycle**, from which **lateral roots** arise.
- **Endoderm**: The vascular cylinder is surrounded by a tightly packed layer of cells called the **endodermis**. Each endoderm cell is wrapped with the **Casparian strip**, a continuous band of **suberin**, a waxy material that is impervious to water and dissolved minerals. The function of the endoderm is to select what minerals enter the vascular cylinder and the body of the plant.

> **STUDY TIP**
>
> The dicot root has a cross or star in the center.

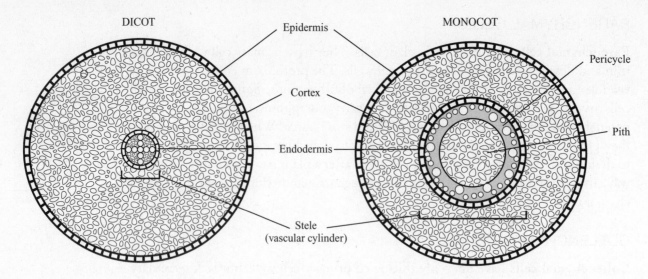

Figure 12.3   Root in Cross Section

Apical meristem, located at the tips of the roots provides **primary growth**, that is, *elongation of the plant down into the soil and up into the air*. Growth in length is concentrated near the root's tip. Three zones of cells at different stages of primary growth are located: the **zone of cell division** called **apical meristem**, the **zone of elongation**, and the **zone of differentiation**; see Figure 12.4. The root tip is protected by a **root cap**, which secretes a substance that helps digest the earth as the root tip grows through the soil.

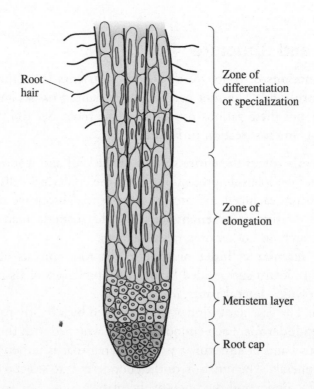

Figure 12.4   Longitudinal View of a Root Tip

- **Zone of cell division**: These are **meristem cells** that are actively dividing and are responsible for producing new cells that grow down into the soil. This is probably the region you observed under the microscope in lab when you were studying cells undergoing mitosis.
- **Zone of elongation**: Here the cells elongate and are responsible for pushing the root cap downward deeper into the soil.
- **Zone of differentiation**: Here cells undergo **specialization** into three primary meristems that give rise to three tissue systems in the plant. The **protoderm** becomes the epidermis, the **ground meristem** becomes the **cortex** (for storage), and the **procambium** becomes the primary xylem and phloem.

Types of Roots

A **taproot** is a single, large root that gives rise to lateral **branch roots**. In many **dicots**, the primary root is the **taproot**. Some taproots tap water deep in the soil. Others, like carrots, beets, and turnips, are modified for storage of food. A **fibrous root system**, common in monocots like grasses, holds the plant firmly in place. As a result, grasses make fine ground cover because they minimize soil erosion. **Adventitious roots** are roots that arise above ground. Trees that grow in swamps or salt marshes like mangroves have **aerial roots** that stick up out of the water and serve to aerate the root cells. English ivy has aerial roots that enable the ivy to cling to the sides of buildings. Some tall plants like corn have **prop roots** that grow above ground out from the base of the stem and help support the plant.

STEMS

Primary Tissue of Stems

Vascular tissue runs the length of the stem in strands called **vascular bundles**. Each vascular bundle contains xylem on the inside, phloem on the outside, and meristem tissue in between the two. In monocots, the vascular bundles are *scattered throughout the stem*; while in dicots, they are *arranged in a ring* around the edge of the stem. The ground tissue of the stem consists of **cortex** and **pith**, parenchymal tissues modified for storage. **Apical meristem**, located at the tips of the shoots and roots, supply cells for the plant to grow in length. See Figure 12.5.

Secondary Growth in Stems

Secondary growth in stems is produced by of **lateral meristem**. Lateral meristem replaces the epidermis with a secondary dermal tissue, such as bark, which is thicker and tougher. A second lateral meristem adds layers of vascular tissue. **Wood** is secondary xylem that accumulates over the years. See Figure 12.5.

REMEMBER

The vascular bundles in a monocot stem are scattered across the stem. In the dicot stem, they are organized in a single ring.

DICOT   MONOCOT

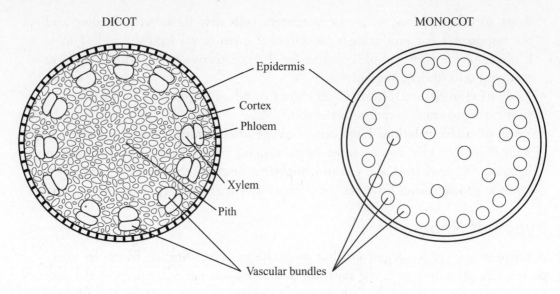

Figure 12.5   Stem

THE LEAF

The leaf is organized to maximize sugar production while minimizing water loss. The **epidermis** is covered by a **waxy cuticle** made of **cutin** to minimize water loss. **Guard cells** are modified epidermal cells that contain chloroplasts, are photosynthetic, and control the opening of the **stomates**. The inner part of the leaf consists of **palisade** and **spongy mesophyll** cells whose function is photosynthesis. The cells in the palisade layer are packed tightly, while the spongy cells are loosely packed to allow for diffusion of gases into and out of these cells. **Vascular bundles** or **veins** are located in the mesophyll and carry water and nutrients from the soil to the leaves and also carry sugar, the product of photosynthesis, from the leaves to the rest of the plant. Specialized mesophyll cells called **bundle sheath cells** surround the veins and separate them from the rest of the mesophyll. Figure 12.6 shows a sketch of a C-3 leaf.

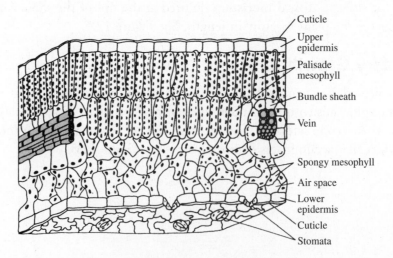

Figure 12.6   The C-3 Leaf

Stomates

About 90 percent of the water a plant loses escapes through the stomates, which account for only about 1 percent of the surface of the leaf. Guard cells are modified epithelium containing chloroplasts that control the opening and closing of the stomates by changing their shape. The cell walls of guard cells are not uniformly thick. Cellulose **microfibrils** are oriented in such a direction (radially) that when the guard cells absorb water by osmosis and become **turgid**, they curve like hot dogs, causing the stomate to open. When guard cells lose water and become **flaccid**, the stomate closes. See Figure 12.7.

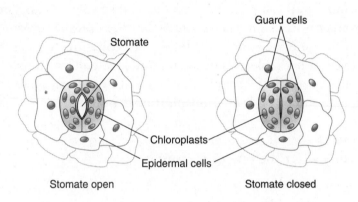

Figure 12.7   Stomates and Guard Cells

Several factors cause stomates to open:

- **Depletion of CO_2 within the air spaces of the leaf**, which occurs when photosynthesis begins, triggers the stomates to open. A plant can be tricked into opening its stomates at night by placing it into a chamber devoid of CO_2.
- An **increase in potassium ions** in the guard cells, which lowers their water potential, causes water to diffuse into them. As they become turgid, the guard cells swell and the stomate opens.
- The stimulation of the **blue light receptor (sensor) in a guard cell** stimulates the activity of ATP-powered proton pumps in the plasma membrane of the guard cells, which, in turn, promote the uptake of K^+ ions. This causes the stomates to open.
- Stomate opening correlates with **active transport of H^+ out of the guard cells** and into the surrounding epithelial cells.

Other factors cause stomates to close:

- **Lack of water** causes the guard cells to lose their turgor, become flaccid, and close the stomate.
- **High temperatures** also close the stomate presumably by stimulating cellular respiration and increasing CO_2 concentration within the air spaces of the leaf.
- **Abscisic acid**, which is produced in the mesophyll cells in response to dehydration, signals guard cells to close the stomates.

TRANSPORT IN PLANTS

Transport of Xylem

Xylem fluid rises in a plant against gravity but requires no energy. The fluid in the xylem can be *pushed up* by root pressure or *pulled up* by **transpirational pull**. **Root pressure** results from water flowing into the stele from the soil as a result of the high mineral content in the root cells. It can push xylem sap upward only a few yard (meters). Droplets of water that appear in the morning on the leaf tips of some herbaceous dicots, like strawberries, is due to root pressure. This is known as **guttation**.

Transpirational pull can carry fluid up the world's tallest trees. **Transpiration**, the **evaporation of water from leaves**, causes negative pressure (tension) to develop in the xylem tissue from the roots to the leaves. The **cohesion** of water due to strong attraction between water molecules makes it possible to pull a column of water from above within the xylem. The absorption of sunlight drives transpiration by causing water to evaporate from the leaf. **Transpirational pull-cohesion tension theory** states that *for each molecule of water that evaporates from a leaf by transpiration, another molecule of water is drawn in at the root to replace it.*

Several factors affect the rate of transpiration:

- High humidity slows down transpiration, while low humidity speeds it up.
- Wind can reduce humidity near the stomates and thereby increase transpiration.
- Increased light intensity will increase photosynthesis and thereby increase the amount of water vapor to be transpired and increase the rate of transpiration.
- Closing stomates stops transpiration.

Absorption of Nutrient and Water

- **Apoplast and symplast**: The movement of water and solutes across a plant, called **lateral movement**, is accomplished along the symplast and apoplast. The **symplast** is a continuous system of cytoplasm of cells interconnected by **plasmodesmata** for long-distance transport. The apoplast is the network of cell walls and intercellular spaces within a plant body that permits short-distance *extracellular movement* of water within a plant. When water reaches the **endodermis**, it can continue to the xylem through the symplast; but water in the apoplast must pass across the endodermis by diffusion.
- **Mycorrhizae**: In mature plants of many species where older regions of roots lack root hairs, mycorrhizae supply the plant with water and minerals. Mycorrhizae are the symbiotic structures consisting of the plant's roots intermingled with the **hyphae** (filaments) of a fungus that greatly increase the quantity of nutrients that a plant can absorb.
- **Rhizobium** is a symbiotic bacterium that lives in the nodules on roots of specific legumes and that fixes nitrogen gas from the air into a form of nitrogen the plant requires.
- **Bulk flow** is how fluids move great distances in plants.

Translocation of Phloem Sap

Phloem sap travels around the plant from sugar **source** to sugar **sink**. This transport is called **translocation**. The source is the structure in which the sugar is being produced by either photosynthesis or the breakdown of starch. Mature leaves are the primary source of sugar. The sugar sink is the structure that stores or consumes the sugar. Growing roots and fruit are sugar sinks. Storage organs like tubers and roots are the sugar sink in the growing season, but they become the source during the spring when starch is broken down to be used as a sugar source by the rest of the plant.

PLANT REPRODUCTION

Asexual Reproduction

Plants can **clone** themselves or reproduce asexually by **vegetative propagation**. In this process, a piece of the **vegetative** part of a plant, the **root**, **stem**, or **leaf**, produces an entirely new plant genetically identical to the parent plant. Examples of asexual reproduction are **grafting**, **cuttings**, **bulbs**, and **runners**.

Sexual Reproduction in Flowering Plants

The flower is the sexual organ of a plant. Fertilization in a flower begins with **pollination** when one pollen grain containing three haploid nuclei, one **tube nucleus**, and two **sperm nuclei** lands on the sticky **stigma** of the flower. The pollen grain absorbs moisture and sprouts, producing a pollen tube that burrows down the style into the ovary. The two sperm nuclei travel down the pollen tube into the ovary. Once inside the ovary, the two remaining sperm nuclei enter the ovule through the **micropyle**. One sperm nucleus fertilizes the egg and becomes the **embryo** (2*n*). The other sperm nucleus fertilizes the **two polar bodies** and becomes the **triploid** (3*n*) **endosperm**, the food for the growing embryo. This process is known as **double fertilization** because two fertilizations occur. After fertilization, the ovule becomes the **seed** and the ripened ovary becomes the **fruit**. In monocots, food reserves remain in the endosperm. In dicots, the food reserves of the **endosperm** are transported to the **cotyledons** and consequently, the mature dicot seed lacks endosperm. In the monocot **coconut**, the endosperm is liquid. See Figure 12.8.

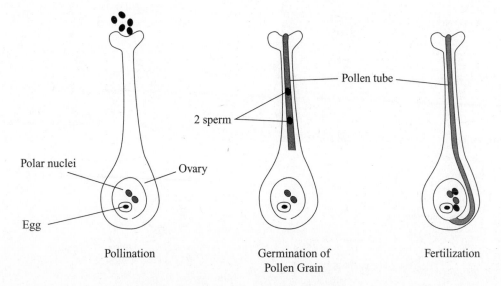

Figure 12.8   Double Fertilization

The Seed

The seed consists of a protective **seed coat**, an **embryo**, and the **cotyledon** or **endosperm**, food for the growing embryo. The embryo consists of the **hypocotyl**, **epicotyl**, and **radicle**. The hypocotyl becomes the lower part of the stem. The epicotyl becomes the upper part of the stem. The radicle, or **embryonic root**, is the first organ to emerge from the germinating seed. Figure 12.9 shows a dicot seed, like a peanut, split in half.

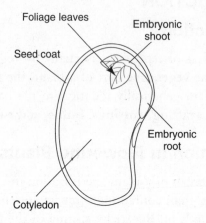

Figure 12.9   The Dicot Seed

ALTERNATION OF GENERATIONS

The sexual life cycle of plants is characterized by an **alternation of generations** in which **haploid** (n) and **diploid** ($2n$) generations alternate with each other. The **gametophyte** (n) produces gametes by mitosis. These gametes fuse during fertilization to yield $2n$ zygotes. Each zygote develops into a **sporophyte** ($2n$), which produces haploid spores (n) by meiosis. Each haploid spore forms a new gametophyte, completing the life cycle; see Figure 12.10.

> **STUDY TIP**
>
> Gametophyte = n
>
> Sporophyte = $2n$

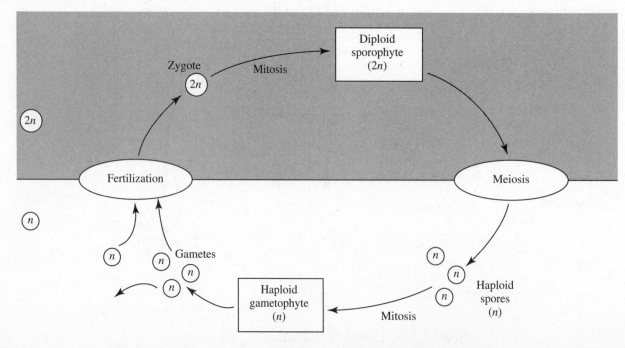

Figure 12.10   Alternation of Generations

Vocabulary for Alternation of Generations

Antheridium	Structure that produces sperm
Archegonium	Structure that produces eggs
Gametophyte	Haploid plant
Heterosporous	Having two kinds of spores, male and female
Homosporous	Producing a single spore that grows into a bisexual gametophyte
Megaspore	Produced by large female cones; develops into female gametophytes
Microspore	Produced by small male cones; develops into male gametophytes or pollen grains
Protonema	Branching one-celled thick filaments produced by germinating moss spores; becomes the gametophyte in moss
Sporangia	Located on mature sporophyte where meiosis occurs, producing haploid spores
Sporophyte	Diploid plant
Sori	Raised spots located on the underside of sporophyte ferns; clusters of sporangia

Mosses and Other Bryophytes

Like all plants, bryophytes exhibit alternation of generations. The **haploid** or **gametophyte** (n) generation is dominant and persists for most of the plant's life. The diploid generation ($2n$) lives only for a short time, and is dependent on the gametophyte for its food.

- *The gametophyte generation dominates.*
- Archegonia and antheridia develop on the tips of the gametophyte.
- The sporophyte grows out of the top of the gametophyte and obtains its nutrients from the gametophyte.
- Haploid spores are formed in mature sporangia.
- See Figure 12.11.

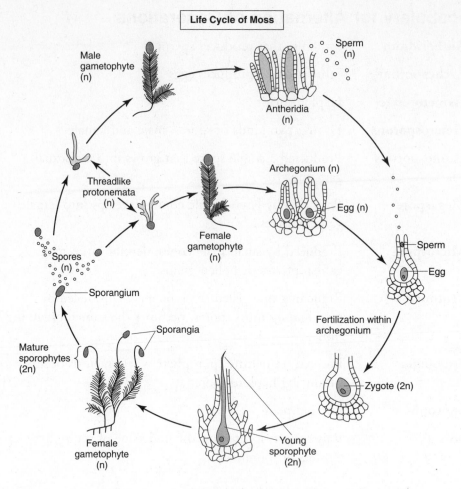

Figure 12.11

Ferns—Seedless Vascular Plants

- *The sporophyte generation is larger than and independent from the gametophyte.*
- Archegonia and antheridia both develop on the underside of the heart-shaped haploid gametophyte.
- Flagellated sperm swim from antheridia to archegonia (of different plants) to form the diploid zygote, which grows into a large diploid sporophyte plant.
- Haploid spores emerge from sori and land on the ground to sprout into the gametophyte generation.
- See Figure 12.12.

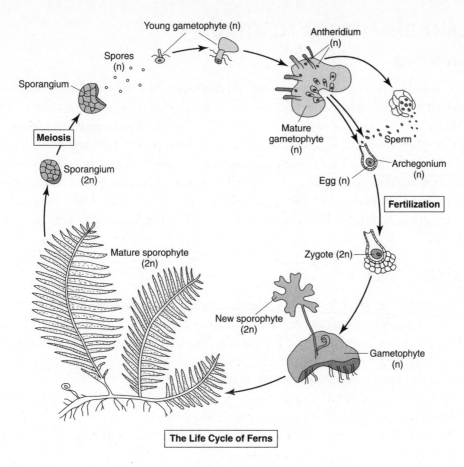

Figure 12.12

Seed Plants

Seed plants are vascular plants that produce seeds. They are divided into two groups: flowering plants and conifers. All are **heterosporous**, meaning there are two kinds of spores: male and female. In the flowering plants, the **angiosperms**, *the gametophyte generation exists inside the sporophyte generation and depends totally on it.* **Meiosis** occurs in the **anthers** and **pistils**. Anthers produce **microspores** that form **male gametophytes**, while the ovules produce **megaspores** that form **female gametophytes**. Fertilization occurs in the ovary, producing zygotes that develop into sporophyte embryos within the ovule. The ovule becomes the seed carrying the embryo and the food necessary for its initial development into another sporophyte.

A gymnosperm like the pine tree is a **sporophyte**. Its **sporangia** are densely packed inside the cones. The gametophyte generation develops from haploid spores retained within the sporangia. Conifers, like all seed plants, are **heterosporous**; male and female gametophytes develop from different types of spores produced by separate cones. Each tree usually has both types of cones. Small pollen cones produce **microspores** that develop into **male gametophytes** or pollen grains. Larger ovulate cones usually develop on separate branches of the tree and produce **megaspores** that develop into **female gametophytes**.

PLANT RESPONSES TO STIMULI
Hormones

Plant hormones help coordinate growth, development, and responses to environmental stimuli. They are produced in very small quantities, but they have a profound effect on the plant because the hormone signal is amplified. **Signal transduction pathways** amplify the hormonal signal and connect it to specific cell responses. (See the section below.) A plant's response to a hormone usually depends not so much on absolute quantities of hormones but on relative amounts. Hormones can have multiple effects on a plant, and they can work synergistically with other hormones or in opposition to them. Here is an overview of plant hormones and what they stimulate.

1. AUXIN

- Responsible for **phototropisms** due to *unequal distribution of auxin.*
- Enhances **apical dominance**, the preferential growth of a plant upward (toward the sun), rather than laterally. (The terminal bud actually suppresses lateral growth by suppressing development of axial buds.)
- Stimulates stem elongation and growth by softening the cell wall.
- **Indoleacetic acid** (**IAA**) is a naturally occurring auxin.
- A humanmade auxin, 2,4-D, is used as a weed killer.
- When used as rooting powder, it causes roots to develop quickly in a plant cutting.
- A synthetic auxin sprayed onto tomato plants will induce fruit production without pollination. This results in seedless tomatoes.

2. CYTOKININS

- *Stimulate cytokinesis and cell division.*
- Delay **senescence** (aging) by inhibiting protein breakdown. (Florists spray cut flowers with cytokinins to keep them fresh.)
- Are produced in roots and travel upward in the plant.

3. GIBBERELLINS

- *Promote stem and leaf elongation.*
- 100 different naturally occurring gibberellins have been identified.
- Work in concert with auxins to promote cell growth.
- Induce **bolting**, the rapid growth of a floral stalk. When a plant, such as broccoli, which normally grows close to the ground, enters the reproductive stage, it sends up a very tall shoot on which the flower and fruit develop. This is a mechanism to ensure pollination and seed dispersal.

4. ABSCISIC ACID (ABA)

- *Inhibits growth.*
- Enables plants to withstand drought.
- Closes stomates during times of water stress.

- Counteracts the breaking of dormancy during a winter thaw.
- *Promotes seed dormancy.* This prevents seeds that have fallen on the ground in the fall from sprouting until the spring when environmental conditions are better. In some desert plants, seeds break dormancy only after heavy rains have washed all ABA out of the cells. (The ground is moist and can support plant growth.)

5. ETHYLENE

- This plant hormone is a gas.
- *Promotes fruit ripening.* This is another example of **positive feedback**. Ethylene gas promotes ripening, which, in turn, triggers increased production of ethylene gas. "One bad apple spoils the whole barrel."
- Commercial fruit sellers pick perishable fruit before they are ripe, while still hard. When they arrive at their destination, they are sprayed with ethylene gas to hasten the ripening. In contrast, apples are kept in an environment of CO_2 to eliminate exposure to ethylene gas and, thus, keep the apples from ripening or rotting. In this way, apples can be stored for long periods of time.
- Is produced in large quantities in times of stress such as drought, flooding, mechanical pressure, injury, and infection.
- *Facilitates **apoptosis***: programmed cell death. Prior to death, cells break down many of their chemical components for the plant to salvage and reuse.
- *Promotes **leaf abscission***—the leaf dies and falls from the plant. Subsequently, a scar forms at the abscission layer to prevent pathogens from entering the plant.

Tropisms

A tropism is the growth of a plant toward or away from a stimulus. Examples are **thigmotropisms** (touch), **geotropisms** or **gravitropisms** (gravity), and **phototropisms** (light). A growth of a plant toward a stimulus is known as a **positive tropism**, while a growth away from a stimulus is a **negative tropism**.

Phototropisms result from an *unequal distribution* of **auxins**, which accumulate on the side of the plant away from the light. Since auxins cause growth, the cells on the shady side of the plant enlarge and the stem bends toward the light.

Geotropisms result from an interaction of auxins and **statoliths**, specialized plastids containing dense starch grains.

Signal Transduction Pathway

Plants, like animals, can respond to stimuli through the **signal transduction pathway**, which occurs in three stages: **reception**, **transduction**, and **response**. To begin, when a **receptor** is stimulated, it undergoes a **conformational change** and activates a second messenger. **Secondary messengers** are cyclic nucleotides, like **cyclic AMP** (**cAMP**) or **cyclic GMP** (**cGMP**), that transfer and amplify the signal from the receptor. The secondary messenger leads directly to a response by altering transcription factors in the nucleus of the target cell and, thus, alters the expression of certain genes.

Photoperiodism

The environmental stimulus a plant uses to detect the time of year is the **photope-riod**, the relative lengths of day and night. Plants have a biological clock set to a 24-hour day, known as a **circadian rhythm**. The physiological response to the photoperiod, such as flowering, is known as **photoperiodism**. Some plants will flower only when the light period is longer than a certain number of hours. These plants are called **long-day plants**. Some plants are **short-day plants**, and some are **day-neutral** and will flower regardless of the length of day. In the 1940s, botanists discovered that plants actually respond to the length of darkness, not the length of light. So a *long-day plant* is actually a *short-night plant*.

The photoreceptor responsible for keeping track of the length of day and night is the pigment **phytochrome**. There are two forms of phytochrome, **Pr (red-light absorbing)** and **Pfr (infrared light absorbing)**. Phytochrome is synthesized in the Pr form. When the plant is exposed to light, Pr converts to Pfr. In the dark, Pfr reverts back to Pr. The conversion from one to the other enables the plant to keep track of time. The plant is able to sense the concentrations of the two phytochromes and respond accordingly.

(daylight)
Red light

$$\text{Pr} \xrightleftharpoons{\hspace{1.5cm}} \text{Pfr (triggers germination)}$$

IR light
(Slow conversion at night)

Multiple-Choice Questions

1. Which is true of mosses?

   (A) They have true xylem but lack phloem.
   (B) They are more advanced than the ferns.
   (C) Their cell walls consist of chitin instead of cellulose.
   (D) They lack vascular tissue.
   (E) They contain photosynthetic pigments other than chlorophyll.

2. The ancestors of land plants were most likely similar to modern

   (A) conifers
   (B) ferns
   (C) green algae
   (D) flowering plants
   (E) bacteria

3. Which is true of seeds?

   (A) They contain the gametophyte.
   (B) They are characteristic of all plants.
   (C) They are a mechanism for dispersal of pollen.
   (D) They are not characteristic of the conifers.
   (E) They contain the cotyledon.

4. Vascular plant tissue includes all of the following EXCEPT

   (A) meristem
   (B) sieve tube cells
   (C) vessels
   (D) tracheids
   (E) companion cells

5. Which is CORRECT about roots?

   (A) Adventitious roots grow below water in certain saltwater plants.
   (B) Only the roots of monocots have a stele.
   (C) The function of the endoderm is to control what water and nutrients enter the body of the plant.
   (D) The Casparian strip is a modification of root hairs in monocots.
   (E) Taproots are a modification for increased absorption.

6. Which of the following is true of ferns?

   (A) They are tracheophytes and have seeds.
   (B) They have transport vessels and flowers.
   (C) Most rely on mychorrizae to increase absorption of nutrients from the soil.
   (D) The sporophyte and gametophyte are independent of each other.
   (E) They are primitive gymnosperms.

7. Which tissue makes up most of the wood of a tree?

   (A) bark
   (B) primary phloem
   (C) secondary phloem
   (D) primary xylem
   (E) secondary xylem

8. Which is CORRECT about monocots?

   (A) Vascular bundles in the stem are in a ring.
   (B) Their floral parts are usually in 3 s.
   (C) They usually have taproots.
   (D) The veins in the leaves are netlike.
   (E) Most common trees, such as maples and oaks, are monocots.

9. Cortex, mesophyll, epidermal cells, and pith all consist of these cells.

    (A) parenchyma
    (B) collenchyma
    (C) sclerenchyma
    (D) All of the above
    (E) None of the above

10. Which of the following causes stomates to close?

    (A) stimulation of the blue light sensor in the guard cells
    (B) active transport of H^+ out of the guard cells into the surrounding cells
    (C) depletion of CO_2 within the air spaces of the leaf
    (D) increase in sunlight
    (E) increase in abscisic acid in the guard cells

11. All of the following evolved in plants in response to water shortage EXCEPT

    (A) mychorrizae
    (B) CAM metabolism
    (C) Kranz anatomy
    (D) cutin
    (E) large surface area of leaves to absorb sunlight

12. You cut your initials in a tree 5 feet above the ground this year. The tree grows 2 feet per year and you return in 8 years. How high will your initials be at that time?

    (A) 5 feet
    (B) 10 feet
    (C) 16 feet
    (D) 21 feet
    (E) 7 feet

13. Which of the following are long, thin cells that overlap, are tapered end to end, and carry water?

    (A) parenchyma
    (B) sieve tube members
    (C) tracheids
    (D) companion cells
    (E) phloem

14. To observe the process of mitosis in plant roots, a student should examine the plant's

    (A) root cap
    (B) zone of maturation
    (C) meristem tissue
    (D) pericycle
    (E) endodermis

15. Which of the following results from the direct expenditure of energy in a plant?

    (A) root pressure
    (B) evaporation of water from the leaves
    (C) water flowing into the root apoplast
    (D) transpirational flow
    (E) sap flowing down from the leaves to the roots

16. Guttation in a plant results directly from

    (A) transpirational pull
    (B) injury to the plant
    (C) condensation of water vapor onto the leaf
    (D) root pressure
    (E) none of the above

17. The primary function of the cortex in a plant stem is

    (A) storage
    (B) transport
    (C) photosynthesis
    (D) absorption
    (E) to maintain proper water potential

18. Phototropisms are controlled by

    (A) excess levels of ethylene gas
    (B) degradation of phytochromes
    (C) circadian rhythm
    (D) an unequal distribution of auxins
    (E) CO_2 levels

19. Which is correct about plants?

    (A) The epicotyl becomes the upper stem.
    (B) The hypocotyl becomes the endosperm.
    (C) Monocots usually have food reserves in the cotyledon.
    (D) Pollen contains the sporophyte.
    (E) The tube nucleus is part of the ovary.

20. Which of the following occurs after fertilization?

    (A) The ovule becomes the seed, the ovary becomes the fruit.
    (B) The ovary becomes the seed, the ovule becomes the fruit.
    (C) The micropyle becomes the seed, the sepals become the fruit.
    (D) The stigma becomes the seed, the ovule becomes the fruit.
    (E) The micropyle becomes the seed, the ovary becomes the fruit.

21. When you pinch off the terminal buds from a young plant to make it grow bushy, which of the following hormones is responsible?

   (A) cytokinins
   (B) auxins
   (C) gibberellins
   (D) abscisic acid
   (E) ethylene

22. "One rotten apple spoils the whole bunch" is the work of which of the following?

   (A) cytokinin
   (B) auxin
   (C) gibberellins
   (D) ethylene
   (E) abscisic acid

Answers to Multiple-Choice Questions

1. **(D)** Mosses are primitive, nonvascular plants, classified as bryophytes. They show an alternation of generations and produce gametes by meiosis and mitosis. The cell walls of fungi consist of chitin.

2. **(C)** The ancestor of the modern multicellular plant is the green algae, Chlorophyta.

3. **(E)** Seeds contain the embryo and the food for the embryo, the endosperm (in monocots) or the cotyledon (in dicots). All plants do not produce seeds; for example, bryophytes and the ferns do not. Conifers do produce seeds. They are located on cones.

4. **(A)** Meristem tissue is actively dividing tissue that gives rise to other tissue, such as xylem and phloem. Xylem consists of vessel and tracheids. Phloem consists of companion cells and sieve tube elements.

5. **(C)** The function of the endoderm is to control what enters the body of the plant. Endoderm cells are wrapped with a waterproof Casparian strip. Adventitious roots are roots that grow above ground. Tap roots are for storage. The roots of both monocots and dicots have a stele, a vascular cylinder. The Casparian strip surrounds the endodermis cells and helps control what enters the stele.

6. **(D)** Ferns are tracheophytes and produce spores, not seeds. They have transport vessels, but they are not flowering plants. They do not have a symbiotic relationship with mychorrizae living in their roots to enhance absorption of nutrients from the soil, as some plants do.

7. **(E)** Wood consists of secondary xylem; the primary xylem was formed first and is located in a small area at the center of a tree.

8. (**B**) Monocots have floral parts oriented in 3 s and have parallel veins in the leaves. Dicots usually have taproots, and the veins in their leaves are netlike. Most trees are dicots. Palm trees are monocots.

9. (**A**) Parenchyma is the most common ground tissue in a plant. Parenchymal cells have primary cell walls that are thin and flexible and look like a traditional plant cell.

10. (**E**) Lack of water, high temperatures, and an increase in abscisic acid cause stomates to close. Things that cause stomates to open are depletion of CO_2 in the air spaces within the leaf, an influx of potassium ions into the guard cells, and the stimulation of the blue-light sensor in the guard cells, which promotes the uptake of K^+ ions. Also, stomate opening correlates with active transport of H^+ out of guard cells.

11. (**E**) All choices describe responses to water shortage except choice E. A large surface area of leaf to maximize the amount of sunlight absorbed would also increase water loss. Leaves in a tropical rain forest, where the environment is very humid, are typically broad. The needles on a conifer evolved to minimize water loss. Conifers are found in cold, dry northern regions.

12. (**A**) A tree grows upward from the apical meristem at the top of the tree, not from the bottom of the tree. Therefore, the initials do not move from their original spot.

13. (**C**) Tracheids and vessels make up the xylem and carry water from the roots to the leaves. Tracheids are long, thin cells that overlap and are tapered at the ends. Vessels are generally wider and shorter, thinner walled, and less tapered than tracheids.

14. (**C**) Meristem tissue is the only tissue of the choices given that is actively dividing tissue. Meristem tissue is growth tissue; it gives rise to other cell types. For example, meristem gives rise to xylem and phloem cells.

15. (**E**) Sap flowing in the phloem requires energy; water moving in the xylem does not. All the other choices occur by passive transport.

16. (**D**) Droplets of water that appear in the morning on the leaf tips of some herbaceous leaves are due to root pressure. This phenomenon is known as guttation.

17. (**A**) The cortex consists of parenchymal cells and stores starch, among other things.

18. (**D**) A phototropism is a plant growth toward or away from a stimulus and is the result of unequal distribution of auxins.

19. (**A**) The epicotyl is part of the embryo and becomes the upper part of the stem and the leaves; the hypocotyl becomes the lower part of the stem and the roots. Monocots store food reserves in the endosperm, not the cotyledon. Pollen consists of three haploid sperm nuclei; it is the gametophyte, not the sporophyte. The tube nucleus is one of the sperm nuclei in pollen. It grows the pollen tube.

20. **(A)** After fertilization, the ovule becomes the seed and the ovary becomes the fruit. The other choices make no sense.

21. **(B)** The growing tip of a young plant produces auxins and enhances apical dominance. Removal of the growing tip, known as pinching back the plant, removes the auxins, and the plant grows laterally, that is, bushier.

22. **(D)** Ethylene gas is given off by plants while they are ripening. To ripen hard fruit quickly, put them into a paper bag with a ripe banana, which gives off large amounts of ethylene gas.

Free-Response Questions

Directions: Answer all questions. You must answer the question in essay—**not** outline—form. You may use labeled diagrams to supplement your essay, but diagrams alone are *not* sufficient. Before you start to write, read each question carefully so that you understand what the question is asking.

1. Discuss the movement of water from the roots to the leaves, naming all the cells and structures water passes through along the way.

2. Discuss the structural and functional strategies that enabled plants to move to land.

Typical Free-Response Answers

Note: The key words are in boldface. Remember, you get credit only when you state a correct fact and use the correct scientific term.

1. Water diffuses into a plant through **root hairs**, which are cytoplasmic extensions of epidermal cells. From the root hairs, water travels into the **parenchymal cells** of the **cortex** in two ways, along the **symplast** and the **apoplast**. The symplast is a continuous system of cytoplasm of cells interconnected by **plasmodesmata**. The apoplast is the network of cell walls and intercellular spaces within the plant body that permits extensive extracellular movement of water.

   From the cortex, water moves to the **endodermis**, a tightly packed layer of cells that surrounds the **vascular cylinder** or **stele**. Each endoderm cell is encircled by a band of wax, the **Casparian strip**, which is not permeable to water. It controls water entering the plant by the apoplast route, which must diffuse into the endoderm cells before continuing. Once water has entered the endodermis, it freely passes into the vascular cylinder and into the xylem, which consists of **tracheids** and **vessel elements**. Once in the xylem, water moves upward toward the leaves by a combination of **transpirational pull** and **cohesion tension**. Water diffuses from the roots, where the **water potential** is highest, to the air spaces in the leaves, where the water potential is the lowest. From the veins in the leaf, water diffuses into **air spaces** within the **spongy mesophyll** and then into the **palisade** and **spongy mesophyll cells** as needed for photosynthesis. In the light reactions of photosynthesis, water molecules are broken down during photolysis, providing electrons for the light reactions and protons for the dark reactions. The **oxygen** from the water molecules is given off into the atmosphere as a waste product.

Note: This question demonstrates a common type of essay question on the AP Exam because it checks for broader understanding by integrating several topics. Key words are boldface to keep you focused on the need for scientific terms. If you need help, review sections in this book on plants, cells, and photosynthesis.

2. Plants began life in the seas and moved to land as **competition** for resources increased. Some of the problems a plant living on land faces are **supporting the plant body**, **absorbing** and **conserving water**, and **reproducing outside of a watery environment**.

Strong stems (trunks) and branches hold leaves up toward the sunlight. Three specialized cells help support plants: parenchyma, collenchyma, and sclerenchyma. Although they lack a secondary cell wall, **parenchymal cells** lend support to a plant when the cell is **turgid**. **Collenchymal cells** have thick primary cells walls and provide flexible support without restraining growth. **Sclerenchymal cells** are very specialized for support. They have rigid and thick walls fortified with lignin. Even after these cells die, their rigid walls provide a skeleton for the plant.

Absorbing and conserving water are two other problems for plants living on land. **Roots**, in addition to anchoring the plant in the soil, absorb nutrients. **Root hairs** are slender cytoplasmic extensions from epidermal cells of the root that greatly increase the absorptive surface area. **Mycorrhizae** are symbiotic fungi that live on mature older regions of roots that lack root hairs. This **symbiont** increases the quantity of water and nutrients plants can absorb. The opening and closing of **stomates** in a leaf limit the loss of water from a plant by transpiration and are controlled by guard cells. A **waxy cuticle** made of **cutin** covers leaves, further minimizing water loss. Some plants have stomates nestled in **stomatal crypts** that minimize exposure of the stomate to air and further minimize transpiration.

Modifications in anatomy and physiology enable plants to be successful on dry land. **C-4 plants** have evolved **Kranz anatomy**, wherein bundle sheath cells sequester CO_2 deep within the leaf and away from stomates. The **Hatch-Slack pathway**, a biochemical pathway, also removes CO_2 from the air spaces and away from the stomates. Both strategies evolved in plants in dry environments to keep their stomates closed as much as possible, thus minimizing excessive water loss through transpiration.

Reproduction on land also offers great challenges for land organisms. In some plants, gametes and zygotes form within a protective jacket of cells called **gametangia** that prevent drying out. **Sporopollenin** is a tough polymer, found in the walls of spores and pollen, that resists harsh terrestrial environments. **Seeds** also have a very tough protective coating to prevent drying out. Seeds have been known to sprout after being dormant for 1,000 years.

Animal Physiology

• Digestion in different animals	• Chemical signals
• Digestion in humans	• Temperature regulation
• Gas exchange in different animals	• Osmoregulation
• Gas exchange in humans	• Excretion
• Circulation in different animals	• Nervous system
• Human circulation	• Muscles

INTRODUCTION

For the most part, this chapter focuses on human physiology. However, some review of other animals is included to gain a broader understanding and better insight into important physiological principles and adaptations common to all animals. For more specifics on animals, see the chapter "Classification." The human immune system, and human reproduction and development are presented as separate chapters with their own questions.

DIGESTION IN DIFFERENT ANIMALS

Hydra

In the **hydra** (cnidarians), digestion occurs in the **gastrovascular cavity**, which has only one opening. Cells of the **gastrodermis** (lining of the gastrovascular cavity) secrete digestive enzymes into the cavity for **extracellular digestion**. Some specialized nutritive cells have flagella that move the food around the gastrovascular cavity, and some have pseudopods that engulf food particles.

Earthworm

The digestive tract of the **earthworm** is a long, straight tube. As the earthworm burrows in the ground, creating tunnels that aerate the soil, the mouth ingests decaying organic matter along with soil. From the mouth, food moves to the esophagus and then to the **crop** where it is stored. Posterior to the crop, the **gizzard**, which consists of thick, muscular walls, grinds up the food with the help of sand

and soil that were ingested along with the organic matter. The rest of the digestive tract consists of the intestines where chemical digestion and absorption occur. Absorption is enhanced by the presence of a large fold in the upper surface of the intestine, called the **typhlosole**, which greatly increases the surface area.

Grasshopper

Like the earthworm, the **grasshopper** has a digestive tract that consists of a long tube consisting of a **crop** and **gizzard**; however, there are several differences. The grasshopper has specialized mouth parts for tasting, biting, and crushing food and has a gizzard that contains **plates** made of **chitin** that help in grinding the food. In addition, in the grasshopper, the digestive tract is also responsible for removing nitrogenous waste (**uric acid**) from the animal.

DIGESTION IN HUMANS

The human digestive system has two important functions: **digestion**—breaking down large food molecules into smaller usable molecules and **absorption**—the diffusion of these smaller molecules in the body's cells. **Fats** get broken down into glycerol and fatty acids, **starch** into monosaccharides, **nucleic acids** into nucleotides, and **proteins** into amino acids. **Vitamins** and **minerals** are small enough to be absorbed without being digested. The digestive tract is about 30 feet (9 m) long and made of **smooth** (**involuntary**) **muscle** that pushes the food along the digestive tract by a process called **peristalsis**.

Mouth

In the mouth, the **tongue** and differently shaped **teeth** work together to break down food mechanically. Form relates to function, and the type of teeth a mammal has reflects its dietary habits. Humans are **omnivores** and have three different types of teeth: **incisors** for cutting, **canines** for tearing, and **molars** for grinding. **Salivary amylase** released by **salivary glands** begins the chemical breakdown of **starch**.

Esophagus

After swallowing, food is directed into the esophagus, and not the windpipe, by the **epiglottis**, a flap of cartilage in the back of the **pharynx** (throat). No digestion occurs in the esophagus.

Stomach

The **stomach** churns food mechanically and secretes **gastric juice**, a mixture of the enzyme pepsinogen and hydrochloric acid, that begins the digestion of **proteins**. The acid environment (pH 2–3) activates **pepsinogen** to become the active enzyme **pepsin** and also kills germs. The stomach of all mammals also contains **rennin** to aid in the digestion of the protein in milk. The **cardiac sphincter** at the top of the stomach keeps food in the stomach from backing up into the esophagus and burning it. The **pyloric sphincter** at the bottom of the stomach keeps the food in the stomach long enough to be digested.

Excessive acid can cause an **ulcer** to form in the esophagus, the stomach, or the **duodenum** (the first 12 inches [30 cm] of the small intestine). Scientists now know that a common cause of ulcers is a particular bacterium, ***Helicobacter pylori***, which can be effectively treated with antibiotics.

Small Intestine

Digestion is completed in the **duodenum**. Intestinal enzymes and pancreatic amylases hydrolyze starch and glycogen into maltose. **Bile**, which is produced in the **liver** and stored in the **gallbladder**, is released into the small intestine as needed and acts as an **emulsifier** to break down fats, creating greater surface area for digestive enzymes. **Peptidases**, such as trypsin and chymotrypsin, continue to break down proteins. Nucleic acids are hydrolyzed by **nucleases**, and **lipases** break down fats. Once digestion is complete, the lower part of the small intestine is the site of **absorption**. Millions of fingerlike projections called **villi** absorb all the nutrients that were previously released from digested food. Each villus contains capillaries, which absorb amino acids, vitamins and monosaccharides, and a **lacteal**, a small vessel of the **lymphatic system**, which absorbs fatty acids and glycerol. Each epithelial cell of the villus has many microscopic cytoplasmic appendages called **microvilli** that greatly increase the rate of nutrient absorption by the villi. Figure 13.1 shows a villus with a lacteal, capillaries, and microvilli.

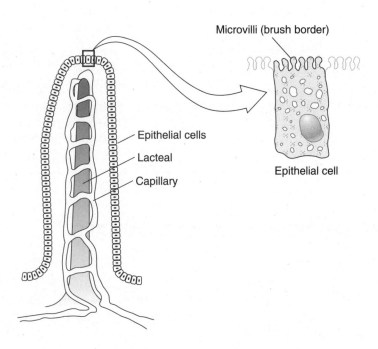

Figure 13.1   Villus

Large Intestine

The **large intestine** or **colon** serves three main functions: **egestion**, the removal of undigested waste; **vitamin production**, from bacteria symbionts living in the colon; and the **removal of excess water**. Together, the small intestine and colon reabsorb 90 percent of the water that entered the alimentary canal. If too much water is removed from the intestine, **constipation** results; if inadequate water is removed, **diarrhea** results. The last 7–8 inches (18–20 cm) of the gastrointestinal tract is called the **rectum**. It stores **feces** until their release. The opening at the end of the digestive tract is called the **anus**.

Hormones that Regulate the Digestive System

Hormones are released as needed as a person sees or smells food or as food moves along the gut. Table 13.1 summarizes the hormones involved in regulating digestion.

TABLE 13.1

Hormones That Regulate Digestion		
Hormone	**Site of Production**	**Effect**
Gastrin	Stomach wall	Stimulates sustained secretion of gastric juice
Secretin	Duodenum wall	Stimulates pancreas to release bicarbonate to neutralize acid in duodenum
Cholecystokinin (CCK)	Duodenum wall	Stimulates pancreas to release pancreatic enzymes and gall bladder to release bile into small intestine

GAS EXCHANGE IN DIFFERENT ANIMALS

Respiration is the exchange of respiratory gases, oxygen and carbon dioxide, between the external environment and the cell or body. It occurs passively by **diffusion**. Therefore, respiratory surfaces must be **thin**, be **moist**, and have **large surface areas**. Although all organisms must exchange respiratory gases, they have evolved different strategies to accomplish it.

- In simple animals, like **sponges** and **hydra**, gas exchange occurs over the entire surface of the organism wherever cells are in direct contact with the environment.
- **Earthworms** and **flatworms** have an **external respiratory surface** because diffusion of O_2 and CO_2 occurs at the skin. Oxygen is carried by **hemoglobin** dissolved in blood.

- The **grasshopper** and other **arthropods** and **crustaceans** have an **internal respiratory surface**. Air enters the body through **spiracles** and travels through a system of **tracheal tubes** into the body, where diffusion occurs in sinuses or hemocoels. In **arthropods** and in some **mollusks**, oxygen is carried by **hemocyanin**, a molecule similar to hemoglobin but with **copper**, instead of iron, as its core atom.
- Aquatic animals like **fish** have **gills** that take advantage of **countercurrent exchange** to maximize the diffusion of respiratory gases.

GAS EXCHANGE IN HUMANS

In **humans**, air enters the nasal cavity and is **moistened**, **warmed**, and **filtered**. From there, air passes through the **larynx** and down the **trachea** and **bronchi** into the tiniest **bronchioles**, which end in microscopic air sacs called **alveoli** where diffusion of respiratory gases occurs; see Figure 13.2. Humans have an **internal respiratory surface**. As the rib cage expands and the **diaphragm** contracts and lowers, the chest cavity expands, making the internal pressure lower than atmospheric pressure. Thus, air is drawn into the lungs by **negative pressure**. The **medulla** in the brain, which contains the breathing control center, sets the rhythm of breathing and monitors **CO_2 levels** in the blood by sensing changes in pH of the blood. CO_2, the by-product of cell respiration, dissolves in blood to form **carbonic acid**. Therefore, the higher the CO_2 concentration in the blood, the lower the pH. Blood pH lower than 7.4 causes the medulla to increase the rate of breathing to rid the body of more CO_2.

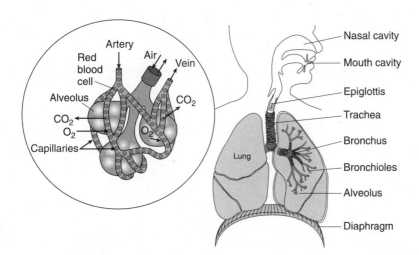

Figure 13.2   Adaptations for Human Respiration

Hemoglobin

Oxygen is carried in the human blood by the respiratory pigment **hemoglobin**, which can combine loosely with four oxygen molecules, forming the molecule **oxy-hemoglobin**. To function in the transport of oxygen, hemoglobin must be able to bind with oxygen in the lungs and unload it at the body cells. The more tightly the hemoglobin binds to oxygen in the lungs, the more difficult it is to unload the cells. Hemoglobin is an **allosteric** molecule and exhibits **cooperativity**. This means that once it binds to one oxygen molecule, hemoglobin undergoes a **conformational change** and binds more easily to the remaining three oxygen molecules. In addition, hemoglobin's conformation is sensitive to pH. A drop in pH lowers the affinity of hemoglobin for oxygen (**Bohr shift**). Because CO_2 dissolves in water to form carbonic acid, actively respiring tissue which releases large quantities of CO_2, will lower the pH of its surroundings and induce hemoglobin to release its oxygen at the cells where needed.

Figure 13.3 contains four graphs showing saturation-dissociation curves for hemoglobin (Hb). *The further to the right the curve is, the less affinity the hemoglobin has for oxygen.*

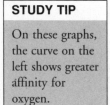

STUDY TIP

On these graphs, the curve on the left shows greater affinity for oxygen.

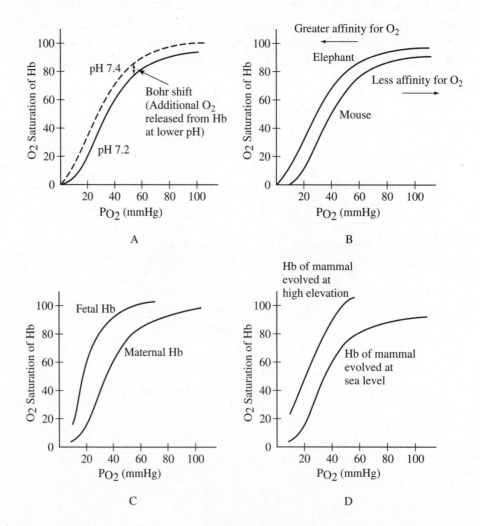

Figure 13.3

- Graph A: Here is a dissociation curve of adult hemoglobin at normal and low blood pH showing the Bohr shift. At a lower pH, the hemoglobin has less affinity for oxygen.
- Graph B: Here is a dissociation curve for hemoglobin of two different mammals, a mouse and an elephant. The mouse has a much higher metabolism, and its body cells have a correspondingly higher oxygen requirement. To accommodate the animal's oxygen needs, the mouse hemoglobin has a dissociation curve located to the right of the elephant hemoglobin. In other words, mouse hemoglobin drops off O_2 at cells more easily than does elephant hemoglobin.
- Graph C: Here is a dissociation curve for fetal and maternal hemoglobin. Fetal hemoglobin has a higher affinity for oxygen than adult hemoglobin so it can take oxygen from the maternal hemoglobin. Also note that the curve of fetal hemoglobin does not have the S-shape common to the other curves. It bonds to each oxygen atom with the same ease; there is no cooperativity.
- Graph D: Here is a dissociation curve for mammals evolved at sea level and mammals evolved at very high altitudes. Since less oxygen is available at high altitudes, mammals that evolved there must have hemoglobin with a greater affinity for oxygen.

Transport of Carbon Dioxide

Very little carbon dioxide is transported by hemoglobin. Most carbon dioxide is carried in the **plasma** as part of the reversible blood buffering **carbonic acid–bicarbonate ion system**, which maintains the blood at a constant pH 7.4. The bicarbonate ion is produced in a two-stage reaction. First, carbon dioxide combines with water to form carbonic acid. This reaction is catalyzed by carbonic acid anhydrase found in red blood cells. Then carbonic acid dissociates into a bicarbonate ion and a proton. The protons can be given up into the plasma, which lowers the blood pH, or taken up by the **bicarbonate ion**, which raises the blood pH. Here is the equation:

$$CO_2 + H_2O \longleftrightarrow H_2CO_3 \longleftrightarrow H_2CO_3^- + H^+$$

carbon dioxide + water   carbonic acid   bicarbonate ion + proton

> **STUDY TIP**
>
> CO_2 is carried in the plasma. O_2 is carried by red blood cells.

CIRCULATION IN DIFFERENT ANIMALS

Primitive animals like the sponge and the hydra have no circulatory systems. All their cells are in direct contact with the environment, and such a system is unnecessary. The earthworm has a closed circulatory system where blood is pumped by the heart through arteries, veins, and capillaries. Oxygen is carried by hemoglobin that is dissolved in the blood. The grasshopper, as a representative animal of the arthropods, has an open **circulatory system**. After blood is pumped by the heart into an artery, it leaves the vessel and seeps through spaces called sinuses or hemocoels as it feeds body cells. The blood then moves back into a vein and circulates back to the heart. This system lacks capillaries. Arthropod blood is colorless and does not carry oxygen.

HUMAN CIRCULATION

Human circulation consists of a **closed circulatory system** with arteries, veins, and capillaries. Table 13.2 shows the components of blood.

TABLE 13.2

Components of Blood

Component	Scientific Name	Properties
Plasma	— —	Liquid portion of the blood
		Contains clotting factors, hormones, antibodies, dissolved gases, nutrients, and wastes
		Maintains proper osmotic potential of blood, 300 mosm/L
Red Blood Cells	Erythrocytes	Carry hemoglobin and oxygen
		Do not have a nucleus and live only about 120 days
		Formed in the bone marrow and recycled in the liver
White Blood Cells	Leukocytes	Fight infection and are formed in the bone marrow
		Die fighting infection and are one component of **pus**
		One type of leukocyte—the **B lymphocyte**—produces antibodies
Platelets	Thrombocytes	These are not cells but cell fragments that are formed in the bone marrow from megakaryocytes
		Clot blood

The Mechanism of Blood Clotting

Blood clotting is a complex mechanism that begins with the release of **clotting factors** from platelets and damaged tissue. It involves a complex set of reactions, including the activation of inactive **plasma proteins**. **Anticlotting factors** normally circulate in the plasma to prevent the formation of a clot or **thrombus**, which can cause serious damage in the absence of injury.

Here is the pathway of normal clot formation.

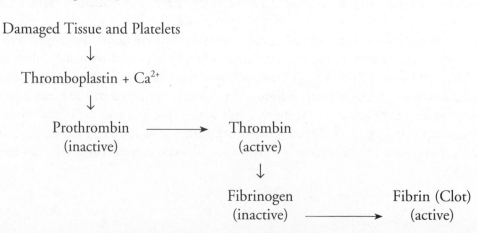

Damaged Tissue and Platelets
↓
Thromboplastin + Ca^{2+}
↓
Prothrombin ⟶ Thrombin
(inactive)   (active)
↓
Fibrinogen ⟶ Fibrin (Clot)
(inactive)   (active)

Structure and Function of Blood Vessels

Table 13.3 describes the various blood vessels.

TABLE 13.3

Blood Vessels		
Vessel	**Function**	**Structure**
Artery and Arteriole	Carry blood away from the heart under enormous pressures	Walls made of thick, elastic, smooth muscle
Vein and venule	Carry blood back to the heart under very little pressure	Thin walls have **valves** to help prevent back flow; veins are located within skeletal muscle, which propels blood upward and back to heart as the body moves
Capillary	Allows for diffusion of nutrients and wastes between cells and blood	Walls are one-cell thick and so small that blood cells travel in single file

The Heart

The heart is located beneath the sternum and is about the size of a clenched fist. It beats about **70 beats per minute** and pumps about 5 quarts (5 L) of blood per minute, or the total volume of blood in the body each minute. Two **atria** receive blood from the body cells, and two **ventricles** pump blood out of the heart. Individual **cardiac muscle cells** have the ability to contract even when removed from the heart. The heart itself has its own innate **pacemaker**, the **sinoatrial (SA) node**, which sets the timing of the contractions of the heart. Located in the wall of the right atrium, it generates and sends electrical signals to the atrioventricular (AV) node. From there, impulses are sent to the bundle of His and Purkinje fibers, which trigger the ventricles to contract. Electrical impulses travel through the cardiac and body tissues to the skin, where they can be detected by an **electrocardiogram** (**EKG**). The heart's pacemaker is influenced by a variety of factors: two sets of nerves that cause it to speed up or slow down, hormones such as adrenalin, and body temperature. **Blood pressure** is lowest in the veins and highest in the arteries when the ventricles contract. Blood pressure for all normal, resting adults is **120/80**. The **systolic number (120)** is a measurement of the pressure when the ventricles contract, while the **diastolic number (80)** is a measure of the pressure when the heart relaxes; see Figure 13.4.

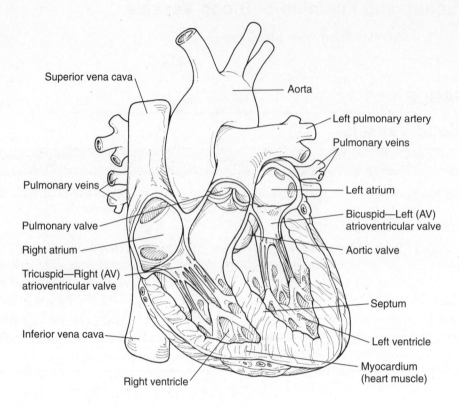

Figure 13.4

Pathway of Blood

Blood enters the heart through the vena cava. From there it continues to the

Right atrium
Right atrioventricular (AV) valve—tricuspid valve
Right ventricle
Pulmonary semilunar valve
Pulmonary artery
Lungs
Pulmonary vein
Left atrium
Bicuspid (left AV) valve
Left ventricle
Aortic semilunar valve
Aorta
To all the cells in the body

Blood circulates through the **coronary circulation** (heart), **renal circulation** (kidneys), and **hepatic circulation** (liver). The **pulmonary circulation** includes the pulmonary artery, lungs, and pulmonary vein. Figure 13.5 is a drawing of the human circulatory system.

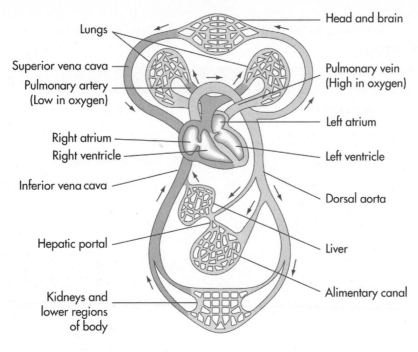

Figure 13.5

CHEMICAL SIGNALS

Animals have two major regulatory systems that release chemicals: the **endocrine system** and the **nervous system**. The endocrine system secretes **hormones**, while the nervous system secretes **neurotransmitters**. Even though the two systems are separate, there is overlap between them, and together they work to regulate the body. **Epinephrine** (adrenaline), for example, functions as the **fight-or-flight** hormone secreted by the adrenal gland and as a neurotransmitter that sends a message from one neuron to another.

Hormones

Hormones are produced in **ductless (endocrine) glands** and move through the blood to a specific **target** cell, tissue, or organ that can be far from the original endocrine gland. They can produce an immediate short-lived response, the way **adrenaline** (epinephrine) speeds up the heart rate, or dramatically alter the development of an entire organism, the way **ecdysone** controls **metamorphosis** in insects. **Tropic hormones** have a far-reaching effect because they stimulate other glands to release hormones. For example, the anterior pituitary releases TSH, which stimulates the thyroid to release thyroxin. Other types of chemical messengers reach their target by special means. **Pheromones** in the urine of a dog carry a message between different individuals of the same species. In vertebrates, **nitric oxide** (NO), a gas, is produced by one cell and diffuses to and affects only neighboring cells before it is broken down. Table 13.4 is an overview of the hormones of the endocrine system.

> **REMEMBER**
>
> Tropic hormones stimulate other glands.

TABLE 13.4

Overview of Hormones

Gland	Hormone	Effect
Anterior pituitary	Growth hormone (GH)	Stimulates growth of bones
	Luteinizing hormone (LH)	Stimulates ovaries and testes
	Thyroid-stimulating hormone (TSH)	Stimulates thyroid gland
	Adrenocorticotropic (ACTH) hormone	Stimulates adrenal cortex to secrete glucocorticoids
	Follicle-stimulating hormone (FSH)	Stimulates gonads to produce sperm and ova
Posterior pituitary	Oxytocin	Stimulates contractions of uterus and milk production by mammary glands
	Antidiuretic hormone (ADH)	Promotes retention of water by kidneys
Thyroid	Thyroxin	Controls metabolic rate
	Calcitonin	Lowers blood calcium levels
Parathyroid	Parathormone	Raises blood calcium levels
Adrenal cortex	Cortisol	Raises blood sugar levels
Adrenal medulla	Epinephrine (adrenaline) Norepinephrine (noradrenaline)	Raises blood sugar level by increasing rate of glycogen breakdown by liver
Pancreas—islets of Langerhans	Insulin—secreted by β cells Glucagon—secreted by α cells	Lowers blood glucose levels Raises blood glucose levels by causing breakdown of glycogen into glucose
Thymus	Thymosin	Stimulates T lymphocytes
Pineal	Melatonin	Involved in biorhythms
Ovaries	Estrogen	Stimulates uterine lining, promotes development and maintenance of primary and secondary characteristics of female
	Progesterone	Promotes uterine lining growth
Testes	Androgens	Support sperm production and promote secondary sex characteristics

The Hypothalamus

The **hypothalamus** plays a special role in the body; it is the *bridge between the endocrine and nervous systems.* The hypothalamus acts as part of the nervous system when, in times of stress, it sends electrical signals to the adrenal gland to release adrenaline. It acts like a nerve when it secretes **gonadotropic-releasing hormone (GnRH)** from neurosecretory cells that stimulate the anterior pituitary to secrete **FSH** and **LH**. It acts as an endocrine gland when it produces **oxytocin** and **antidiuretic hormone** that it stores in the posterior pituitary. The hypothalamus also contains the body's thermostat and centers for regulating hunger and thirst.

Feedback Mechanisms

A feedback mechanism is a self-regulating mechanism that increases or decreases the level of a particular substance. **Positive feedback** *enhances an already existing response*. During childbirth, for example, the pressure of the baby's head against sensors near the opening of the uterus stimulates more uterine contractions, which causes increased pressure against the uterine opening, which causes yet more contractions. This positive-feedback loop brings labor *to an end* and the birth of a baby. This is very different from negative feedback. **Negative feedback** is a common mechanism in the endocrine system (and elsewhere) that *maintains homeostasis*. A good example is how the body maintains proper levels of thyroxin. When the level of thyroxin in the blood is too low, the hypothalamus stimulates the anterior pituitary to release a hormone, thyroid-stimulating hormone (TSH), which stimulates the thyroid to release more thyroxin. When the level of thyroxin is adequate, the hypothalamus stops stimulating the pituitary. (See "Positive and Negative Feedback of Menstrual Cycle" on page 347.)

How Hormones Trigger a Response in Target Cells

There are two types of hormones, and they stimulate target cells in different ways. These are illustrated in Figure 13.6.

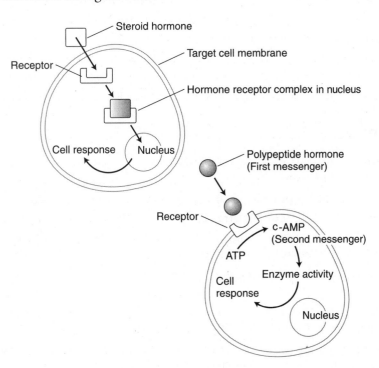

Figure 13.6

1. **Lipid** or **steroid hormones** diffuse directly through the plasma membrane and bind to a receptor inside the nucleus that triggers the cell's response.
2. **Protein** or **peptide hormones** (nonsteroidal) cannot dissolve in the plasma membrane, so they bind to a receptor on the surface of the cell. Once the hormone (the first messenger) binds to a receptor on the surface of the cell, it triggers a secondary messenger, such as c-AMP inside the cell, which converts the extracellular chemical signal to a specific response.

TEMPERATURE REGULATION

Most life exists within a fairly narrow range, from 0°C (the temperature at which water freezes) to about 50°C. Temperatures on land fluctuate enormously. Therefore, temperature regulation, like water conservation, became a problem for animals when they moved to the land. To stay alive, animals must generate their own body heat, seek out a more suitable climate, and/or change behavior. Here are some examples of behavioral changes that alter body temperature:

- A snake can warm itself in the sun or cool off by hiding under a rock.
- Animals on a cold winter prairie huddle together to decrease heat loss.
- Bees swarming in a hive raise the temperature inside the hive.
- Dogs pant and sweat through their tongues to cool themselves off.
- Elephants lack sweat glands, but they cool themselves off by squirting water onto their skins and flapping their ears like fans.
- Humans shiver, jump around, and develop goose bumps to keep warm.
- Birds migrate to a warmer climate and a better source of food to feed themselves and their young.

Ectotherms

Ectotherms are animals that gain most of their body heat from their environment. They have such a low metabolic rate that the amount of heat they can generate is too small to have any effect on body temperature. They must maintain adequate body temperature through behavioral means, like those previously described. Ectotherms include fish, amphibians, and reptiles. The term ectotherm is closest in meaning to the common term cold-blooded, which has no real scientific meaning.

Endotherms

Endotherms are animals that use metabolic processes (oxidizing sugar) to produce body heat. In a cold environment, an endotherm can keep its body warmer than the environment. The temperature of the human body fluctuates during the course of a day, but the average temperature is 37°C (98.6°F.). All mammals and birds are endotherms. Some scientists believe that certain dinosaurs may have been endotherms. The term endotherm is closest in meaning to the common term warm-blooded, which has no real scientific meaning.

In terms of energy consumption, endothermy is very costly. Humans can use 60% of what we eat to maintain our body heat. The metabolic rate of a mammal is much higher than that of a reptile of similar size. As a consequence, mammals must take in many more calories than a reptile of similar size. Flying birds have an even higher metabolic requirement than do mammals.

Despite the fact that the requirement for being an endotherm is so great, it may have given birds and mammals a critical advantage during the time when ancient Earth was dominated by reptiles. Perhaps being able to maintain a high body temperature made it possible for mammals and birds to invade and colonize colder environments that reptiles found uninhabitable.

Poikilotherms and Homeotherms

You may see the terms **poikilotherm** (having a body temperature that varies with the environment) and **homeotherm** (having a constant body temperature despite fluctuations in environmental temperature). However, these terms are not always meaningful. When discussing thermoregulation, the real issue is not whether an animal has variable or constant body temperature but, rather, the *source of heat* used to maintain its body temperature. For example, some scientists classify a fish as a poikilotherm. However, many fish inhabit water that has such a constant temperature that the fish also have a constant body temperature, just like a homeotherm. Also, some may classify mammals as homeotherms because they maintain a constant body temperature. However, that is not the case during **torpor**, a condition that enables animals to save energy by drastically decreasing metabolic rate, or during **hibernation** (extended torpor) for animals like chipmunks.

Problems of Living on Land

Through natural selection, animals have evolved various anatomical and physiological adaptations for life in different environments. The size of the ears in a jackrabbit can be correlated to the climate in which it lives. Jackrabbits living in cold northern regions have small ears to minimize heat loss. In contrast, rabbits in warm, southern regions have long ears to dissipate heat from the many capillaries that make their ears appear pink. This anatomical difference across a geographic range is called a **north-south cline**.

Countercurrent heat exchange is a mechanism that has evolved in a variety of organisms. It helps to warm or cool extremities. See Figure 13.7. This can be seen in arctic animals such as the polar bear that must reach into icy waters to catch fish to eat. While its arm is in the frigid water, warm core blood is flowing out to the paw in the arteries, warming the chilled blood returning to the heart in veins, which lie directly next to the arteries.

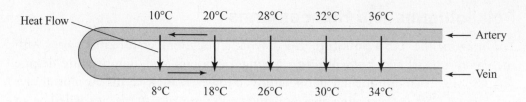

Figure 13.7 Countercurrent Heat Exchangers

OSMOREGULATION

Osmoregulation is the management of the body's water and solute concentration. Organisms in different environments face different problems maintaining the proper concentration of body fluids.

For **marine vertebrates**, like bony fish, the ocean is a strongly dehydrating environment because it is very hypertonic to the organisms living in it. Fish constantly lose water through their gills and skin to the surrounding environment. To counteract the problem, they produce very little urine and drink large amounts of seawater. The extra salt that is taken in with the seawater is actively transported out through the gills.

The osmoregulation problems faced by **freshwater organisms** are opposite those living in salt water. The environment is hypotonic to the organisms, and they are constantly gaining water and losing salt. Freshwater fish excrete copious amounts of dilute urine.

Terrestrial organisms face entirely different problems. They have had to evolve mechanisms and structures that enable them to rid themselves of metabolic wastes while retaining as much water as possible.

EXCRETION

Different excretory mechanisms have evolved in various organisms for the purpose of osmoregulation and removal of metabolic wastes. Here are some organisms matched with their excretory structures.

Organism	**Structures**
Protista	Contractile vacuole
Platyhelminthes (planaria)	Flame cells
Earthworm	Nephridia (metanephridia)
Insects	Malpighian tubules
Humans	Nephrons

Excretion is the removal of metabolic wastes, which include **carbon dioxide** and **water** from cell respiration and **nitrogenous wastes** from protein metabolism. The organs of excretion in humans are the **skin**, **lungs**, **kidney**, and the **liver** (where urea is produced). There are three nitrogenous wastes: **ammonia**, **urea**, and **uric acid**. Which waste an organism excretes is the result of the environment it evolved and lives in. Here are some characteristics of each nitrogenous waste.

Nitrogenous Wastes

1. **Ammonia**

   - Very soluble in water and highly toxic
   - Excreted generally by organisms that live in water, including the hydra and fish

2. **Urea**

   - Not as toxic as ammonia
   - Excreted by earthworms and humans
   - In mammals, it is formed in the liver from ammonia

3. **Uric acid**

   - Pastelike substance that is not soluble in water and therefore not very toxic
   - Excreted by insects, many reptiles, and birds, with a minimum of water loss

The Human Kidney

The kidney functions as both an **osmoregulator** (regulates blood volume and concentration) and an organ of **excretion**. Humans have two kidneys supplied by blood from the **renal artery** and **renal vein**. The kidneys filter about 1,000–2,000 liters of blood per day and produce, on average, about 1.5 liters of urine (700–2,000 mL). As terrestrial animals, humans need to conserve as much water as possible. However, people must balance the need to conserve water against the need to rid the body of poisons. The kidney must adjust both the volume and the concentration of urine depending on the animal's intake of water and salt and the production of urea. If fluid intake is high and salt intake is low, the kidney will produce large volumes of **dilute (hyposmotic) urine**. In periods of high salt intake where water is unavailable, the kidney can produce **concentrated (hyperosmotic) urine**.

The Nephron

The functional unit of the kidney is the **nephron**; see Figure 13.8. The nephron consists of a cluster of capillaries, the **glomerulus**, which sits inside a cuplike structure called **Bowman's capsule** and connects to a long narrow tube, the **renal tubule**. Each human kidney contains about 1 million nephrons. The nephron carries out its job in four steps: **filtration**, **secretion**, **reabsorption**, and **excretion**.

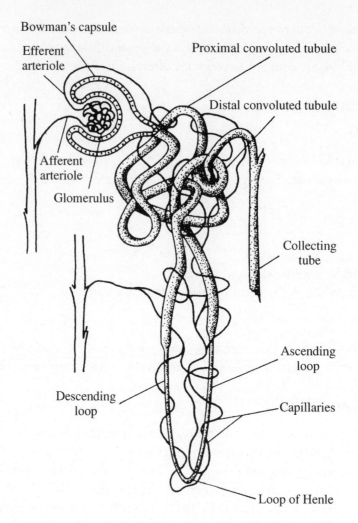

Figure 13.8 The Nephron

Filtration occurs as blood pressure (about twice that of capillaries outside the kidney) forces fluid from the blood in the **glomerulus** into Bowman's capsule. Specialized cells of **Bowman's capsule** are modified into **podocytes**, which, along with **slit pores**, increase the rate of filtration. Filtration occurs by *diffusion* and is *passive* and *nonselective*. The filtrate contains everything small enough to diffuse out of the glomerulus and into Bowman's capsule, including glucose, salts, vitamins, waste such as urea, and other small molecules. From Bowman's capsule, the filtrate travels into the **proximal tubule**.

Secretion occurs in the **proximal** and **distal tubules**. It is the *active, selective* uptake of certain drugs and toxic molecules that did not get filtered into Bowman's capsule. The proximal tubule also secretes ammonia to neutralize the acidic filtrate.

Reabsorption is the process by which most of the water and solutes (glucose, amino acids, and vitamins) that initially entered the tubule during filtration are transported back into the peritubular capillaries and, thus, back to the body. This process begins in the proximal convoluted tubule and continues in the loop of Henle and collecting tubule. The main function of the loop of Henle is to move salts from the filtrate and accumulate them in the medulla surrounding the loop of Henle and the collecting tubule. In this way, the loop of Henle acts as a **countercurrent exchange mechanism** (see the chapter "The Cell"), maintaining a steep salt gradient surrounding the loop. This gradient ensures that water molecules will continue to flow out of the collecting tubule of the nephron, thus creating hypertonic urine and conserving water. *The longer the loop of Henle, the greater is the reabsorption of water.*

Excretion is the removal of metabolic wastes, for example, nitrogenous wastes. Everything that passes into the collecting tubule is **excreted** from the body. From the collecting tubule or duct, urine passes through the **ureter** to the **urinary bladder**. Urine is temporarily stored in the urinary bladder until it passes out of the body via the **urethra**. See Figure 13.9. [attached]

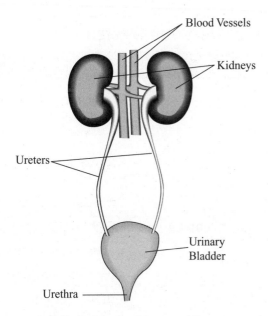

Figure 13.9   Urinary Tract

Hormone Control of the Kidneys

The kidney is able to respond quickly to maintain **homeostasis** of the body because it is under the control of the endocrine system. Two hormones, **antidiuretic hormone (ADH)** and **aldosterone**, respond to different osmoregulatory problems.

ADH is produced by the hypothalamus and both stored and released by the posterior pituitary. It is released *in response to dehydration* due to excessive sweating or inadequate water intake, which causes the blood to become too concentrated. ADH increases the permeability of the collecting tubules to water so that more water can be reabsorbed and urine volume reduced.

Aldosterone is a hormone released by the adrenal glands in *response to a decrease in blood pressure or volume.* Aldosterone acts on the distal tubules of the nephron to reabsorb more sodium (Na^+) and water, thus increasing blood volume and pressure.

NERVOUS SYSTEM

The vertebrate nervous system consists of central and peripheral components.

1. The **central nervous system (CNS)** consists of the brain and spinal cord.
2. The **peripheral nervous system (PNS)** consists of all nerves outside the CNS.

The peripheral nervous system is then further divided and subdivided into various systems. Table 13.5 gives an overview of the peripheral nervous system.

TABLE 13.5

Components of the Peripheral Nervous System

System	Function
SENSORY	Conveys information from sensory receptors or nerve endings
MOTOR	
■ Somatic System	Controls the voluntary muscles
■ Autonomic System	Controls involuntary muscles
Sympathetic	• **Fight or Flight** response
	• Increase heart and breathing rate
	• Liver converts glycogen to glucose
	• Bronchi of lungs dilate and increase gas exchange
	• Adrenalin raises blood glucose levels
Parasympathetic	• **Opposes the sympathetic system**
	• Calms the body
	• Decreases heart/breathing rate
	• Enhances digestion

The Neuron

The neuron consists of a cell body, which contains the **nucleus** and other organelles, and two types of cytoplasmic extensions called **dendrites** and **axons**. **Dendrites** are **sensory**; they receive incoming messages from other cells and carry the electrical signal to the cell body. A neuron can have hundreds of dendrites. A neuron has only one axon, which can be several feet long in large mammals. **Axons** transmit an impulse from the cell body outward to another cell. Many axons are wrapped in a fatty **myelin sheath** that is formed by **Schwann cells**. Figure 13.10 shows a sketch of a neuron.

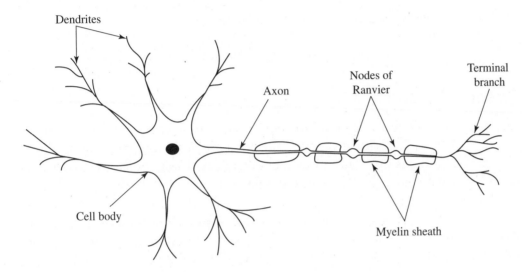

Figure 13.10   The Neuron

There are three types of neurons.

- **Sensory neurons** receive an initial stimulus from a sense organ, such as the eyes and ears, or from another neuron.
- A **motor neuron** stimulates **effectors** (**muscles** or **glands**.) A motor neuron, for example, can stimulate a digestive gland to release a digestive enzyme or to stimulate a muscle to contract.
- The **interneuron** or **association neuron** resides within the spinal cord and brain, receives sensory stimuli, and transfers the information directly to a motor neuron or to the brain for processing.

The Reflex Arc

The simplest nerve response is a **reflex arc**. It is *inborn, automatic, and protective.* An example is the **knee-jerk reflex**, which consists of only two neurons: sensory and motor. A stimulus, a tap from a hammer, is felt in the sensory neuron of the kneecap, which sends an impulse to the motor neuron, which directs the thigh muscle to contract. A more complex reflex arc consists of three neurons, sensory, motor, and interneuron or association neurons; see Figure 13.11. A sensory neuron transmits an impulse to the interneuron in the **spinal cord**, which sends one impulse to the brain for processing and also one to the motor neuron to effect change immediately (at the muscle). This is the type of response that quickly jerks your hand away from a hot iron before your brain has figured out what occurred.

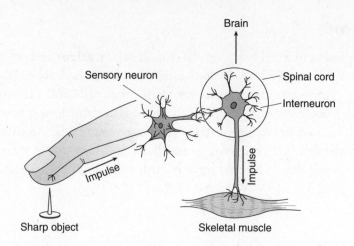

Figure 13.11 The Three-Neuron Reflex Arc

Resting Potential

All living cells exhibit a **membrane potential**, a difference in electrical charge between the cytoplasm (negative ions) and extracellular fluid (positive ions). Physiologists measure this difference in membrane potential using microelectrodes connected to a voltmeter. This potential should be between −50 mV and −100 mV. The negative sign indicates that the inside of the cell is negative relative to the outside of the cell. A neuron in its unstimulated or **polarized state** (**resting potential**) has a membrane potential of about **−70 mV**. The **sodium-potassium pump** maintains the polarization by actively pumping ions that leak across the membrane. In order for the nerve to fire, a stimulus must be strong enough to overcome the **resting threshold** or **resting potential**. *The larger the membrane potential, the stronger the stimulus must be to cause the nerve to fire.*

Gated Channels

Neurons have **gated-ion channels** that open or close in response to a stimulus and play an essential role in transmission of electrical impulses. These channels allow only one kind of ion, such as sodium or potassium, to flow through them. If a stimulus triggers a **sodium ion-gated channel** to open, sodium flows into the cytoplasm, resulting in a decrease in polarization to about −60 mV. The membrane becomes somewhat **depolarized**, so it is easier for the nerve to fire. In contrast, if a **potassium ion-gated channel** is stimulated, the membrane potential increases and the membrane becomes **hyperpolarized**, to about −75 mV, so that it is harder for the neuron to fire.

Action Potential

An **action potential**, or **impulse**, can be generated only in the **axon** of a neuron. When an axon is stimulated sufficiently to overcome the threshold, the permeability of a region of the membrane suddenly changes and the impulse can pass. **Sodium channels** open and sodium floods into the cell, down the concentration gradient. In response, **potassium channels open** and potassium floods out of the cell. This rapid movement of ions or **wave of depolarization** reverses the polarity of the membrane and is called an **action potential**. The action potential is localized and lasts a very short time. The **sodium-potassium pump** restores the membrane to its original polarized condition by pumping sodium and potassium ions back to their original position. This period of repolarization, which lasts a few milliseconds, is called the **refractory period**, during which the neuron cannot respond to another stimulus. The refractory period ensures that an impulse moves along an axon in one direction only since the impulse can move only to a region where the membrane is polarized. Figure 13.12 shows the axon of a neuron as an impulse action potential passes from left to right, depolarizing the membrane in front of it.

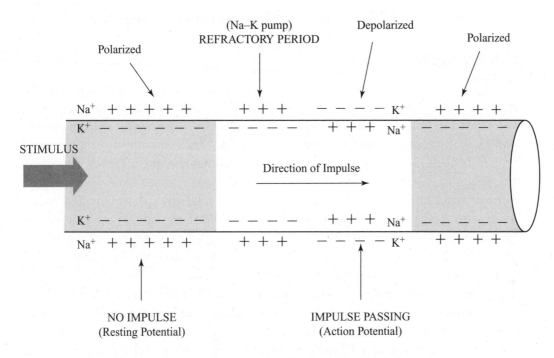

Figure 13.12   An Impulse Passing Along an Axon

The action potential is like a row of dominoes falling in order after the first one is knocked over. The first action potential generates a second action potential, which generates a third, and so on. The impulse moves along the axon propagating itself *without losing any strength*. If the axon is **myelinated**, the impulse travels faster because it leaps from **node to node** (Ranvier) in saltatory (jumping) fashion.

The action potential is an *all-or-none* event; either the stimulus is strong enough to cause an action potential or it is not. The body distinguishes between a strong stimulus and a weak one by the **frequency** of action potentials. A strong stimulus sets up more action potentials than a weak one does.

The graph in Figure 13.13 traces the events of a membrane experiencing sufficient stimulation to undergo an action potential.

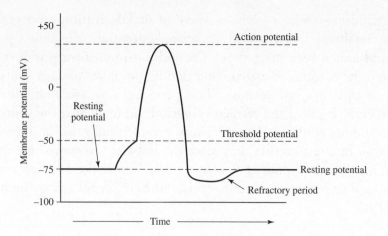

Figure 13.13   Axon Membrane Undergoing an Action Potential

The Synapse

Although an impulse travels along an axon electrically, it crosses a synapse **chemically**. The cytoplasm at the **terminal branch** of the **presynaptic neuron** contains many **vesicles**, each containing thousands of molecules of **neurotransmitter**. Depolarization of the presynaptic membrane causes **Ca^{++} ions** to rush into the terminal branch through **calcium-gated channels**. This sudden rise in Ca^{++} levels stimulates the vesicles to fuse with the presynaptic membrane and release the neurotransmitter by **exocytosis** into the synaptic cleft. Once in the synapse, the neurotransmitter bonds with receptors on the **postsynaptic side**, altering the membrane potential of the postsynaptic cell. Depending on the type of receptors and the ion channels they control, the postsynaptic cell will be inhibited or excited. In either case, shortly after the neurotransmitter is released into the synapse, it is destroyed by an enzyme called an **esterase** and recycled by the presynaptic neuron. One common neurotransmitter is **acetylcholine**. Others are **serotonin**, **epinephrine**, **norepinephrine**, and **dopamine**. In addition, many cells release gas molecules in response to chemical signals. The neurotransmitter acetylcholine stimulates some cells to release the gas **nitric oxide (NO)**, which, in turn, stimulates other cells.

Figure 13.14 shows the terminal branch of the neuronal axon and the synapse.

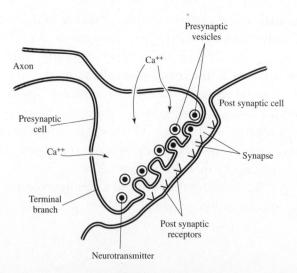

Figure 13.14   Terminal Branch of Neuron at Synapse

MUSCLE

There are three types of muscle: smooth, cardiac, and skeletal. **Smooth** or **involuntary** muscle makes up the walls of the **blood vessels** and the **digestive tract**. Because of the arrangement of its actin and myosin filaments, it does not have a striated appearance. It is under the control of the autonomic nervous system. **Cardiac muscle** makes up the **heart** and appears striated because of the presence of dark bands called intercalated discs that separate adjacent cells. It generates its own action potential; individual heart cells will beat on their own in a saline solution. **Skeletal** or **voluntary muscles** are very large and multinucleate. They work in pairs, one muscle contracts while the other relaxes. The biceps and triceps are one example. This discussion of muscle focuses on skeletal muscles; see Figure 13.15.

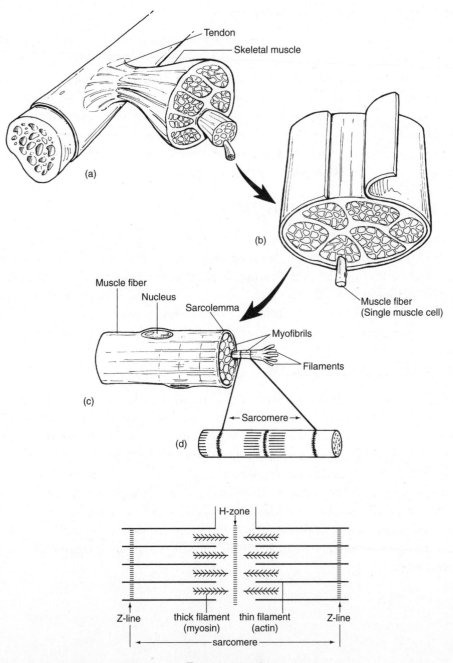

Figure 13.15

Every muscle consists of bundles of thousands of **muscle fibers** that are individual cylindrical **muscle cells**. Each cell is exceptionally **large** and **multinucleate**. Skeletal muscle consists of modified structures that enable the cell to contract.

- The **sarcolemma** is modified **plasma membrane** that surrounds each muscle fiber and can propagate an **action potential**.
- The **sarcoplasmic reticulum (SR)** is modified **endoplasmic reticulum** that contains sacs of Ca^{++} necessary for normal muscle contraction.
- The **T system** is a system of tubules that runs perpendicular to the SR and that connects the SR to the extracellular fluid.
- The **sarcomere** is the functional unit of the muscle fiber (cell). Its boundaries are the **Z lines**, which give skeletal muscle its characteristic striated appearance.

The Sliding Filament Theory

Within the cytoplasm of each muscle cell are thousands of fibers called **myofibrils** that run parallel to the length of the cell. Myofibrils consist of **thick** and **thin filaments**. Each **thin filament** consists of two strands of **actin proteins** wound around one another. Each **thick filament** is composed of two long chains of **myosin** molecules each with a globular head at one end. The heads serve two functions. The contraction of the sarcomere depends on two other molecules: **troponin** and **tropomyosin**, in addition to Ca^{++} **ions** to form and break cross-bridges. Muscle contracts as thick and thin filaments slide over each other.

The Neuromuscular Junction

The axon of a motor neuron synapses on a skeletal muscle at the **neuromuscular junction**. The neurotransmitter **acetylcholine**, released by vesicles from the axon, binds to **receptors** on the **sarcolemma**, depolarize the muscle cell membrane, and set up an **action potential**. The impulse moves along the **sarcolemma**, into the **T system**, and stimulates the **sarcoplasmic reticulum** to release Ca^{++}. The Ca^{++} **ions** then alter the troponin-tropomyosin relationship and the muscle contracts.

Summation and Tetanus

A single action potential in a muscle will cause the muscle to contract locally and minutely for a few milliseconds and then to relax. This brief contraction is called a **twitch**. If a second action potential arrives before the first response is over, there will be a **summation effect** and the contraction will be larger. If a muscle receives a series of overlapping action potentials, even further summation will occur. If the rate of stimulation is fast enough, the twitches will blur into one smooth, sustained contraction called **tetanus** (not related to the disease of the same name). Tetanus is what occurs when a large muscle such as the biceps muscle contracts. If the muscle continues to be stimulated without respite, it will eventually **fatigue** and relax; see Figure 13.16.

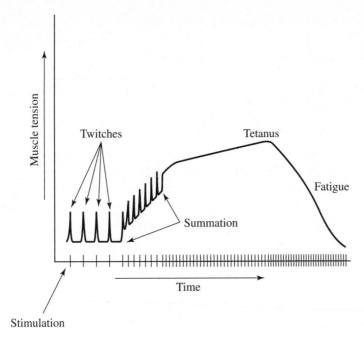

Figure 13.16

Multiple-Choice Questions

DIGESTION

1. Gastric enzyme works best at a pH of

   (A) 2
   (B) 6
   (C) 8
   (D) 11
   (E) 14

Questions 2–6

Questions 2–6 refer to the parts of the human digestive system listed below.

   (A) Mouth
   (B) Esophagus
   (C) Stomach
   (D) Small intestine
   (E) Large intestine

2. Where does the digestion of fats occur?

3. Where does the reabsorption of water used during digestion occur?

4. In which structure is there no digestion?

5. In which structure is digestion completed?

6. Which structure contains the microvilli?

---

7. Which is TRUE of the stomach?

   (A) The pyloric sphincter is at the top of the stomach.
   (B) The stomach lining releases lipases to begin fat digestion.
   (C) Hydrochloric acid activates the enzyme pepsinogen.
   (D) The pH of the stomach varies from acid to basic depending on what must be digested.
   (E) Digestion in the stomach is free of influences by hormones.

8. The hormone gastrin is released by the _____ and has its effect on the _____.

   (A) duodenum; stomach
   (B) duodenum; pancreas
   (C) stomach; gastric lining
   (D) stomach; small intestine
   (E) stomach; pancreas

9. The lacteal is found in the _____, and is involved with _____.

   (A) stomach; the release of hormones
   (B) duodenum; the hydrolysis of lipids
   (C) small intestine; the absorption of fatty acids
   (D) colon; the reabsorption of water
   (E) liver; production of hydrolytic enzymes

10. Absorption of nutrients occurs in the

   (A) duodenum of the colon
   (B) duodenum of the small intestine
   (C) latter part of the small intestine
   (D) latter part of the large intestine
   (E) small and large intestine

GAS EXCHANGE

11. Which is CORRECT about gas exchange in humans?

   (A) The diaphragm rises and air is pulled into the lungs.
   (B) Air is forced down the windpipe when a person inhales.
   (C) The breathing rate is controlled by the hypothalamus in the brain.
   (D) Hemoglobin carries carbon dioxide and oxygen in fairly equal amounts.
   (E) As humans inhale, the pressure in the chest cavity decreases and air is drawn into the lungs.

12. Tracheal tubes are found in

    (A) earthworms
    (B) hydra
    (C) fish
    (D) insects
    (E) birds

13. Breathing in humans is usually regulated by

    (A) the number of red blood cells
    (B) the amount of hemoglobin in the blood
    (C) inherent genetic control
    (D) CO_2 levels and pH sensors
    (E) the pituitary gland

14. All of the following statements about the normal direction of the flow of blood are correct EXCEPT

    (A) lungs $\rightarrow$ pulmonary artery
    (B) right ventricle $\rightarrow$ tricuspid valve
    (C) aorta $\rightarrow$ aortic semilunar valve
    (D) vena cava $\rightarrow$ right atrium
    (E) pulmonary vein $\rightarrow$ left atrium

15. The pacemaker of the heart is

    (A) the bundle of His
    (B) the atrioventricular node
    (C) the diastolic node
    (D) the semilunar node
    (E) sinoatrial node

16. The Bohr effect on the oxyhemoglobin dissociation curve is produced directly by changes in

    (A) temperature
    (B) pH
    (C) CO_2 levels
    (D) oxygen concentration
    (E) it is a constant and not affected by environmental conditions

17. Which is TRUE of the human circulatory system?

    (A) The right ventricle of the heart has the thickest wall.
    (B) Veins have thick walls consisting of smooth muscle cells to assist in returning blood to the heart.
    (C) Blood flow is slowest in capillaries to maximize the diffusion of nutrients and wastes.
    (D) The left and right ventricles contract alternately, which is responsible for the pulse sound.
    (E) The heartbeat for men is usually faster than the heartbeat for women.

18. In humans, the largest amount of the carbon dioxide produced by body cells is carried to the lungs as

    (A) CO_2 attached to hemoglobin in the red blood cells
    (B) attached to hemoglobin circulating in the plasma
    (C) the bicarbonate ion attached to hemoglobin
    (D) CO_2 gas in solution in the plasma
    (E) the bicarbonate ion dissolved in the plasma

19. During ventricular systole, the _____ valve(s) _____.

    (A) semilunar; close
    (B) semilunar; open
    (C) AV; open
    (D) AV and semilunar; close
    (E) heart; all close

20. All of the following are true about blood EXCEPT

    (A) red blood cells live for about 120 days
    (B) white blood cells are formed in the bone marrow
    (C) platelets are not cells but are cell fragments
    (D) platelets derive from specialized cells known as neutrophils
    (E) the liquid portion of the blood is called plasma

CHEMICAL SIGNALS

21. Which hormone acts opposite parathormone?

    (A) calcitonin
    (B) glucagon
    (C) insulin
    (D) adrenaline
    (E) thyroxin

22. The main target of antidiuretic hormone is the

    (A) heart
    (B) kidney
    (C) liver
    (D) spleen
    (E) thyroid

23. The hormone secreted by the adrenal cortex

    (A) reduces inflammatory response
    (B) is an amino acid
    (C) stimulates the adrenal medulla
    (D) lowers blood sugar
    (E) is oxytocin

The following questions refer to the list of chemical messengers below.

Questions 24–29
Questions 24–29 refer to the list of hormones below. You may use each one once, more than once, or not at all.

(A) Glucagon
(B) Adrenocorticotropic hormone
(C) Oxytocin
(D) Thyroxin
(E) Follicle-stimulating hormone

24. Induces labor

25. Released by the posterior pituitary

26. Stimulates the adrenal gland

27. Controls metabolic rate

28. Produced in the pancreas

29. Causes blood sugar levels to increase

---

30. Tropic hormones

(A) are those that are secreted by the liver
(B) are those that are secreted by the thyroid
(C) stimulate other glands
(D) are released only in females
(E) are released only by the hypothalamus

EXCRETION

31. Which of the following processes of the kidney is the LEAST selective?

(A) secretion
(B) filtration
(C) reabsorption
(D) the target of ADH
(E) the pumping of sodium ions out of the tubule

32. In humans, urea is produced in the

(A) kidneys
(B) urinary bladder
(C) urethra
(D) liver
(E) ureter

33. The main nitrogenous waste excreted by birds is

    (A) ammonia
    (B) urea
    (C) uric acid
    (D) nitrite
    (E) All of the above are correct.

34. Which of the following is incorrectly paired?

    (A) nephridia-earthworm
    (B) flame cell-bird
    (C) Malpighian tubules-insect
    (D) nephron-ape
    (E) paramecium-contractile vacuole

35. Following a company picnic on a hot summer day where you ate traditional fare of hot dogs and hamburgers, ice cold beer, and played softball for an hour, you become very thirsty. Which would probably contribute LEAST to your thirst?

    (A) eating salty food
    (B) drinking an ice cold beer
    (C) drinking lots of ice water
    (D) sweating
    (E) None of the above

36. Which nitrogenous waste requires the least water for its excretion?

    (A) ammonia
    (B) urea
    (C) nitrites
    (D) uric acid
    (E) All of the above require water for their safe removal from the body.

Questions 37–40
Questions 37–40 refer to the parts of the kidney listed below.

    (A) Descending loop of Henle
    (B) Ascending loop of Henle
    (C) Bowman's capsule
    (D) Collecting tubule
    (E) Proximal and distal tubules

37. Where does filtration occur?

    (A) A
    (B) B
    (C) C
    (D) D
    (E) E

38. Identify where secretion occurs.

    (A) A
    (B) B
    (C) C
    (D) D
    (E) E

39. Identify the area that is impermeable to the diffusion of water.

    (A) A
    (B) B
    (C) C
    (D) D
    (E) E

40. Identify the target structure of ADH.

    (A) A
    (B) B
    (C) C
    (D) D
    (E) E

NERVOUS SYSTEM

41. Which of the following sequences describes the passage of a nerve impulse through a simple reflex arc in humans?

    (A) receptor → effector → interneuron → motor neuron → sensory neuron
    (B) receptor → sensory neuron → interneuron → effector → motor neuron
    (C) sensory neuron → effector → motor neuron → interneuron → receptor
    (D) receptor → sensory neuron → interneuron → motor neuron → effector
    (E) effector → receptor → sensory neuron → interneuron → motor neuron

42. The function of a Schwann cell is to

    (A) produce esterases to break down neurotransmitters
    (B) form the myelin sheath around the axon of a neuron
    (C) act as an interneuron in the spinal cord
    (D) receive impulses and send them to the neuron
    (E) act as an effector for a neuron

43. What is the function of cholinesterase in the transmission of an impulse?

    (A) It binds with postsynaptic receptors and blocks transmission of the impulse.
    (B) It prevents the release of any more neurotransmitter from presynaptic vesicles.
    (C) It enhances neurotransmission across the synapse.
    (D) It breaks down acetylcholine, preventing continuous neurotransmission.
    (E) It prevents calcium ions from flowing into the terminal branch.

44. Which would be associated with the parasympathetic system?

    (A) increase in blood sugar
    (B) increase in adrenaline
    (C) increase in breathing rate
    (D) increase in digestion
    (E) increase in epinephrine

45. The threshold potential of a particular membrane measures −70 mV at time zero. After 10 minutes, it measures −90 mV. What is the best explanation for what has happened to the membrane?

    (A) It became depolarized.
    (B) The concentrations of Na^+ and K^+ became balanced.
    (C) The membrane became hyperpolarized.
    (D) The membrane became hypopolarized.
    (E) The membrane is more likely to pass an impulse at 90 mV.

46. Which is an example of an effector?

    (A) hormone
    (B) gland
    (C) interneuron
    (D) brain
    (E) any organ in the digestive system

47. If the threshold potential were measured at −75 mV, what would be the value if the polarization across the membrane were increased?

    (A) −70 mV
    (B) −80 mV
    (C) zero
    (D) There would be no change because the membrane potential of a particular neuron can never change.
    (E) The answer cannot be determined.

48. How do action potentials relay different intensities of information?

    (A) by changing the amplitude of the action potential
    (B) by changing the speed with which the impulse passes
    (C) by changing the frequency of the action potential
    (D) by changing the duration of the action potential
    (E) by reversing the direction of the action potential

MUSCLE

49. Tendons connect

    (A) muscle to muscle
    (B) muscle to bone
    (C) ligaments to bones
    (D) ligaments to ligaments
    (E) muscle to ligaments

50. The walls of arteries consist of

    (A) striated muscle and are under voluntary control
    (B) striated muscle and are not under voluntary control
    (C) smooth muscle and are controlled by the somatic nervous system
    (D) smooth muscle and are controlled by the autonomic nervous system
    (E) a mixture of striated and smooth muscle under control of the autonomic nervous system

51. What is the basic unit of function of a skeletal muscle fiber?

    (A) myosin filaments
    (B) actin filaments
    (C) the sarcomere
    (D) Z line
    (E) myofibril

52. What neurotransmitter at the synapse of a neuromuscular junction causes a muscle to contract?

    (A) GABA
    (B) norepinephrine
    (C) dopamine
    (D) acetylcholine
    (E) serotonin

53. Which of the following is the name of a single muscle cell?

    (A) sarcomere
    (B) myofibril
    (C) troponin
    (D) sarcolemma
    (E) muscle fiber

54. All of the following are true about the contracting of skeletal muscle cells
   EXCEPT

   (A) thick filaments are composed of actin
   (B) Ca⁺⁺ ions are necessary for normal muscle contraction
   (C) the sarcolemma can propagate an action potential
   (D) the T-system connects the sarcolemma to the sarcoplasmic reticulum
   (E) Z lines are responsible for the striated appearance of skeletal muscle

Answers to Multiple-Choice Questions

NUTRITION

1. (**A**) Gastric enzymes are strongly acidic and activate the gastric enzyme pepsinogen. Other enzymes, such as intestinal enzymes, are activated in an alkaline pH.

2. (**D**) Fats are digested by lipases in the small intestine.

3. (**E**) The large intestine is the site of egestion of undigested wastes, vitamin production, and water reabsorption.

4. (**B**) There is no digestion in the esophagus.

5. (**D**) All digestion and absorption of nutrients is completed in the small intestine.

6. (**D**) The microvilli are cytoplasmic extensions of the villi. Millions of villi line the endothelium of latter sections of the small intestine.

7. (**C**) The cardiac sphincter is at the top of the stomach; the pyloric sphincter is at the bottom of the stomach. The stomach releases inactive pepsinogen, which is activated by the acid environment, and digests proteins. The pH of the stomach is very acidic. The stomach is stimulated by the hormone gastrin.

8. (**C**) This is a statement of fact. See Table 13.1, "Hormones that Regulate Digestion."

9. (**C**) The lacteal is inside the villi that line the inside of the small intestine. The duodenum is the first 12 inches (30 cm) of the small intestine and is the site of digestion, not absorption. "Colon" is another name for large intestine.

10. (**C**) Digestion occurs in the duodenum—the first part of the small intestine, and absorption occurs in the latter part.

GAS EXCHANGE

11. (**E**) Humans breathe by negative pressure. When we inhale, the chest cavity expands as the diaphragm lowers. This increase in volume causes a decrease in internal pressure, and air is drawn into the lungs because the internal pressure is less than the external pressure.

12. (**D**) Spiracles, openings in the exoskeleton of the insects, connect to tracheal tubes that lead to the hemocoels where diffusion of respiratory gases occur. The respiratory surface is internal.

13. (**D**) The breathing rate is controlled by the medulla in the brain, which is primarily sensitive to CO_2 levels in the blood. The other choices are false.

14. (**A**) From the lungs, blood flows into the left atrium via the pulmonary vein.

15. (**E**) The pacemaker of the heart is the sinoatrial node. From the sinoatrial node, the impulse passes to the atrioventricular node, then to the bundle of His.

16. (**B**) When the pH becomes more acidic, hemoglobin has less affinity for oxygen, so the hemoglobin will drop off some oxygen at the cells that have become slightly acidic due to an accumulation of CO_2 from cell respiration.

17. (**C**) Diffusion occurs in the thin-walled capillaries where the blood circulates slowly. All other choices are false.

18. (**E**) Although a small amount of CO_2 is carried by the red blood cells, most is carried as the bicarbonate ion.

19. (**B**) Systole is the contraction of the ventricles of the heart. When the ventricles contract, blood is pushed out of the arteries through the semilunar valves while the bicuspid and tricuspid valves remain closed.

20. (**D**) Platelets are fragments of cells known as megakaryocytes. All the other choices are correct.

CHEMICAL SIGNALS

21. (**A**) Parathormone is released by the parathyroid and raises calcium levels in the blood; calcitonin is released by the thyroid and lowers calcium levels.

22. (**B**) The target of antidiuretic hormone is the collecting duct of the nephron. If ADH is released, less urine is excreted.

23. (**A**) The hormone released by the adrenal cortex is cortisol, which reduces inflammation. The pharmaceutical version of cortisol is cortisone, which has the same function.

24. (**C**) Oxytocin induces labor and the production of milk by the mammary glands.

25. (**C**) Oxytocin is produced in the hypothalamus and stored and released from the posterior pituitary.

26. (**B**) ACTH is a tropic hormone. A tropic hormone is one that stimulates another endocrine gland to release a substance. ACTH stimulates the adrenal cortex to release a glucocorticoid-like cortisol.

27. **(D)** The thyroid gland releases thyroxin, which controls the rate of metabolism.

28. **(A)** Glucagon is produced in the α cells of the pancreas and raises blood sugar by breaking down glycogen that is stored in the liver.

29. **(A)** See # 28.

30. **(C)** Tropic hormones are released by one gland and stimulate another gland to release its hormone. For example, TSH is released by the anterior pituitary and stimulates the thyroid to release thyroxin.

EXCRETION

31. **(B)** During filtration, all substances small enough to diffuse out of the glomerulus will do so. It is the least-selective process occurring in the nephron.

32. **(D)** Urea is a nitrogenous waste that is produced in the liver; urine is produced in the kidney.

33. **(C)** Uric acid is excreted as a crystal to conserve water by many terrestrial animals. However, humans and earthworms release nitrogenous waste in the form of urea.

34. **(B)** The flame cell is a primitive excretory structure in Platyhelminthes like planaria. All freshwater Protista have contractile vacuoles that pump out water that continually leaks in. The other statements are all true.

35. **(C)** All of the other options would contribute to thirst. Alcohol, particularly, is dehydrating because it blocks production of antidiuretic hormone, thus causing the release of extra water in the urine.

36. **(D)** Uric acid is not soluble in water, therefore, it does not require water in order to be excreted from the body.

37. **(C)** Filtration occurs in Bowman's capsule.

38. **(E)** Secretion occurs in the proximal and distal tubules.

39. **(B)** The ascending loop of Henle is impermeable to water.

40. **(D)** The target structure of ADH is the collecting tubule within the kidney.

NERVOUS SYSTEM

41. **(D)** An effector is a muscle or a gland. The interneuron is located in the spinal cord.

42. **(B)** The Schwann cell forms the myelin sheath around the axon and nourishes the neuron. There are more Schwann cells in the nervous system than there are neurons.

43. **(D)** There are several classes of esterases. They are all specific in action for a particular neurotransmitter.

44. (**D**) The parasympathetic system is calming. It is active when you are relaxing or digesting food and it slows down when you are excited.

45. (**C**) The gradient is steeper at −90 mV than it was at −75 mV. The membrane is now hyperpolarized. The threshold is higher. Depolarizing the membrane and passing an impulse are more difficult.

46. (**B**) An effector is a gland or muscle.

47. (**B**) The membrane potential reads zero when there is no differential between one side of the membrane and the other. The greater the differential, the more negative the potential.

48. (**C**) The only way a neuron distinguishes the intensity of the stimulus is by changing the frequency of the action potentials. A nerve response is all or none.

MUSCLE

49. (**B**) This is a fact.

50. (**D**) The autonomic nervous system controls those things that are not consciously controlled. It controls smooth and cardiac muscles.

51. (**C**) The sarcomere is the basic unit of contraction of skeletal muscle.

52. (**D**) Acetylcholine is the neurotransmitter at a neuromuscular junction.

53. (**E**) A single muscle cell is called a muscle fiber. It is a very large, multinucleated cell.

54. (**A**) Thick filaments are composed of myosin, not actin. The other statements are correct about muscles.

Free-Response Questions

Directions: Answer all questions. You must answer the question in essay—**not** outline—form. You may use labeled diagrams to supplement your essay, but diagrams alone are *not* sufficient. Before you start to write, read each question carefully so that you understand what the question is asking.

1. Beginning at an axon at rest, describe the conduction of an impulse through a neuromuscular junction to the contraction of a skeletal muscle.

2. Regulation and homeostasis are critical to living things.
   a. Explain what a feedback mechanism is and how it helps an organism to maintain homeostasis.
   b. Explain the difference between positive feedback, and negative feedback and give examples.

Typical Free-Response Answers

Note: The answer to this first essay question has three parts: what happens at the axon, what happens at the synapse, and what happens at the muscle. That means each section is worth from two to four points. You cannot get full credit unless you cover all three parts to some degree. If you write a masterpiece on the axon and the synapse and do not have time to discuss the muscle, you will not get the full ten points. Key words are all in boldface. This essay question is broad, and there is so much to write about. However, you can get full credit without writing an essay as detailed as this one.

1. An axon at rest is polarized and has a **resting potential** of about −70 mV. The **sodium-potassium pump** maintains this level of **polarization** by actively pumping ions that leak across the membrane. In order for the nerve to fire and an impulse to pass, a stimulus must be strong enough to overcome the resting threshold. The larger the resting potential, the stronger the stimulus must be to cause the nerve to fire.

   When an axon is stimulated sufficiently to overcome the **threshold**, the permeability of a region of the membrane suddenly changes. Sodium channels open and sodium floods into the cell, down the concentration gradient. In response, potassium channels open and potassium floods out of the cell. This **wave of depolarization** reverses the polarity of the membrane and is called an **action potential**. The action potential is localized and lasts a very short time. The sodium-potassium pump restores the membrane to its original polarized condition by pumping sodium and potassium ions back to their original positions. This period of repolarization, which lasts a few milliseconds, is called the **refractory period**, during which time the neuron cannot respond to another stimulus. The refractory period ensures that an impulse moves along an axon in **one direction only** since the impulse can move only to a region where the membrane is polarized. The impulse moves along the axon propagating itself without losing any strength. The action potential is an **all-or-none event**; either the stimulus is strong enough to cause an action potential or it is not. The body distinguishes between a strong stimulus and a weak one by the **frequency of action potentials**. A strong stimulus sets up more action potentials than a weak one does.

   The action potentials end at the terminal branch of the axon where the cytoplasm contains many **vesicles**, each containing thousands of molecules of **neurotransmitter**. Depolarization of the **presynaptic membrane** causes Ca⁺⁺ ions to rush into the terminal branch through **calcium-gated channels**. This sudden rise in Ca⁺⁺ levels stimulates the vesicles to fuse with the presynaptic membrane and release neurotransmitter by **exocytosis** into the synapse. Once in the synapse, the neurotransmitter bonds with **receptors** on the postsynaptic side, altering the membrane potential of the postsynaptic cell. Shortly after the neurotransmitter is released into the synapse, it is destroyed by an enzyme called an **esterase** and recycled by the presynaptic neuron. The neurotransmitter at a neuromuscular junction is **acetylcholine**.

   Neurotransmitter binds to receptors on the **sarcolemma** of the skeletal muscle membrane, depolarizes it, and sets up an action potential. This impulse moves along the sarcolemma, into the **T system**, and stimulates the **sarcoplasmic reticulum**, the modified endoplasmic reticulum, to release Ca^{++}. The Ca^{++} ions cause the thick **myosin** filaments and the thin **actin** filaments of the **sarcomere**, the basic functional unit of muscle, to slide over one another, and the muscle contracts.

2a. The body maintains homeostasis by acting through complex feedback mechanisms. A real-life example of a feedback system is the thermostat and a home heating system. You set the thermostat for a desired temperature. As soon as the temperature deviates from that set point, either the furnace or air conditioner turns on. In this way, the temperature in the building is maintained at a steady point.

2b. **Positive feedback** enhances an already existing response to complete something. During childbirth, the pressure of the baby's head against sensors near the opening of the uterus stimulates more uterine contractions, which causes increased pressure against the uterine opening, which causes yet more contractions. This positive feedback loop brings labor *to an end* and the birth of a baby. Another example of positive feedback can be seen in the helper T cells (T_h) in the human immune system. When a specific T_h cell is selected and activated by the MHC-antigen complex, it proliferates and makes thousands of copies of itself called clones. These clones secrete cytokines, which further stimulate T_h cell activity. In this case, an army of helper T cells is produced to fight a specific antigen. When the invading antigen is destroyed, the helper T population is no longer needed. Only a small number of memory helper cells remain circulating in the blood. This is very different from negative feedback.

    Negative feedback is a common mechanism in the endocrine system (and elsewhere) that maintains homeostasis. A good example is how the body maintains proper levels of thyroxin. When the level of thyroxin in the blood is too low, the hypothalamus stimulates the anterior pituitary to release a hormone, thyroid-stimulating hormone (TSH), which stimulates the thyroid to release more thyroxin. When the level of thyroxin in the bloodstream is adequate again, the hypothalamus stops stimulating the pituitary.

The Human Immune System

- Nonspecific defense mechanisms
- Specific defense mechanisms
- Types of immunity
- Blood groups and transfusion
- AIDS
- Positive feedback in the immune system
- Other topics in immunity

INTRODUCTION

We live in a sea of germs. They are in the air we breathe, the food we eat, and the water we drink and swim in. Our body has three cooperating lines of defense that are able to respond to new threats as quickly as they appear. This chapter covers this complex topic in detail.

NONSPECIFIC DEFENSE MECHANISMS

First Line of Defense

The **first line of nonspecific defense** is a **barrier** that helps prevent **pathogens** (things that cause disease) from entering the body. These include

- Skin
- Mucous membranes, which release mucus that contains antimicrobial substances including **lysozyme**
- Cilia in the respiratory system to sweep out mucus with its trapped microbes
- Stomach acid

Second Line of Defense

Microbes that get into the body encounter the **second line of nonspecific defense**, which is meant to limit the spread of invaders in advance of specific immune responses. This second line includes the following.

- **Inflammatory response**

  ✔ **Histamine** triggers **vasodilation** (enlargement of blood vessels), which increases blood supply to the area, bringing more **phagocytes**. It is secreted by **basophils** (a type of circulating white blood cell) and **mast cells**, found in the connective tissue. Histamine is also responsible for the symptoms of the common cold: sneezing, coughing, redness, itchy eyes, and runny nose.

  ✔ **Prostaglandins** further promote blood flow to the area.

  ✔ **Chemokines** secreted by blood vessel endothelium and monocytes also attract phagocytes to the area.

  ✔ **Pyrogens**, released by certain leukocytes, increase body temperature to speed up the immune system and make it more difficult for microbes to function.

- **Phagocytes** ingest invading microbes. Two types, called **neutrophils** and **monocytes**, migrate to an infected site in response to local chemical attractants. This response is called **chemotaxis**. **Neutrophils** engulf microbes and die within a few days. **Monocytes** transform into **macrophages** ("giant eaters"), extend **pseudopods**, and engulf huge numbers of microbes over a long period of time. They digest the microbes with a combination of **lysozyme** and two toxic forms of oxygen: **superoxide anion** and **nitric oxide**.

- **Complement**, a group of proteins, that leads to the lysis (bursting) of invading cells.

- **Interferons** block against cell-to-cell viral infections.

- **Natural killer (NK) cells** destroy virus-infected body cells (as well as cancerous cells). They attack the cell membrane, causing it to **lyse** (burst open) and die.

SPECIFIC DEFENSE MECHANISMS

Third Line of Defense

The **third line of defense** is **specific** and relies on **B lympocytes** and **T lymphocytes**, which originate in the **bone marrow**. Once mature, both cell types circulate in the **blood, lymph,** and **lymphatic tissue:** the **spleen, lymph nodes, tonsils,** and **adenoids.** Both recognize different specific **antigens** (substances that cause the production of antibodies).

- **B lymphocytes** mature in the bone marrow and fight disease in what is referred to as a **humoral response** by producing **antibodies.** Each activated B lymphocyte is estimated to secrete 2,000 antibodies per second over the cell's 4–5 day life span. They become stimulated by T lymphocytes or by free viruses and bacterial toxins floating in the blood.

- **T lymphocytes** mature in the **thymus gland** and fight pathogens by what is called **cell-mediated response.** They are stimulated by body cells that have been infected with bacteria, viruses, and parasites and by **antigen-presenting cells** (**APCs**), which display foreign antigens on their surface. There are two main types of T cells, **cytotoxic T cells** and **helper T cells.** Each responds to a different class of major histocompatability complex (MHC) molecule.

  ✔ **Cytotoxic T (T_c) cells** kill body cells infected with viruses or other pathogens and cancer cells. They have antigen receptors on their cell surface that bind to protein fragments displayed by **class I MHC** molecules. When a T_c cell binds to a class I MHC molecule, a **CD8** surface protein from the T_c cell holds the two together until the T_c cell is activated. When a T_c cell is activated, it proliferates and differentiates into **plasma cells** and **memory cells.** (See "Clonal Selection" on page 333.) Because T_c cells have the CD8 proteins on their cell surface, they are also referred to as **CD8 cells.** Activated T_c cells attack and kill infected cells by releasing **perforin** (a protein that forms pores in the target cell's membrane), which causes the cell to lyse and die. The infecting microbes are now free in the blood or tissue and are disposed of by circulating antibodies.

  ✔ **Helper T (T_h) cells** announce to the immune system, with the help of macrophages and some B cells, that foreign antigens have entered the body. T_h cells have receptors on their cell surface that bind to protein fragments displayed by the body's **Class II MHC** molecules. When a T_h cell binds to a class II MHC molecule, a CD4 surface protein from the T_h cell holds the two cells together until the T_h cell is activated. When a T_h cell is activated, it proliferates and differentiates into **plasma cells** and **memory cells** just as do T_c cells. (See "Clonal Selection" on page 333.) Because T_h cells have CD4 proteins on their cell surface, they are referred to as **CD4** cells. T_h cells stimulate T_c cells, B cells, and other T_h cells by releasing the **cytokines, interleukin-1 (Il-1),** and **interleukin-2 (Il-2).**

Major Histocompatibility Complex (MHC) Molecules

Major histocompatibility complex (MHC) molecules or markers, also known as **HLA (human leukocyte antigens)**, are a collection of cell surface markers that identify the cells as **self**. No two people, except identical twins have the same MHC markers. There are two main classes of MHC markers, **class I** and **class II**.

- **Class I MHC molecules** are found on almost every body cell.

- **Class II MHC molecules** are found on specialized cells, including **macrophages**, **B cells**, and **activated T cells**; see Figure 14.1.

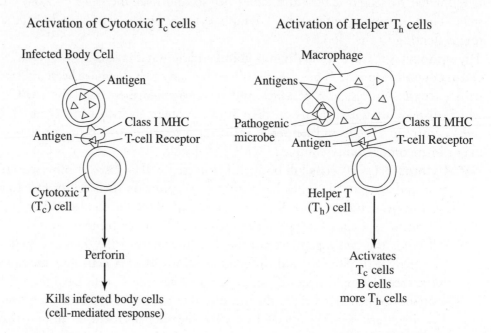

Figure 14.1   Class I and Class II Major Histocompatibility (MHC) Molecules

Clonal Selection

Clonal selection is a fundamental mechanism in the development of immunity; see Figure 14.2. Antigenic molecules select or bind to specific **B** or **T lymphocytes**. Once a lymphocyte is **selected**, it becomes metabolically active, **proliferates** (**clones** thousands of copies of itself in lymphatic tissue), and **differentiates** into **plasma cells** and **memory cells**.

- **Plasma cells** are short-lived and fight the antigen immediately in what is called the **primary immune response**. This takes about 10–17 days.
- **Memory cells** are long-lived cells bearing receptors specific to the same antigen as the plasma cells and that remain circulating in the blood in small numbers for a **lifetime**. If the body is attacked or *challenged* in the future by the same antigen, the memory cells rapidly reactivate. This response, called the **secondary immune response**, is faster (2–7 days) and of greater magnitude than the primary immune response. The capacity of the immune system to generate a secondary immune response is called **immunological memory**. The immunological memory is the mechanism that prevents you from getting any specific viral infection, such as chicken pox, more than once.

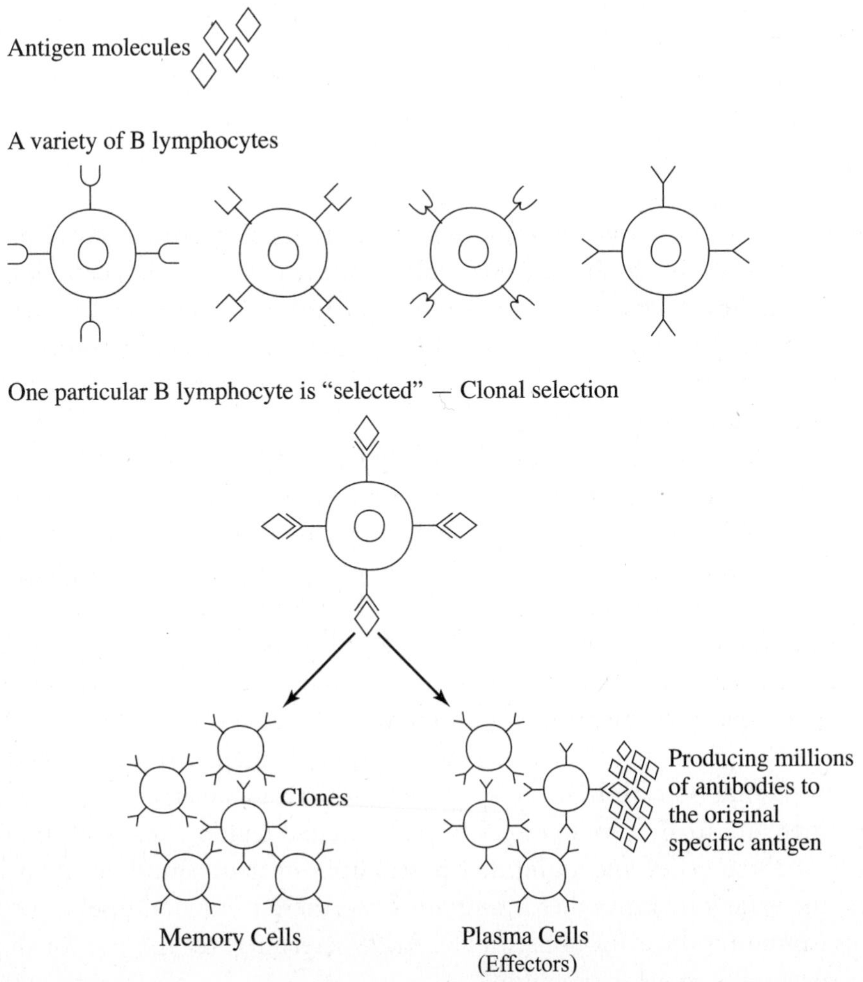

Antigen molecules

A variety of B lymphocytes

One particular B lymphocyte is "selected" — Clonal selection

Clones

Memory Cells

Plasma Cells (Effectors)

Producing millions of antibodies to the original specific antigen

Figure 14.2   Clonal Selection

How Macrophages Activate Helper T Cells and the Immune System

- A **macrophage**, acting as an **APC** (**antigen-presenting cell**), engulfs a bacterium and presents a fragment of it to the cell surface by an MHC II molecule.
- A specific T_h **cell** is activated by binding to the **MHC-antigen complex**. The activation is enhanced by **interleukin-1** and **CD4 protein** from the T_h cell.
- The activated T_h cell proliferates, cloning itself into thousands of copies, each with the same MHC-antigen combination. These clones secrete **cytokines**.
- The **cytokines** further stimulate T_h cells, B cells and T_c cells.
- See Figure 14.3.

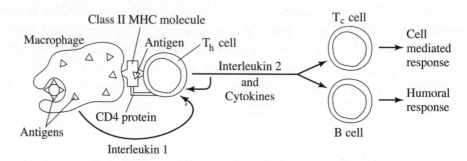

Figure 14.3   How Macrophages Activate the Immune System

Antibodies

Antibodies, also called **immunoglobins,** are a group of **globular proteins**. Each antibody molecule is a **Y-shaped molecule** consisting of four polypeptide chains, **two identical heavy chains** and **two identical light chains** joined by **disulfide bridges**. The molecule consists of four unchanging or **constant regions (C)** and four **variable regions (V)**. The tips of the Y have specific shapes and are the binding sites for different antigens.

There are five classes of immunoglobins: IgM, IgA, IgD, IgE, and IgG.

Antibodies destroy free-floating antigens in two major ways. One way is for antibodies to neutralize the antigen by causing them to clump, forming an **antigen-antibody complex**. This large complex is then phagocytosed by **macrophages**. The other way antibodies destroy antigen is through **complement fixation**, the activation of **complement**, a group of about twenty proteins that assist in lysing cells.

Despite the fact that there are only four variable regions on one antibody, it is now accepted that the blueprints for antibodies are made early in life, *prior* to any exposure to antigenic agents. When you are exposed to an antigen, antibodies are chosen by **clonal selection** from a limitless variety as needed. This is possible because there are many genes coding for the variable regions, many alleles coding for each of these genes and unlimited possibilities of gene shuffling. As a consequence, the variety in antibodies is *unlimited* and there is no viral disease for which humans cannot produce antibodies. (The AIDS screening test is a test for the presence of antibodies, not for the virus.)

TYPES OF IMMUNITY

Table 14.1 describes the two types of immunity.

TABLE 14.1

Types of Immunity

Passive Immunity	Active Immunity
Temporary Antibodies are transferred to an individual from someone else. Examples are maternal antibodies that pass through the placenta to the developing fetus or through breast milk to the baby. Also, a person with a weak immune system often receives an injection of gamma globulin (IgG), which are antibodies culled from many people, to boost the weak immune system.	**Permanent** The individual makes his or her own antibodies after being ill and recovering or after being given an **immunization** or **vaccine**. A vaccine contains dead or live viruses or enough of the outer coat of a virus to stimulate a full immune response and to impart lifelong immunity.

BLOOD GROUPS AND TRANSFUSION

ABO antibodies circulate in the plasma of the blood and bind with ABO antigens in the event of an improper transfusion. *The certain danger in a transfusion comes if the recipient has antibodies to the donor's antigens.* However, before someone receives a transfusion of blood, two samples of donor and recipient blood must be mixed to determine compatibility. This procedure is called a **cross-match**.

Blood Type	Antigens Present on the Surface of the Red Blood Cells	Antibodies Present Circulating in the Plasma
A	A	B
B	B	A
O	None	A and B
AB	A and B	None

> **REMEMBER**
>
> The danger in transfusion is if the recipient has antibodies to the donor's antigens.

Blood type O is known as the **universal donor** because it has no blood cell antigens to be clumped by the recipient's blood. Blood type AB is known as the **universal recipient** because there are no antibodies to clump the donor's blood.

Rh Factor

Rh factor is another antigen located on the surface of red blood cells. Most of the population—85 percent—has the antigen and are called Rh$^+$. Those without the antigen (15 percent) are Rh$^-$.

AIDS

AIDS stands for acquired immune deficiency disease. People with AIDS are highly susceptible to opportunistic diseases, infections, and cancers that take advantage of a collapsed immune system. The virus that causes AIDS, **HIV (human immuno-deficiency virus)**, attacks cells that bear CD4 molecules on their surface, mainly helper T cells. HIV is a retrovirus. Once inside a cell, it reverse transcribes itself, using the enzyme **reverse transcriptase**, and integrates the newly formed DNA into the host cell genome. It remains in the nucleus as a provirus, directing the production of new viruses.

POSITIVE FEEDBACK IN THE IMMUNE SYSTEM

Positive feedback enhances an already existing process until some endpoint or maximum rate is reached. An example in the immune system can be seen in the activity of helper T cells. When a T_h cell becomes activated by a class II MHC molecule, it releases two cytokines, interleukin-I and interleukin-II. Il-2 stimulates B cells and other T cells into action. However, interleukin-I enhances the activity of the already activated T_h cells, stimulating them more until they are activated to a maximum. This is in contrast to negative feedback, which is a means to achieve stability.

OTHER TOPICS IN IMMUNITY

- Allergies are hypersensitive immune responses to certain substances called **allergens**. They involve the release of histamine, an anti-inflammatory agent, which causes blood vessels to dilate. A normal allergic reaction involves redness, a runny nose, and itchy eyes. Antihistamines can normally counteract these symptoms. However, sometimes an acute allergic response can result in a life-threatening response called **anaphylactic shock**, which can result in death within minutes.
- **Antibiotics** are medicines that kill bacteria or fungi. Where vaccines are given to prevent illness caused by viruses, antibiotics are administered after a person is sick.
- **Autoimmune diseases** such as **multiple sclerosis, lupus, arthritis,** and **juvenile diabetes** are caused by a terrible mistake of the immune system. The system cannot properly distinguish between **self** and **nonself**. It perceives certain structures in the body as nonself and attacks them. In the case of multiple sclerosis, the immune system attacks the **myelin sheath** surrounding certain neurons in the CNS.
- **Monoclonal antibodies** are antibodies produced by a single B cell that have been selected for a specific antigen. They are important in research and in the treatment and diagnosis of certain diseases.

Multiple-Choice Questions

1. Which of the following is directly responsible for humoral immunity?

   (A) cytotoxic T cells
   (B) helper T cells
   (C) macrophages
   (D) neutrophils
   (E) B cells

2. Which of the following is correct about blood type?

   (A) Blood type O has O antigens on the surface of the red blood cells.
   (B) Blood type A has A antibodies circulating in the plasma.
   (C) The danger in a transfusion is if the donor has antibodies to the recipient.
   (D) A and B antigens are on the surface of the red blood cells.
   (E) Type O blood is the universal donor because type O blood has no circulating antibodies.

3. All of the following are true of white blood cells EXCEPT

   (A) they are also called leukocytes
   (B) examples are neutrophils and thrombocytes
   (C) Lymphocytes are of two varieties: B type and T type
   (D) T lymphocytes engage in cell-mediated immunity
   (E) B lymphocytes are involved with humoral immunity

4. All of the following are true about major histocompatibility molecules EXCEPT

   (A) there are two major classes
   (B) MHC molecules are found on virtually every normal cell in the body
   (C) class II MHC molecules are found on macrophages
   (D) everybody except identical twins has different MHC molecules on their body cells
   (E) MHC molecules work in opposition to HLA markers

5. Which of the following is true about immunity?

   (A) T cells produce antibodies.
   (B) Another name for antibody is antibiotic.
   (C) Juvenile diabetes is an autoimmune disease.
   (D) Macrophages are one type of B lymphocyte.
   (E) Antibodies kill infected cells by lysing them.

6. Which of the following would be considered a safe transfusion?

    (A) type A donor, type B recipient
    (B) type AB donor, type O recipient
    (C) type O donor, type AB recipient
    (D) type B donor, type A recipient
    (E) All can be made safe by modern medical technology.

Questions 7–11

For the questions 7–11 choose from the choices below. Use each one once and only once.

    (A) B cells
    (B) Macrophages
    (C) Helper T cells
    (D) Cytotoxic T cells
    (E) Natural killer cells

7. Act as antigen-presenting cells (APC)

8. Responsible for humoral immunity

9. Bonds to class I MHC molecules

10. Secrete cytokines to stimulate other lymphocytes

11. Nonspecific cell that kills virus-infected cells by lysing them, not phagocytosing them

---

12. Which is NOT part of the lymphatic system?

    (A) spleen
    (B) liver
    (C) tonsils
    (D) adenoids
    (E) lymph nodes

13. Which is NOT part of the nonspecific immune defense?

    (A) macrophages
    (B) skin
    (C) immunoglobins
    (D) natural killer cells
    (E) histamine

14. Which is NOT part of the nonspecific immune defense?

   (A) histamine
   (B) prostaglandins
   (C) pyrogens
   (D) cytokines
   (E) neutrophils

Question 15

Use the graph below to answer the following question.

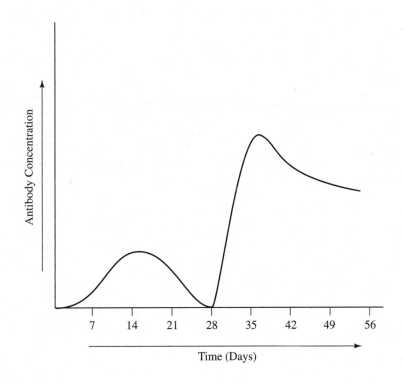

15. What is happening at day 28?

   (A) The person begins to feel ill.
   (B) Antibody concentration begins to decline.
   (C) This is the first exposure to an antigen.
   (D) This is the second exposure to an antigen.
   (E) The person quickly spikes a fever.

Answers to Multiple-Choice Questions

1. **(E)** Humoral immunity refers to the production of antibodies. Choice (E), B cells, is the only choice given that produces antibodies. Neutrophils are part of nonspecific immunity. Macrophages are antigen-presenting cells, and helper T cells activate B cells to produce antibodies.

2. **(D)** Blood type O has no antigens on the red blood cells. Blood type A has A antigens on the surface of the red blood cells and B antibodies circulating in the plasma. Type O is the universal donor because it has no antigens on the surface of the red blood cells to be clumped by the recipient's antibodies.

3. **(B)** is the only false statement because thrombocytes is another name for platelets.

4. **(E)** HLA, human leukocytic antigen, is another name for MHC. All other statements are correct.

5. **(C)** Antibodies are produced by B cells, not T cells. Another name for antibody is immunoglobin. Gamma globulin is the most common immunoglobin or antibody. Antibodies kill infected cells by bonding with the antibodies on the surface of the pathogens and clumping them. This makes them visible and easy for phagocytosing cells to gobble them up.

6. **(C)** The rule for safe transfusion is that the recipient must not have antibodies to the donor's blood. Type O blood is the universal donor because type O blood has no antigens on the surface of the red blood cells. Type AB blood is the universal recipient because type AB blood has no ABO antibodies circulating in the blood.

7. **(B)**

8. **(A)**

9. **(D)**

10. **(C)**

11. **(E)**

12. **(B)** The liver has many functions. It is part of the digestive system and is responsible for detoxifying the body. However, it does not function in the immune system.

13. **(C)** Immunoglobin is another name for antibody, and antibodies are key to the specificity of the immune system. Macrophages are part of both systems, nonspecific and specific immunity.

14. **(D)** Cytokines are part of the body's specific defenses because they are released by helper T cells, which activate T cells and B cells.

15. **(D)** This is the second exposure to the antigen that stimulates the secondary immune response. Antibody production is much more rapid than during the first immune response because antibodies are being produced from circulating memory B cells.

Free-Response Question

Directions: Answer all questions. You must answer the question in essay—**not** outline—form. You may use labeled diagrams to supplement your essay, but diagrams alone are *not* sufficient. Before you start to write, read each question carefully so that you understand what the question is asking.

Humans live in a sea of germs protected by three lines of defense.

a. Describe the three lines of defense and how each protects us.

b. Compare and contrast the role of T and B cells.

Typical Free-Response Answer

a. The immune system consists of three lines of defense. The first and second lines of defense are **nonspecific**, while the third line of defense is **specific**. The first line of defense is a barrier that prevents germs from entering the body. These include the **skin**, **mucous membranes** (which contain the hydrolytic enzymes, lysozyme), **cilia**, and **stomach acid**. The second line of defense, which is more complex, is engaged when microbes breach the first line of defense. The second line of defense is characterized by the **inflammatory response**, redness, swelling, sneezing, and coughing. This is brought about primarily by **histamine**, which triggers vasodilation and thus increases blood supply to the area. In addition, two types of phagocytes, **neutrophils** and **monocytes**, ingest invading microbes. Also, **natural killer cells** destroy virus-infected body cells, **interferons** block the advance of invading viruses, and complement helps destroy cells directly in a number of ways.

   The third line of defense is characterized by specificity and diversity. It is responsible for attacking a broad spectrum of microbes with a variety of specific responses. **B lymphocytes** produce **antibodies**, thousands per second, that fight germs in what is known as humoral response. **T lymphocytes** fight pathogens by what is called **cell-mediated response**, a kind of hand-to-hand combat. There are two types of T cells: cytotoxic T cells or killer T cells, and helper T cells. Both are activated by macrophages that present antigens on their surface for identification. T_c cells destroy body cells that have been infected with viruses by releasing perforin and lysing the cell. **Helper T** cells tell the immune system, with the help of macrophages and some B cells, that foreign antigens have entered the body. T_h cells stimulate T_c cells, B cells, and other T_h cells by releasing the **cytokines, interleukin-1 (Il-1)** and **interleukin-2 (Il-2)**.

b. Once a T or B lymphocyte is activated, it becomes metabolically active, proliferating and differentiating into plasma cells and memory cells. **Plasma cells** are short-lived and fight antigens immediately in what is called the primary immune response. This takes about 10–17 days. Memory cells are long-lived cells, bearing receptors specific to the same antigen as the plasma cells. **Memory cells** remain circulating in the blood in small numbers for a lifetime.

Animal Reproduction and Development

CHAPTER 15

- Asexual reproduction
- Sexual reproduction
- Embryonic development
- Factors that influence embryonic development

INTRODUCTION

From an evolutionary standpoint, reproduction and passing one's genes to the next generation is the goal of life. Nutrition, transport, excretion, and the other life functions are the processes that enable an organism to live long enough to reproduce. This chapter is primarily about how animals reproduce and develop, although initially there is a brief review of asexual reproduction. A small section covers spermatogenesis and oogenesis; the rest of meiosis is in the chapter entitled "Cell Division." Reflecting the AP Exam, the material on sexual reproduction is at a basic high school level. However, the material on embryonic development is at a college level.

ASEXUAL REPRODUCTION

Asexual reproduction produces offspring genetically identical to the parent. It has several advantages over sexual reproduction:

- It enables animals living in isolation to reproduce without a mate.
- It can create numerous offspring quickly.
- There is no expenditure of energy-maintaining reproductive systems or hormonal cycles.
- Because offspring are **clones** of the parent, it is advantageous when the environment is stable.

The following list shows the types of asexual reproduction in sample organisms.

- **Fission** is the separation of an organism into two new cells. (Amoeba)
- **Budding** involves the splitting off of new individuals from existing ones. (Hydra)
- **Fragmentation** and **regeneration** occur when a single parent breaks into parts that regenerate into new individuals. (Sponges, planaria, starfish)
- **Parthenogenesis** involves the development of an egg without fertilization. The resulting adult is haploid. (Honeybees, some lizards)

SEXUAL REPRODUCTION

Sexual reproduction has one major advantage over asexual reproduction: **variation**. Each offspring is the product of both parents and may be better able to survive than either parent, especially in an environment that is changing.

The Human Male Reproductive System

- **Testes** (*testis,* singular) are the male gonads, where sperm are produced.
- **Seminiferous tubules** are the site of sperm formation in the testes.
- **Epididymis** is the tube in the testes where sperm gain mobility.
- The **vas deferens** is the muscular duct that carries sperm during ejaculation from the epididymis to the urethra in the penis.
- **Seminal vesicles** secrete mucus, fructose sugar (which provides energy for the sperm), and the hormone **prostaglandin** (which stimulates uterine contractions) during ejaculation.
- The **prostate gland** is the large gland that secretes **semen** directly into the urethra.
- The **scrotum** is the sac outside the abdominal cavity that holds the testes. The cooler temperature there enables sperm to survive.
- The **urethra** is the tube that carries semen and urine.
- See Figure 15.1.

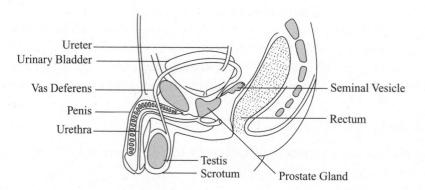

Figure 15.1   The Human Male Reproductive System

The Human Female Reproductive System

- The ovaries (**ovary,** singular) are where meiosis occurs and where the secondary oocyte forms prior to birth.
- The **oviducts** or **Fallopian tubes** are where fertilization occurs. After ovulation, the egg moves through the oviduct to the uterus.
- The **uterus** is where the **blastocyst** will implant and where the embryo will develop during the nine-month **gestation** if fertilization occurs.
- The **endometrium** is the lining of the uterus that thickens monthly in preparation for implantation of the blastocyst.
- The **vagina** is the birth canal. During labor and delivery, the baby passes through the **cervix** (the mouth of the uterus) and into the vagina.
- See Figure 15.2.

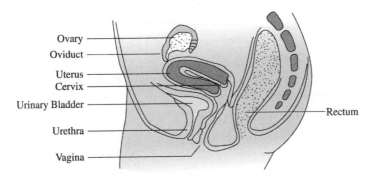

Figure 15.2   The Human Female Reproductive System

The Menstrual Cycle of the Human Female

Follicular phase	Several follicles in the ovaries grow and secrete increasing amounts of estrogen in response to **follicle-stimulating hormone (FSH)** from the **anterior pituitary**.
Ovulation	The **secondary oocyte** ruptures out of the ovaries in response to **luteinizing hormone**.
Luteal phase	The **corpus luteum** forms in response to luteinizing hormone. It is the follicle left behind after ovulation and secretes **estrogen** and **progesterone**, which thicken the **endometrium** of the **uterus**.
Menstruation	The monthly shedding of the lining of the uterus when implantation of an embryo does not occur.

Hormonal Control of the Menstrual Cycle

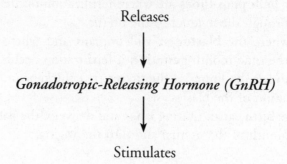

Hypothalamus

Releases

↓

Gonadotropic-Releasing Hormone (GnRH)

↓

Stimulates

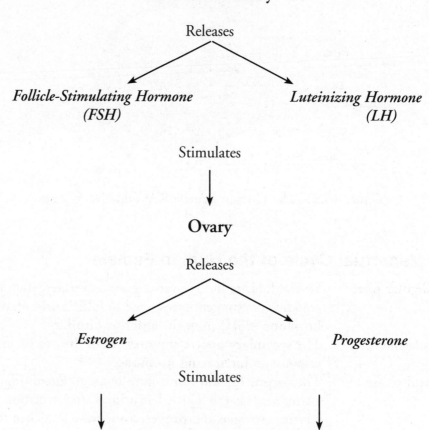

Anterior Pituitary

Releases

Follicle-Stimulating Hormone (FSH)   *Luteinizing Hormone (LH)*

Stimulates

↓

Ovary

Releases

Estrogen   *Progesterone*

Stimulates

↓   ↓

Thickening of the lining of the uterus

Positive and Negative Feedback of Menstrual Cycle

- **Positive feedback in the menstrual cycle**: *Positive feedback enhances a process until it is completed.* During the **folliclular phase**, estrogen released from the follicle stimulates the release of **LH** from the **anterior pituitary**. The increase in LH stimulates the follicle to release *even more* estrogen. The hormone levels continue to rise until the follicle matures and ovulation occurs.
- **Negative feedback in the menstrual cycle**: *Negative feedback stops a process once homeostasis is reached.* During the **luteal phase, LH** stimulates the **corpus luteum** to secrete **estrogen** and **progesterone**. Once the levels of estrogen and progesterone reach sufficiently high levels, they trigger the **hypothalamus** and **pituitary** to *shut off*, thereby **inhibiting** the secretion of LH and FSH.

STUDY TIP

Understanding positive and negative feedback may help you answer essay questions.

Spermatogenesis

Spermatogenesis, the process of sperm production, is a continuous process that starts at puberty. It begins as **luteinizing hormone** (**LH**) induces the **interstitial cells** of the testes to produce **testosterone**. Together with **FSH**, **testosterone** induces maturation of the **seminiferous tubules** and stimulates the beginning of sperm production. In the seminiferous tubules, each **spermatogonium cell** ($2n$) divides by mitosis to produce two **primary spermatocytes** ($2n$), which each can undergo meiosis I to produce 2 **secondary spermatocytes** (n). Each secondary spermatocyte then undergoes meiosis II, which yields 4 **spermatids** (n). These spermatids **differentiate** and move to the **epididymis** where they become motile. See Figure 15.3.

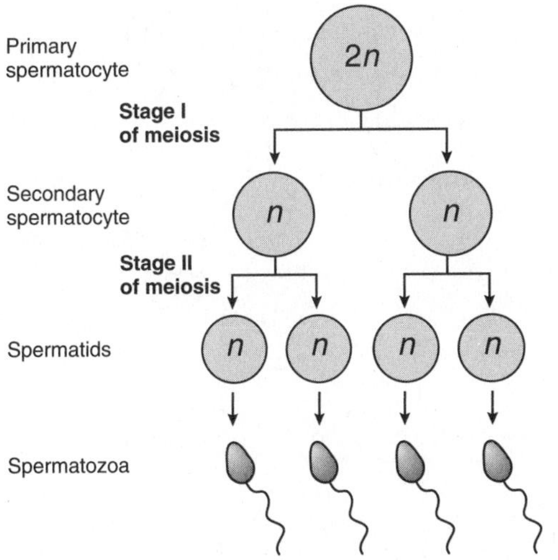

Figure 15.3 Spermatogenesis

Oogenesis

Oogenesis, the production of ova, begins prior to birth. Within the embryo, an **oogonium cell** ($2n$) undergoes mitosis to produce **primary oocytes** ($2n$). These remain quiescent within small follicles in the ovaries until puberty, when they become reactivated by hormones. **FSH** periodically stimulates the follicles to complete meiosis I, producing **secondary oocytes** (n), which are released at ovulation. Meiosis II then stops again and does not continue until **fertilization,** when a sperm penetrates the secondary oocyte. See Figure 15.4. Oogensis differs from sperm formation in three ways. First, it is a **stop-start process**. It begins prior to birth and is completed after fertilization. Second, cytokinesis divides the cytoplasm of the cell unequally, producing one large cell and *two small polar bodies* which will disintegrate. Third, one primary oogonium cell produces only *one active egg cell*.

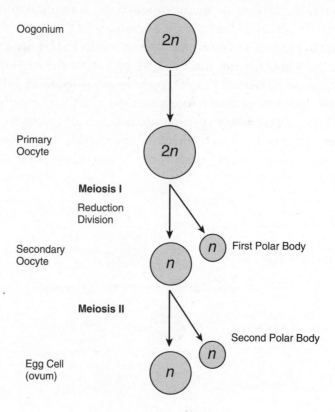

Figure 15.4   Oogenesis

Fertilization

Fertilization, the fusion of sperm and ovum nuclei, is a complex process. It begins with the **acrosome reaction**, when the head of the sperm, the **acrosome**, releases hydrolytic enzymes that penetrate the jelly coat of the egg. Specific molecules from the sperm bind with receptor molecules on the vitelline membrane of the egg. This specific recognition ensures that the egg will be fertilized by only sperm from the same species. Once a sperm binds to receptors on the egg, the membrane is dramatically depolarized and no other sperm can penetrate the egg membrane.

Although the fusion of the sperm and egg triggers this activation, an unfertilized egg can be activated artificially by electrical stimulation or by injection with Ca^{++}. The development of an unfertilized egg is called **parthenogenesis**. The adult that results is haploid. **Drone honeybees** develop by natural parthenogenesis from unfertilized eggs and are haploid males.

EMBRYONIC DEVELOPMENT

Embryonic development consists of three stages: **cleavage**, **gastrulation**, and **organogenesis**. The stages below generally describe the development of all animal eggs but are typical of the **sea urchin** because it has almost no yolk. In eggs with more yolk, such as those of the **frog**, cleavage is unequal, with very little cell division in the yolky region. In eggs with a great deal of yolk, such as a **bird egg**, cleavage is limited to a small, nonyolky disc at the top of the egg.

Cleavage *is the rapid mitotic cell division of the zygote that occurs immediately after fertilization.* When the embryo begins to divide, the early cell divisions follow one of two patterns. In protostomes, mollusks, annelids, and arthropods, cleavage is spiral and determinant, which means that the future of each cell has been assigned by the four-ball stage. At this time, if one cell is separated from the rest, it will not develop into a complete embryo. In deuterostomes, echinoderms, and chordates, cleavage is radial and indeterminate, which means that each cell retains the capacity to develop into a complete and normal embryo. In both groups, however, cleavage produces a fluid-filled ball of cells called a **blastula**. The cells of the blastula are called **blastomeres**, and the fluid-filled center is called the **blastocoel**. Figure 15.5 shows both radial and spiral cleavage.

> **STUDY TIP**
>
> Cleavage and gastrulation are important topics.

> **STUDY TIP**
>
> Gastrulation produces three germ layers: ectoderm, mesoderm, and endoderm.

Spiral Cleavage    Radial Cleavage

Figure 15.5

Gastrulation is a process that involves rearrangement of the blastula and begins with the formation of the **blastopore**, an opening into the blastula. In some animals (protostomes), the blastopore becomes the mouth; in other animals (deuterostomes), the blastopore becomes the anus. Some of the cells on the surface of the embryo migrate into the blastopore to form a new cavity called the **archenteron** or primitive gut. As a result of this cell movement, a **three-layered** embryo called a **gastrula** is formed. In most animals, the gastrula consists of three differentiated layers called **embryonic germ layers**. They are the **ectoderm**, **endoderm**, and **mesoderm**. These three layers will develop into all the parts of the adult animal.

- The **ectoderm** will become the **skin** and the **nervous system**.
- The **endoderm** will form the **viscera** including the lungs, liver, and digestive organs, and so on.
- The **mesoderm** will give rise to the **muscle, blood,** and **bones**. Some primitive animals (sponges and cnidarians) develop a noncellular layer, the **mesoglea,** instead of the mesoderm. See Figure 15.6.

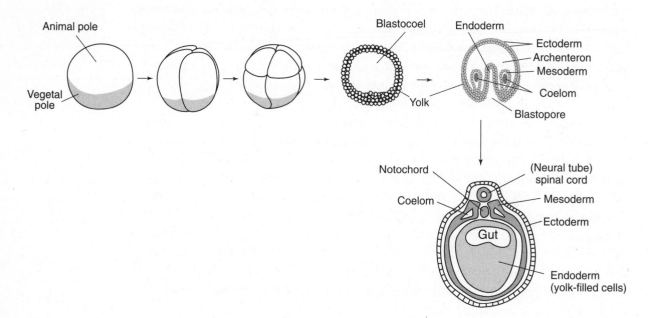

Figure 15.6

Organogenesis is organ building. It is the process by which cells continue to **differentiate**, producing organs from the three embryonic germ layers. Three kinds of morphogenetic changes—folds, splits, and dense clustering (**condensation**)—are the first evidence of organ building. Once all the organ systems have been developed, the embryo simply increases in size.

The Frog Embryo

Fertilization: One third of the frog egg is yolk, which is massed in the lower portion of the egg. This yolky portion of the egg is called the **vegetal pole**. The top half of the egg is called the **animal pole** and has a **pigmented cap**. Eggs are laid directly into water and fertilization is external. When the sperm penetrates the egg, the pigmented cap rotates toward the point of penetration and a **gray crescent** appears on the side opposite the point of entry of the sperm. The gray crescent is critical to normal development of the growing embryo. (See "The Gray Crescent" on page 352.)

Cleavage and gastrulation: Because of the presence of yolk, cleavage is uneven. The **blastopore** forms at the border of the gray crescent and the vegetal pole. Cells at the **dorsal lip** above (dorsal to) the blastopore begin to stream over the dorsal lip and into the blastopore in a process called involution. As these ectoderm cells stream inward by what is called epibolic movement, the blastocoel disappears and is replaced by another cavity called the **archenteron**. The region of mesoderm lining the archenteron that formed opposite the blastopore is called the **dorsal mesoderm**.

Organogenesis: In chordates, the organs to form first are the **notochord**, the skeletal rod characteristic of all chordate embryos, and the **neural tube**, which will become the **central nervous system**. The neural tube forms from the **dorsal ectoderm** just above the notochord. Both form by **embryonic induction** (see "Embryonic Induction" below). After the blueprints of the organs are laid down, the embryo develops into a larval stage, the tadpole. Later, **metamorphosis** will transform the tadpole into a frog.

The Bird Embryo

Cleavage and gastrulation: The bird's egg has so much yolk that development of the embryo occurs in a flat disc or **blastodisc** that sits on top of the yolk. A **primitive streak** forms instead of a gray crescent. Cells migrate over the primitive streak and flow inward to form the **archenteron**. As cleavage and gastrulation occur, the yolk gets smaller.

Extraembryonic membranes: Tissue outside the embryo forms four **extraembryonic membranes** necessary to support the growing embryo inside the shell. They are the **yolk sac, amnion, chorion**, and **allantois**. The **yolk sac** encloses the yolk, food for the growing embryo. The **amnion** encloses the embryo in protective **amniotic fluid**. The **chorion** lies under the shell and allows for the diffusion of respiratory gases between the outside and the growing embryo. The **allantois** is analogous to the placenta in mammals. It is a conduit for respiratory gases between the environment and the embryo. It is also the repository for **uric acid**, the **nitrogenous waste** from the embryo that accumulates until the chick hatches. See Figure 15.7.

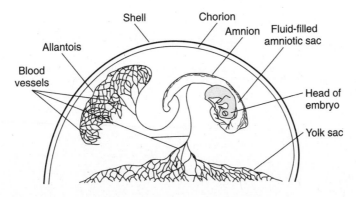

Figure 15.7   The bird embryo

FACTORS THAT INFLUENCE EMBRYONIC DEVELOPMENT

Cytoplasmic Determinants

When an eight-ball sea urchin embryo is separated into two halves, the future development of the two halves depends on the plane in which they are cut. If the dissection is longitudinal, producing embryos containing cells from both animal and vegetal poles, subsequent development is normal. If, however, the plane of dissection is horizontal, the result is four abnormally developing embryos. This demonstrates that the cytoplasm surrounding the nucleus has profound effects on embryonic development. The importance of the cytoplasm in the development of the embryo is known as **cytoplasmic determinants**; see Figure 15.8.

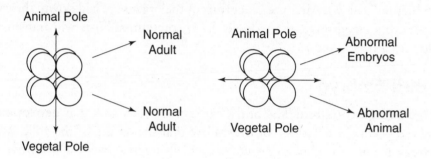

Figure 15.8

The Gray Crescent

Hans Spemann, in his now-famous experiment, demonstrated the importance of the **cytoplasm** associated with the **gray crescent** in the normal development of the animal. He dissected embryos in the two-ball stage in different ways. Only the cell containing the gray crescent developed normally. In addition, these experiments provide more proof that the cytoplasm plays a major role in determining the course of embryonic development; see Figure 15.9.

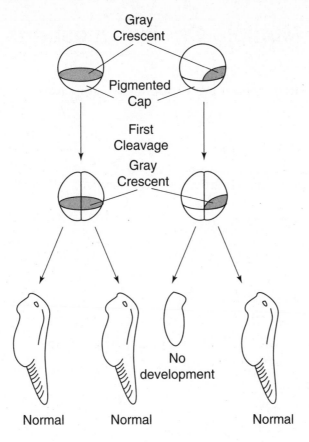

Figure 15.9   Development of the Frog Egg

Embryonic Induction

Embryonic induction is the ability of one group of embryonic cells to influence the development of another group of embryonic cells. Spemann proved that the **dorsal lip** of the blastopore normally initiates a chain of inductions that results in the formation of a **neural tube**. In the experiment, he grafted a piece of **dorsal lip** from one amphibian embryo onto the ventral side of a second amphibian embryo. What developed on the recipient was a second notochord and neural tube at the location of the graft. *The dorsal lip induced the abdomen tissue above it to become neural tissue.* Because it plays a crucial role in development, Spemann named the dorsal lip the **primary organizer**.

Homeotic, Homeobox, or Hox Genes

Homeotic, **homeobox**, or **Hox genes** are **master regulatory genes** that control the expression of genes that regulate the placement of specific anatomical structures. They play a critical role in normal embryonic development. A **homeotic gene** might give the instruction "place legs here" in the developing embryo.

Multiple-Choice Questions

Questions 1–4
Questions 1–4 refer to the list below of primary germ layers.
  (A) Ectoderm
  (B) Endoderm
  (C) Mesoderm

1. Gives rise to the lining of the digestive tract

2. Gives rise to the brain and eye

3. Gives rise to the blood

4. Gives rise to the bone

---

5. In human females, fertilization normally occurs in the _____ and implantation occurs in the _____.

   (A) ovary; uterus
   (B) Fallopian tube; uterus
   (C) ovary; oviduct
   (D) oviduct; vagina
   (E) Fallopian tube; cervix

Questions 6–10
Questions 6–10 refer to the list of terms below about hormones. Choose the best answer that fits each of the following descriptions.

   (A) Follicle-stimulating hormones (FSH)
   (B) Luteinizing hormone
   (C) Gonadotropic-releasing hormone (GnRH)
   (D) Estrogen
   (E) Oxytocin

6. Causes labor

7. Released by the hypothalamus and stimulates the anterior pituitary

8. Stimulates the ovary to mature a secondary oocyte

9. Responsible for thickening the endometrial lining of the uterus

10. Stimulates sperm production

Questions 11–13

How many chromosomes are in each of the following human cells?

11. Primary spermatocyte?

    (A) 23
    (B) 46
    (C) 96

12. Spermatogonium cell?

    (A) 23
    (B) 46
    (C) 96

13. Spermatid?

    (A) 23
    (B) 46
    (C) 96

---

14. During the menstrual cycle, what is the main source of progesterone in human females?

    (A) anterior pituitary
    (B) posterior pituitary
    (C) hypothalamus
    (D) oviduct
    (E) corpus luteum

15. Sperm gain motility in the

    (A) vas deferens
    (B) epididymis
    (C) seminiferous tubules
    (D) interstitial cells
    (E) all of the above

16. In vertebrate animals, one primary oogonium develops into _____ active egg cell(s).

    (A) 1
    (B) 2
    (C) 4
    (D) 8
    (E) 16

17. Which is FALSE about embryonic development?

    (A) Early embryonic division in deuterostomes is spiral.
    (B) The hollow ball stage is called the blastula.
    (C) The end of gastrulation is defined by the formation of primary germ layers.
    (D) The archenteron is the primary gut.
    (E) The opening in the gastrula is called a blastopore.

Questions 18–21

For questions 18–21 match the descriptions with the extraembryonic membranes below.

    (A) yolk sac
    (B) allantois
    (C) amnion
    (D) chorion

18. It is analogous to the placenta in mammals. It is for the diffusion of nutrients and wastes.

19. It provides food for the growing embryo.

20. It lies beneath the shell and allows for the exchange of O_2 and CO_2 between the egg and the outside.

21. It protects the developing embryo from physical trauma.

---

22. What does gastrulation accomplish?

    (A) It changes a solid ball into a hollow ball.
    (B) It changes a blastula into a morula.
    (C) It creates the neural tube.
    (D) It creates a hollow embryo with three tissue layers.
    (E) Fertilization

23. Spemann bisected a two-ball stage of a frog embryo and found that only the cell containing the gray crescent developed normally. This demonstrated that

    (A) the gray crescent contains the DNA necessary for normal development
    (B) the gray crescent is really the nucleus
    (C) the gray crescent is necessary for proper development under only certain conditions
    (D) the cytoplasm plays a major role in determining the course of embryonic development
    (E) frog cleavage is unique in the animal kingdom

24. Spemann grafted a piece of dorsal lip from one amphibian embryo onto the ventral side of a second embryo. A second notochord and neural tube developed at the location of the graft. This experiment proved that

    (A) embryonic development does not follow any particular developmental pathway and can be easily altered
    (B) the dorsal lip can transform into the archenteron
    (C) the dorsal lip can transform into any organ or structure
    (D) the dorsal lip is an inducer that causes adjacent tissue to transform into some structure
    (E) any embryonic tissue can be grafted onto another animal and divert the second animal's normal development

25. Master genes that control the expression of other genes responsible for anatomical structures are called

    (A) mesoglea
    (B) acrosomes
    (C) Hox genes
    (D) cortical genes
    (E) parthenogenetic genes

26. Which of the following is TRUE about parthenogenesis?

    (A) It is the form of asexual reproduction carried out by Protista.
    (B) It is a primitive form of sexual reproduction.
    (C) It involves the development of the egg into an adult without fertilization.
    (D) The adult that results from parthenogenesis is diploid.
    (E) The adult that results from parthenogenesis can be either haploid or diploid.

27. Embryonic induction is best illustrated by which of the following?

    (A) development of the chorion in a developing chick embryo
    (B) replacement of cartilage with bone in a developing human embryo
    (C) development of the ectoderm into skin
    (D) development of the neural tube after contact with the dorsal mesoderm
    (E) None of the above is correct.

Answers to Multiple-Choice Questions

1. (**B**) The endoderm gives rise to the viscera, the internal organs.

2. (**A**) The ectoderm gives rise to the skin and the nervous system. The eye is part of the peripheral nervous system.

3. (**C**) The mesoderm gives rise to the blood, bones, and muscle.

4. (**C**) The mesoderm gives rise to the blood, bones, and muscle.

5. (**B**) Fact.

6. **(E)** Oxytocin is produced in the hypothalamus and released from the posterior pituitary.

7. **(C)** Gonadotropic-releasing hormone from the hypothalamus stimulates the anterior pituitary to release hormones, such as FSH and TSH.

8. **(A)** FSH is released by the anterior pituitary and stimulates the follicle in the ovary to mature a secondary oocyte.

9. **(D)** Estrogen and progesterone are responsible for thickening the lining of the uterine wall in preparation for implantation of an embryo.

10. **(A)** FSH is active in males as well as females. In males, it stimulates sperm production; in females, it stimulates the maturation of a secondary oocyte in the ovary.

11. **(B)** See #13 below.

12. **(B)** See #13 below.

13. **(A)** The spermatogonium cell has 46 chromosomes. It undergoes mitosis and produces two primary spermatocytes, each containing 46 chromosomes. Each primary spermatocyte can undergo meiosis I, producing two secondary spermatocytes each containing 23 chromosomes. Each secondary spermatocyte then undergoes meiosis II, which yields 4 spermatids (n). Each spermatid undergoes differentiation and becomes an active sperm.

14. **(E)** The corpus luteum produces progesterone, which thickens the uterine wall.

15. **(B)** Sperm gain motility in the epididymis.

16. **(A)** In vertebrate animals, one primary oogonium develops into one active egg cell and two polar bodies. The polar bodies disintegrate.

17. **(A)** Early embryonic division in deuterostomes is radial. All the other choices are correct statements.

18. **(B)**

19. **(A)**

20. **(D)**

21. **(C)**

22. **(D)** The end of gastrulation is marked by the formation of the three embryonic layers, the ectoderm, endoderm, and mesoderm.

23. **(D)** During embryonic development, the cytoplasm of the egg plays a major role in influencing the development of the embryo.

24. **(D)** Embryonic induction is the ability of one group of cells to influence the development of another group of embryonic cells. The dorsal lip has the ability to induce tissue to which it is adjacent to become a neural tube.

25. **(C)** Hox genes are master genes that control the expression of other genes that regulate the placement of specific anatomical parts. Hox stands for homeotic genes.

26. (**C**) In honeybees, the queen bee fertilizes some eggs and does not fertilize others. The ones she fertilizes become female (2*n*) worker bees. The ones she does not fertilize become male (*n*) drones.

27. (**D**) Embryonic induction is the ability of one group of embryonic cells to influence the development of another group of embryonic cells.

Free-Response Questions

Directions: Answer all questions. You must answer the question in essay—**not** outline—form. You may use labeled diagrams to supplement your essay, but diagrams alone are *not* sufficient. Before you start to write, read each question carefully so that you understand what the question is asking.

1. Discuss the embryonic development of a frog egg.
   a. Describe the embryonic stages and explain what is accomplished at each stage.
   b. Give an example of embryonic induction.
   c. Describe one experiment in embryonic induction, and explain what it demonstrated.

2. Compare and contrast the formation of sperm and eggs.

Typical Free-Response Answers

Note: Once again, to help you study, terms that should be included in the essay are in bold.

1a. One-third of the frog egg is yolk, which is massed in the lower portion of the egg and is called the **vegetal pole**. The top half of the egg, the **animal pole**, has a **pigmented cap** and is where most of the cell division will occur. When the sperm penetrates the egg, the pigmented cap rotates toward the point of penetration and a **gray crescent** appears on the side opposite the point of entry of the sperm. The gray crescent exerts major influence on the normal development of the growing embryo. Because the frog is a **deuterostome**, the symmetry during early cleavage is **radial** and **indeterminate**, meaning the fate of the individual cells is not yet decided. Cleavage ends with the formation of the hollow-ball (blastula) stage. The cells of the blastula are called **blastomeres**, and the fluid-filled center of the ball is called the **blastocoel**. A blastopore (the first opening) forms at the border of the gray crescent and the vegetal pole. Because the frog is a deuterostome, the blastula will become the anus and the mouth will develop elsewhere.

   Gastrulation is the process that begins with the formation of the blastopore and ends with the formation of three germ layers: **ectoderm**, **endoderm**, and **mesoderm**. The ectoderm will become the skin and the nervous system. The endoderm will form the viscera, including the lungs, liver, digestive organs, and so on. The mesoderm will give rise to the muscles, blood, and bones.

1b. After the three germ layers are formed, organs begin to form in a process known as **organogenesis**. In chordates, like the frog, the organs to form first are the **notochord**, the skeletal rod characteristic of all chordates, and the neural tube, which will become the central nervous system. The neural tube is formed by **embryonic induction** by the endoderm layer lying under it called the dorsal endoderm. After the blueprints of the organs are laid down, the embryo develops into a larval stage, the tadpole. Later, metamorphosis will transform the tadpole into a frog.

1c. In an experiment, **Spemann** grafted a piece of **dorsal lip** from one amphibian embryo onto the ventral side of a second embryo. A second notochord and neural tube developed at the location of the graft. This experiment proved the mechanics of embryonic induction. It showed that the dorsal lip is an **inducer** that causes adjacent tissue to transform into some structure that, in this case, it was not originally destined to be.

> *Note: Compare means discuss those things that are similar. Contrast means discuss what is different about the two processes.*

2. There are two stages in meiosis: **meiosis I** (**reduction division**), in which homologous chromosomes separate, and meiosis II, which is like **mitosis**. In meiosis I, each chromosome pairs up precisely with its homologue into a synaptonemal complex by a process called **synapsis** and forms a structure known as a tetrad or bivalent. Synapsis is important for two reasons. First, it ensures that each daughter cell will receive one homologue from each parent. Second, it makes possible the process of crossing-over by which homologous chromatids exchange genetic material. The two stages of meiosis are further divided into phases. At the beginning of meiosis, cells have the diploid chromosome number ($2n$). By the end of meiosis, cells contain the haploid or haploid chromosome number (n). Each meiotic cell division consists of the same four stages as mitosis: prophase, metaphase, anaphase, and telophase.

In spermatogenesis, each spermatogonium cell undergoes meiosis to produce four active sperm cells. A **diploid spermatogonium** ($2n$) divides by mitosis to produce **primary spermatocytes** ($2n$) of equal size, which undergo meiosis I and yield two secondary spermatocytes (n), also of equal size. These spermatocytes undergo meiosis II yielding four **spermatids** (n), which differentiate in the epididymis to form four active motile sperm.

During **oogenesis**, a **primary oogonium cell** ($2n$) undergoes mitosis to produce **primary oocytes** ($2n$). These remain quiescent within small follicles in the ovaries until puberty, when they become reactivated by hormones. **FSH** periodically stimulates the follicles to complete meiosis I, producing secondary oocytes (n), which are released at ovulation.

Oogenesis differs from sperm formation in three ways. First, oogenesis is a stop-start process. The first part occurs prior to birth, the second part after fertilization. Second, cytokinesis divides the cell unequally. Almost all the cytoplasm remains in the egg, the other cells produced, polar bodies, have very little cytoplasm and disintegrate. Third, one primary oogonium cell produces only one active egg cell.

Ecology

CHAPTER 16

- Properties of populations
- Population growth
- Community structure and population interactions
- The food chain
- Ecological succession
- Biomes
- Chemical cycles
- Humans and the biosphere

INTRODUCTION

Ecology is the study of the interactions of organisms with their physical environment and with each other. Here is some introductory vocabulary for the topic.

1. A **population** is a group of individuals of one species living in one area who have the ability of interbreeding and interacting with each other.
2. A **community** consists of all the organisms living in one area.
3. An **ecosystem** includes all the organisms in a given area as well as the abiotic (nonliving) factors with which they interact.
4. **Abiotic factors** are nonliving and include temperature, water, sunlight, wind, rocks, and soil.
5. **Biosphere** is the global ecosystem.

PROPERTIES OF POPULATIONS

Here are 5 properties of populations you should know.

1. Size

Size is the total number of individuals in a population and is represented by N, below.

2. Density

Density is the number of individuals per unit area or volume. Counting the number of organisms inhabiting a certain area is often very difficult, if not impossible. For example, imagine trying to count the number of ants in 1 acre (0.5 ha) of land. Instead, scientists use **sampling techniques** to estimate the number of organisms living in one area. One sampling technique commonly used to estimate the size of a population is called **mark and recapture**. In this technique, organisms are captured, tagged, and then released. Some time later, the same process is repeated and the following formula is used for the collected data.

$$N = \frac{\text{(number marked in first catch)} \cdot \text{(total number in second catch)}}{\text{number of recaptures in second catch}}$$

Suppose 50 zebra mussels are captured, marked, and released. One week later, 100 zebra mussels are captured and 10 are found to have markings already. When using the formula, the total population would be about 500 zebra mussels.

3. Dispersion

Dispersion is the pattern of spacing of individuals within the area the population inhabits, see Figure 16.1. The most common pattern of dispersion is **clumped**. Fish travel this way in schools because there is safety in numbers. Some populations are spread in a **uniform** pattern. For example, certain plants may secrete toxins that keep away other plants that would compete for limited resources. **Random** spacing occurs in the absence of any special attractions or repulsions. Trees can be spaced randomly in a forest.

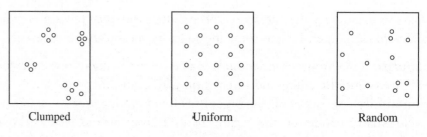

Clumped   Uniform   Random

Figure 16.1

4. Survivorship Curves

Survivorship or **mortality curves** show the size and composition of a population. There are three types of survivorship curves.

- **Type 1** curves show organisms with low death rates in young and middle age and high mortality in old age. There is a great deal of parenting, which accounts for the high survival rates of the young. This is characteristic of humans.
- **Type 2** curves describe a species with a death rate that is constant over the life span. This describes the hydra, reptiles, and rodents.
- **Type 3** curves show a very high death rate among the young but then shows that death rates decline for those few individuals that have survived to a certain age. This is characteristic of fish and invertebrates that release thousands of eggs, have external fertilization, and have no parenting; see Figure 16.2.

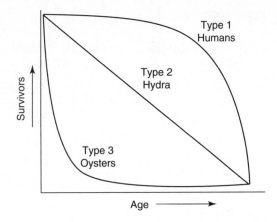

Figure 16.2   Survivorship Curves

5. Age Structure Diagrams

Another important parameter of populations is age structure. An age structure diagram shows the relative numbers of individuals at each age. Figure 16.3 shows two age structure diagrams. Country I shows the age structure of the human population of India; the pyramidal shape is characteristic of developing nations with half the population under the age of 20. Even after taking into account the disease, famine, natural disaster and emigration that will occur, the population in 20 years will be enormous. Country II shows an age structure for a developed nation like the United States with a stable population, **zero population growth**, where the number of people at each age group is about the same and the birth rates and the death rates are about equal.

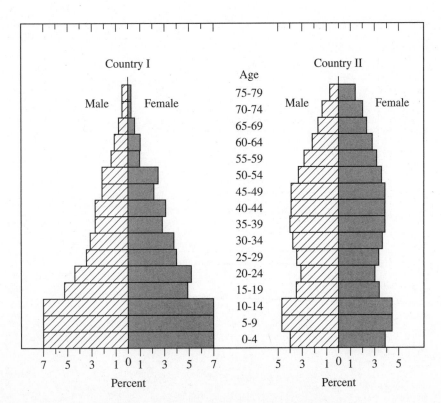

Figure 16.3   Age Structure Diagrams

POPULATION GROWTH

Every population has a characteristic **biotic potential**, the maximum rate at which a population could increase under ideal conditions. Different populations have different biotic potentials, which are influenced by several factors. These factors include *age at which reproduction begins, the life span during which the organisms are capable of reproducing, the number of reproductive periods in the lifetime, and the number of offspring the organism is capable of reproducing.* Regardless of whether a population has a large or small biotic potential, certain characteristics about growth are common to all organisms.

Exponential Growth

The simplest model for population growth is one with unrestrained or **exponential growth**. This population has no predation, parasitism, or competition. It has no immigration or emigration and is in an environment with unlimited resources. This is characteristic of a population that has been recently introduced into an area, such as a sample of bacteria newly inoculated onto a petri dish. Although exponential growth is usually short-lived in nature, the human population has been in the exponential growth phase for over 300 years.

Carrying Capacity

Ultimately, there is a limit to the number of individuals that can occupy one area at a particular time. That limit is called the **carrying capacity (K)**. Each particular environment has its own carrying capacity around which the population size oscillates; see Figure 16.4.

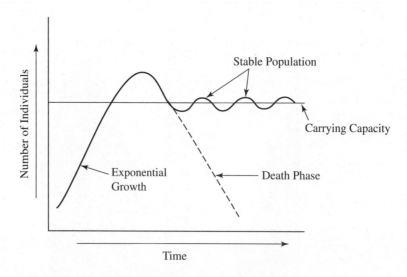

Figure 16.4

In addition, the carrying capacity changes as the environmental conditions change. Perhaps a fire destroyed several acres of forest habitat. See Figure 16.5.

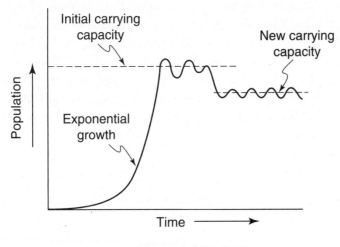

Figure 16.5

Limiting Factors

Limiting factors are those factors that limit population growth. They are divided into two categories: **density-dependent** and **density-independent** factors.

- **Density-dependent factors** are those factors that increase directly as the population density increases. They include competition for food, the buildup of wastes, predation, and disease.
- **Density-independent factors** are those factors whose occurrence is unrelated to the population density. These include earthquakes, storms, and naturally occurring fires and floods.

Growth Patterns

Some species are opportunistic; they reproduce rapidly when the environment is uncrowded and resources are vast. They are referred to as **r-strategists**. Other organisms, the **K-strategists**, live at a density near the carrying capacity (K). Table 16.1 is a chart comparing the two life strategies.

TABLE 16.1

Comparison of Two Life Strategies

r-strategists	K-strategists
Many young	Few young
Little or no parenting	Intensive parenting
Rapid maturation	Slow maturation
Small young	Large young
Reproduce once	Reproduce many times
Example: insect	Example: mammals

A Case Study—The Hare and the Lynx

A perfect study in population growth involves the populations of **snowshoe hare** and **lynx** at the Hudson Bay Company, which kept records of the pelts sold by trappers from 1850–1930. The data reveal fluctuations in the populations of both animals. The hare feeds on the grass, and the lynx feeds on the hare. So the cycles in the lynx population are probably caused by cyclic fluctuations in the hare population. The hare population experiences cycles of **exponential growth** and **crashes**. Additionally, cycles in the hare population are probably due to a **limited food supply** for the hare due to a combination of malnutrition from cyclical overcrowding and overgrazing and of predation by the lynx; see Figure 16.6.

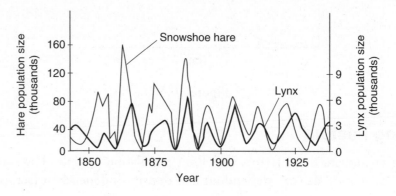

Figure 16.6

COMMUNITY STRUCTURE AND POPULATION INTERACTIONS

Communities are made of populations that interact with the environment and with each other. These interactions are very complex but can be divided into five categories: **competition, predation, parasitism, mutualism,** and **commensalism.**

1. Competition

The Russian scientist **G. F. Gause** developed the **competitive exclusion principle** after studying the effects of interspecific competition in a laboratory setting. He worked with two very similar species, *Paramecium caudatum* and *Paramecium aurelia.* When he cultured them separately, each population grew rapidly and then leveled off at the carrying capacity. However, when he put the two cultures together, *P. aurelia* had the advantage and drove the other species to extinction. His principle states that *two species cannot coexist in a community if they share a **niche**, that is, if they use the same resources.*

In nature, there are two related outcomes, besides extinction, if two species *inhabit the same niche and therefore compete for resources.* One of the species will evolve through natural selection to exploit different resources. This is called **resource partitioning.** Another possibility is what occurred on the **Galapagos Islands.** Finches evolved different beak sizes through natural selection and were able to eat different kinds of seeds and avoid competition. This divergence in body structure is called **character displacement.**

2. Predation

Predation can refer to one animal eating another animal, or it can also refer to animals eating plants. For their protection, animals and plants have evolved defenses against predation.

- **Plants** have evolved **spines** and **thorns** and chemical **poisons** such as **strychnine, mescaline, morphine**, and **nicotine** to fend off attack by animals.
- **Animals** have evolved **active defenses** such as **hiding, fleeing,** or **defending** themselves. These, however, can be very costly in terms of energy. Animals have also evolved **passive defenses** such as **cryptic coloration** or **camouflage** that make the prey difficult to spot. Here are three examples.

  - ✔ **Aposematic coloration** is the very bright, often red or orange, coloration of poisonous animals as a warning that possible predators should avoid them.
  - ✔ **Batesian mimicry** is copycat coloration where one harmless animal mimics the coloration of one that is poisonous. One example is the **viceroy butterfly** which is harmless but looks very similar to the **monarch butterfly,** which stores poisons in its body from the milkweed plant.
  - ✔ In **Müllerian mimicry**, two or more poisonous species, such as the cuckoo bee and the yellow jacket, resemble each other and gain an advantage from their combined numbers. Predators learn more quickly to avoid any prey with that appearance.

3. Mutualism

Mutualism is a symbiotic relationship where both organisms benefit (+/+). An example is the bacteria that live in the human intestine and produce vitamins.

4. Commensalism

Commensalism is a symbiotic relationship where one organism benefits and one is unaware of the other organism (+/o). Barnacles that attach themselves to the underside of a whale benefit by gaining access to a variety of food sources as the whale swims into different areas. The whale is unaware of the barnacles.

5. Parasitism

Parasitism is a symbiotic relationship (+/−) where one organism, the parasite, benefits while the host is harmed. A tapeworm in the human intestine is an example.

ENERGY FLOW AND PRIMARY PRODUCTION

Every day, Earth is bombarded with enough sunlight to supply the needs of the entire human population for the next 25 years. Most solar radiation, though, is absorbed, scattered, or reflected by the atmosphere. Only a small fraction actually reaches green plants, and less than 1% is actually converted to chemical bond energy by photosynthesis. However, that energy is the basis for almost all of Earth's food chains and fuels all life on Earth. (An example of a food chain that does not rely on solar energy is one located around deep-ocean thermal vents.) Ecologists use two terms when they discuss energy flow on Earth: gross primary productivity and net primary productivity. **Gross primary productivity (GPP)** is the amount of light energy that is converted to chemical energy by photosynthesis per unit time. **Net primary productivity (NPP)** is equal to the GPP minus the energy used by producers for their own cellular respiration.

Different ecosystems vary in their NPP as well as what they contribute to the total or **global NPP** of Earth. Tropical rain forests are among the most productive terrestrial ecosystems and contribute a large portion of Earth's overall net primary production. (Unfortunately, that number is shrinking as we cut down rain forests to make way for farming.) Coral reefs, on the other hand, have a very high NPP but contribute relatively little to the global NPP because they occupy such a tiny part of the planet. The open oceans are just the opposite of coral reefs. Their NPP is very low per unit area. Because they occupy three-fourths of the globe, their global PNN is higher than that of any other biome.

ENERGY FLOW AND THE FOOD CHAIN

The **food chain** is the pathway along which food is transferred from one **trophic** or feeding **level** to another. Energy, in the form of food, moves from the **producers** to the **herbivores** to the **carnivores**. Only about **10 percent** of the energy stored in any trophic level is converted to organic matter at the next trophic level. This means that if you begin with 1,000 g of plant matter, the food chain can support 100 g of herbivores (primary consumers), 10 g of secondary consumers, and only 1 g of tertiary consumers. As a result of the loss of energy from one trophic level to the next, food chains are rather short. They never have more than four or five trophic levels. As you might expect, long food chains are less stable than short ones. This is because population fluctuations at lower trophic levels are magnified at higher levels, causing local extinction of top predators. A good model to demonstrate the interaction of organisms in the food chain and the loss of energy is the **food pyramid,** see Figure 16.7.

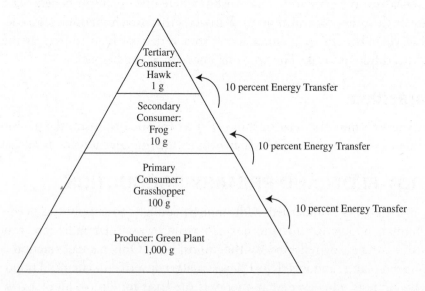

Figure 16.7   Food Pyramid

Food chains are not isolated; they are interwoven with other food chains into a **food web**. An animal can occupy one trophic level in one food chain and a different trophic level in another food chain. Humans, for example, can be primary consumers when eating vegetables but are tertiary consumers when eating a steak. Here are two sample food chains, each with four trophic levels.

Producers → Primary Consumers → Secondary Consumers → Tertiary Consumers

Terrestrial Food Chain
Green Plant → Grasshopper → Frog → Hawk

Marine Food Chain
Phytoplankton → Zooplankton → Small Fish → Shark

- **Producer**
  - ✔ **Autotrophs**
  - ✔ **Green plants**
  - ✔ Convert light energy to chemical bond energy
  - ✔ Have the greatest biomass of any trophic level
  - ✔ Examples: diatoms and phytoplankton

- **Primary consumers**
  - ✔ **Heterotrophs**
  - ✔ **Herbivores**
  - ✔ Eat the producers
  - ✔ Examples: grasshoppers, zooplankton

- **Secondary consumers**
  - ✔ **Heterotrophs**
  - ✔ **Carnivores**
  - ✔ Eat the primary consumers
  - ✔ Examples: frogs, small fish

- **Tertiary consumers**
  - ✔ **Heterotrophs**
  - ✔ **Carnivores**
  - ✔ Eat the secondary consumers
  - ✔ Top of the food chain
  - ✔ Have the **least biomass** of any other trophic level in the food chain.
  - ✔ Least stable trophic level and most sensitive to fluctuations in populations of the other trophic levels
  - ✔ Example: hawk

Biological Magnification

Organisms at higher trophic levels have greater concentrations of accumulated toxins stored in their bodies than those at lower trophic levels. This phenomenon is called **biological magnification**. The bald eagle almost became extinct because Americans sprayed heavily with the pesticide DDT in the 1950s, which entered the food chain and accumulated in the bald eagle at the top of the food chain. Because DDT interferes with the deposition of calcium in eggshells, the thin-shelled eggs were broken easily and few eaglets hatched. DDT is now outlawed, and the bald eagle was saved from extinction by human intervention.

Decomposers

Decomposers—bacteria and **fungi**—are usually not depicted in any diagram of a food chain. However, without decomposers to recycle nutrients back to the soil to nourish plants, there would be no food chain and no life.

ECOLOGICAL SUCCESSION

Most communities are dynamic, not stable. The size of a population increases and decreases around the carrying capacity. Migration of a new species into a habitat can alter the entire food chain. Major disturbances, natural and human-made, like volcanic eruptions, strip mining, clear-cutting a forest, and forest fires, can suddenly and drastically destroy a community or an entire ecosystem. What follows this destruction is the process of sequential rebuilding of the ecosystem called **ecological succession**.

If the rebuilding begins in a lifeless area where even soil has been removed, the process is called **primary ecological succession**. *The essential and dominant characteristic of primary succession is soil building.* After an ecosystem is destroyed, the first organisms to inhabit a barren area are **pioneer organisms** like **lichens** (a symbiont consisting of algae and fungi) and **mosses**, which are introduced into the area as spores by the wind. Soil develops gradually as rocks weather and organic matter accumulates from the decomposed remains of the pioneer organisms. Once soil is present, pioneer organisms are overrun by other larger organisms: grasses, bushes, and then trees. The final stable community that remains is called the **climax community**. It remains until the ecosystem is once again destroyed by a **blowout**, a disaster that destroys the ecosystem once again.

One example of **primary succession** that was studied in detail is at the southern edge of Lake Michigan. As the lakeshore gradually receded northward after the last ice age (10,000 years ago), it left a series of new beaches and sand dunes exposed. Today, someone who begins at the water's edge and walks south for several miles will pass through a series of communities that were formed in the last 10,000 years. These communities represent the various stages beginning with bare, sandy beach and ending with a climax community of old, well-established forests. In some cases, the climax community is a beech-sugar maple forest, in other areas the forest is a mix of hickory and oak.

The process known as **secondary succession** occurs when an existing community has been cleared by some disturbance that leaves the soil intact. This is what happened in 1988 in Yellowstone National Park when fires destroyed all the old growth that was dominated by lodgepole pine but left the soil intact. Within one year, the burned areas in Yellowstone were covered with new vegetation.

BIOMES

Biomes are very large regions of the earth whose distribution depends on the amount of **rainfall** and the **temperature** in an area. Each biome is characterized by **different vegetation** and **animal** life. There are many biomes, including freshwater, marine, and terrestrial. In the northern hemisphere, from the equator to the most northerly climes, there is a trend in terrestrial biomes: from tropical rain forest, desert, grasslands, temperate deciduous forest, taiga, and finally, tundra in the north. Changes in altitude produce effects similar to changes in latitudes. On the slopes of the Appalachian Mountains in the east and the Rockies and coastal ranges in the west, there is a similar trend in biomes. As elevation increases and temperatures and humidity decrease, one passes through temperate deciduous forest to taiga to tundra. Here is an overview of the major biomes of the world.

STUDY TIP

Know the characteristics of each biome.

Marine

- The largest biome, covering three-fourths of the earth's surface
- The most stable biome with temperatures that vary little because water has a high heat capacity and there is such enormous volume of water.
- Provides most of the earth's food and oxygen
- The marine biome is itself divided into different regions classified by the amount of sunlight they receive, the distance from the shore, and the water depth and whether it is open water or ocean bottom.

Tropical Rain Forest

- Found near the equator with abundant rainfall, stable temperatures, and high humidity.
- Although these forests cover only 4 percent of the earth's land surface, they account for more than 20 percent of the earth's net carbon fixation (food production).
- The most diversity of species of any biome on earth. May have as many as 50 times the number of species of trees as a temperate forest.
- Dominant trees are very tall with interlacing tops that form a dense canopy, keeping the floor of the forest dimly lit even at midday. The canopy also prevents rain from falling directly onto the forest floor, but leaves drip rain constantly.
- Many trees are covered with **epiphytes**, photosynthetic plants that grow on other trees rather than supporting themselves. They are not parasites but may kill the trees inadvertently by blocking the light.
- The most diverse animal species of any biome, including birds, reptiles, mammals, and amphibians.
- Some are biodiversity **hotspots**, meaning that many species are endangered.

Desert

- Less than 10 inches (25 cm) of rainfall per year; not even grasses can survive.
- Experiences the most extreme temperature fluctuations of any biome. Daytime *surface* temperatures can be as high as 158°F (70°C). With no moderating influence of vegetation, heat is lost rapidly at night. Shortly after sundown, temperatures drop drastically.
- Characteristic plants are the drought-resistant cactus with shallow roots to capture as much rain as possible during hard and short rains, which are characteristic of the desert.
- Other plants include sagebrush, creosote bush, and mesquite.
- There are many small annual plants that germinate only after a hard rain, send up shoots and flowers, produce seeds, and die, all within a few weeks.

- Most animals are active at night or during a brief early morning period or late afternoon, when heat is not so intense. During the day, they remain cool by burrowing underground or hiding in the shade.
- Cacti can expand to hold extra water and have modified leaves called spines, that protect against animals attacking the cactus for its water.
- As an example of how severe conditions in a desert can be, in the Sahara Desert are regions hundreds of miles across that are completely barren of any vegetation.
- Characteristic animals include rodents, kangaroo rat, snakes, lizards, arachnids, insects, and a few birds.

Temperate Grasslands

- Covers huge areas in both the temperate and tropical regions of the world.
- Characterized by low total annual rainfall or uneven seasonal occurrence of rainfall, makes conditions inhospitable for forests.
- Principal grazing mammals include bison and pronghorn antelope in the United States and wildebeest and gazelle in Africa. Also, burrowing mammals, such as prairie dogs and other rodents, are common.

Temperate Deciduous Forest

- Found in the northeast of North America, south of the taiga, and characterized by trees that drop their leaves in winter.
- Includes many more plant species than does the taiga.
- Shows **vertical stratification** of plants and animals, that is, there are species that live on the ground, the low branches, and the treetops.
- Soil is rich due to decomposition of leaf litter.
- Principal mammals include squirrels, deer, foxes, and bears, which are dormant or hibernate through the cold winter.

Conifer Forest—Taiga

- Located in northern Canada and much of the world's northern regions.
- Dominated by conifer (evergreens) forests, like spruce and fir.
- Landscape is dotted with lakes, ponds, and bogs.
- Very cold winters.
- This is the largest terrestrial biome.
- Characterized by heavy snowfall, trees are shaped with branches directed downward to prevent heavy accumulations of snow from breaking their branches.
- Principal large mammals include moose, black bear, lynx, elk, wolverines, martens, and porcupines.
- Flying insects and birds are prevalent in summer.
- Has greater variety in species of animals than does the tundra.

Tundra

- Located in the far northern parts of North America, Europe, and Asia.
- **Permafrost**, permanently frozen subsoil found in the farthest point north including Alaska.

- Commonly referred to as the **frozen desert** because it gets very little rainfall, which cannot penetrate the frozen ground.
- Has the appearance of gently rolling plains with many lakes, ponds, and bogs in depressions.
- Insects, particularly flies, are abundant. As a result, vast numbers of birds nest in the tundra in the summer and migrate south in the winter.
- Principal mammals include reindeer, caribou, Arctic wolves, Arctic foxes, Arctic hares, lemmings, and polar bears.
- Though the number of individual organisms in the tundra is high, the number of species is small.

CHEMICAL CYCLES

Although the earth receives a constant supply of energy from the sun, chemicals must be recycled. You must know several chemical cycles: **carbon**, **nitrogen**, **water cycles**.

The Water Cycle

Water evaporates from the earth, forms clouds, and rains over the oceans and land. Some rain percolates through the soil and makes its way back to the seas. Some evaporates directly from the land, but most evaporates from plants by **transpiration**.

The oceans contain 97% of the water in the biosphere. About 2% is locked in glaciers and polar ice caps, and the remaining 1% is in lakes, rivers, and ground water. A negligible amount is in the atmosphere.

The Carbon Cycle

The basis of this are the reciprocal processes of **photosynthesis** and **respiration**.

- Cell respiration by animals and bacterial decomposers adds CO_2 to the air and removes O_2.
- Burning of fossil fuels adds CO_2 to the air.
- Photosynthesis removes CO_2 from the air and adds O_2.

The major reservoir of carbon is fossil fuels, plant and animal biomass. Carbon is also found in the soil, in dissolved carbon compounds in the oceans, in sediments in aquatic ecosystems, and in the atmosphere as CO_2 (carbon dioxide) and CO (carbon monoxide).

The Nitrogen Cycle

Very little nitrogen enters ecosystems directly from the air. Most of it enters ecosystems by way of bacterial processes.

- **Nitrogen-fixing bacteria** live in the nodules in the roots of legumes and convert **free nitrogen** into the **ammonium ion** (NH_4^+).
- **Nitrifying bacteria** convert the ammonium ion into **nitrites** and then into **nitrates**.
- **Denitrifying bacteria** convert **nitrates** (NO_3) into **free** atmospheric **nitrogen**.

The main reservoir of nitrogen is the atmosphere, which contains about 79% nitrogen gas (N₂). Nitrogen is also found bound in the soil and in lake, river, and ocean sediments. It is also fixed into animal and plant biomass and in groundwater. See Figure 16.8.

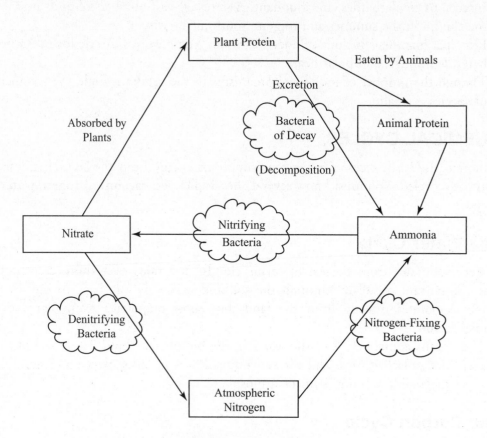

Figure 16.8   Nitrogen Cycle

HUMANS AND THE BIOSPHERE

Humans threaten to make the earth uninhabitable as the population increases exponentially and as people waste natural resources and pollute the air and water. Here are several examples.

Eutrophication of the Lakes

Humans have disrupted freshwater ecosystems, causing a process called **eutrophication**. Runoff from sewage and manure from pastures increase nutrients in lakes and cause excessive growth of algae and other plants. Shallow areas become choked with weeds, and swimming and boating become impossible. As these large populations of photosynthetic organisms die, two things happen. First, organic material accumulates on the lake bottom and reduces the depth of the lake. Second, **detrivores** use up oxygen as they decompose the dead organic matter. Lower oxygen levels make it impossible for some fish to live. As fish die, decomposers expand their activity and oxygen levels continue to decrease. The process continues, more organisms die, the oxygen levels decrease, more decomposing matter accumulates on the lake bottom, and ultimately, the lake disappears. See Figure 16.9.

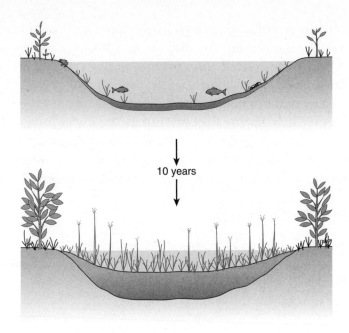

Figure 16.9

Acid Rain

Acid rain is caused by pollutants in the air from **combustion of fossil fuels**. Nitrogen and sulfur pollutants in the air turn into **nitric**, **nitrous**, **sulfurous**, and **sulfuric acids**, which cause the pH of the rain to be less than 5.6. This kills the organisms in lakes and damages ancient stone architecture.

Toxins

Toxins from industry have gotten into the **food chain**. Most cattle and chicken feed contain **antibiotics** and **hormones** to accelerate animal growth but may have serious ill affects on humans who eat the chicken and beef. Any **carcinogens** or **teratogens** (causing birth defects) that get into the food chain accumulate and remain in the human body's fatty tissues because we occupy the top of the food chain. This process is called biological magnification.

Global Warming

To understand global warming, we first must talk about the **greenhouse effect**. CO_2 and water vapor in the atmosphere absorb and retain much of the light and heat that comes to Earth from the sun. If the greenhouse effect did not occur, the average temperature on the surface of Earth would be much colder and life as we know it would not exist. However, atmospheric CO_2 levels have markedly increased during the last 150 years, which scientists link to **global warming**. According to NASA, the top four warmest years since the 1890s were the last few years and global temperatures continue to climb.

> **STUDY TIP**
>
> One of the four essays will most likely be a mix of ecology and evolution.

This increase in temperature is called global warming and could have disastrous effects for the world's population. An increase of 1.0°C on average temperature worldwide would cause the polar ice caps to melt, raising the level of the seas. As a result, major coastal cities in the United States, including New York, Los Angeles, and Miami, would be under water.

Coral reefs like the Great Barrier Reef in Australia are under increased physiological stress due to an increase in global warming. This stress makes it more difficult for coral to build their skeletons. Oysters and sea urchins are also suffering because of increased acidification of the oceans due to an increase in CO_2 dissolved in the oceans.

Depleting the Ozone Layer

The accumulations in the air of **chlorofluorocarbons**, chemicals used for refrigerants and aerosol cans, have caused the formation of a hole in the protective **ozone layer**. This allows more ultraviolet (UV) light to reach the earth, which is responsible for an increase in the incidence of **skin cancer** (melanoma) worldwide.

Introducing New Species

Humans have moved species from one area to another with serious consequences. Two examples are the "killer" honeybees and the zebra mussel.

- The **"killer" honeybee**: The African honeybee is a very aggressive subspecies of honeybee that was brought to Brazil in 1956 to breed a variety of bee that would produce more honey in the tropics than the Italian honeybee. The African honeybees escaped by accident and have been spreading throughout the Americas. By the year 2000, ten people were killed by these bees in the United States.
- The **zebra mussel**: In 1988, the zebra mussel, a fingernail-sized mollusk native to Asia, was discovered in a lake near Detroit. No one knows how the mussel got transplanted there, but scientists assume it was accidentally carried by a ship from a freshwater port in Europe to the Great Lakes. Without any local natural predator to limit its growth, the mussel population exploded. They were first discovered when they were found to have clogged the water intake pipes of those cities whose water is supplied by Lake Erie. To date, the zebra mussel has caused millions of dollars of damage. In addition, the influx of the zebra mussel threatens several native species with extinction by outcompeting indigenous species.

Pesticides vs. Biological Control

Scientists have developed a variety of pesticides, chemicals that kill organisms that we consider to be undesirable. These include insecticides, herbicides, fungicides, and mice and rat killers. On the one hand, these pesticides save lives by increasing food production and by killing animals that carry and cause diseases like bubonic plague (diseased rats) and malaria (anopheles mosquitoes). On the other hand, exposure to pesticides can cause cancer in humans. Moreover, spraying with pesticides ensures the development of resistant strains of pests through natural selection. The pests come back stronger than before. This problem requires that we spray more and more, which means more people will be exposed to these toxic chemicals.

An alternative to widescale spraying with pesticides is called biological control. The following are some biological solutions to get rid of pests without using dangerous chemicals.

1. Use crop rotation—change the crop planted in a field.
2. Introduce natural enemies of the pests—you must be careful, however, that you do not disrupt a delicate ecological balance by introducing an invasive species.
3. Use natural plant toxins instead of synthetic ones.
4. Use insect birth control—male insect pests can be sterilized by exposing them to radiation and then releasing them into the environment to mate unsuccessfully with females.

Multiple-Choice Questions

1. The high level of pesticides in birds of prey is an example of

   (A) the principle of exclusion
   (B) cycling of nutrients by decomposers
   (C) exponential growth
   (D) biological magnification
   (E) ecological succession

2. Which of the following best explains why there are usually no more than five trophic levels in a food chain?

   (A) There are not enough organisms to fill more than five levels.
   (B) There is too much competition among the organisms at the lower levels to support more animals at higher levels.
   (C) The statement is not true; there can be unlimited trophic levels.
   (D) Energy is lost at each trophic level.
   (E) There are not enough decomposers to support more than five trophic levels.

3. Which of the following is NOT an important characteristic of the marine biome?

   (A) the largest biome
   (B) provides most of the earth's food
   (C) temperatures vary tremendously
   (D) the largest source of oxygen
   (E) covers most of the earth

4. The most important factors affecting the distribution of biomes are

   (A) temperature and rainfall
   (B) amount of sunlight and human population size
   (C) latitude and longitude
   (D) altitude and water supply
   (E) amount of sun and length of day

5. Which of the following is NOT an abiotic factor?

   (A) air
   (B) water
   (C) wind
   (D) temperature
   (E) decomposers

6. Which of the following encompasses all the others?

   (A) ecosystem
   (B) community
   (C) population
   (D) individual
   (E) species

7. Which of the following lists the biomes as they appear as you move from the equator to the North Pole in North America?

   (A) tropical rain forest—taiga—tundra—desert—temperate deciduous forest
   (B) desert—tundra—taiga—temperate deciduous forest—tropical rain forest
   (C) taiga—temperate deciduous forest—tundra—desert—tropical rain forest
   (D) tundra—taiga—temperate deciduous forest—desert—tropical rain forest
   (E) tropical rain forest—desert—temperate deciduous forest—taiga—tundra

Questions 8–11
Questions 8–11 refer to the survivorship curve shown below.

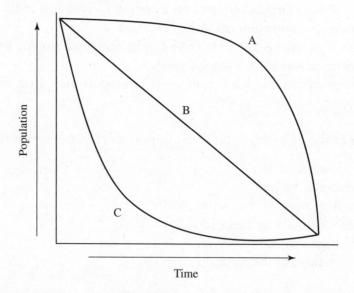

8. Which best describes a K-strategist?

   (A) A
   (B) B
   (C) C
   (D) None of the above
   (E) All of the above at different times

9. Which best describes starfish?

   (A) A
   (B) B
   (C) C
   (D) None of the above
   (E) All of the above at different times

10. Which best describes an organism that fertilizes externally?

    (A) A
    (B) B
    (C) C
    (D) None of the above
    (E) All of the above at different times

11. Which curve best describes an organism that invests a lot of energy
    in parenting?

    (A) A
    (B) B
    (C) C
    (D) None of the above
    (E) All of the above at different times

---

12. A species described as r-strategist would definitely NOT have which of the
    following characteristics?

    (A) clumped population pattern
    (B) much parenting
    (C) many offspring
    (D) random population pattern
    (E) high mortality of young

13. Which of the following is a density-independent factor limiting human
    population growth?

    (A) famine
    (B) disease
    (C) competition for food
    (D) overpopulation
    (E) naturally occurring fires

14. What would most likely be the cause of bushes of one species growing in
    one area in a uniform spacing pattern?

    (A) random distribution of seeds
    (B) interactions among individuals in the population
    (C) chance
    (D) the varied nutrient supplies in that area
    (E) variation in sunlight

15. Animals from two different species utilize the same source of nutrition in one area. It is most accurate to say that the animals

   (A) will learn to get along
   (B) will compete for food
   (C) will die because there will not be enough food for both of them
   (D) will learn to eat different foods
   (E) any of the above is possible

16. Two animals live together in close association. One benefits, while the other is unaware of the first animal. The relationship is best described as

   (A) parasitism
   (B) mutualism
   (C) commensalism
   (D) predation
   (E) None of the above is correct.

Questions 17–20
Questions 17–20 refer to the following depiction of a food web for a terrestrial ecosystem.

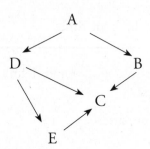

17. Which species is the producer?

   (A) A
   (B) B
   (C) C
   (D) D
   (E) E

18. A toxic pollutant would be found in highest concentrations in which species?

   (A) A
   (B) B
   (C) C
   (D) D
   (E) E

19. Which would have the greatest biomass?

    (A) A
    (B) B
    (C) C
    (D) D
    (E) E

20. Which would have the smallest biomass?

    (A) A
    (B) B
    (C) C
    (D) D
    (E) E

---

21. Eutrophication in lakes results from

    (A) an increase in ambient temperatures
    (B) a decrease in temperatures
    (C) an increase in carbon dioxide in the air
    (D) pollution in the air
    (E) an increase in nutrients in the lake

Questions 22–24

Questions 22–24 refer to the graph below that shows changes in population over time.

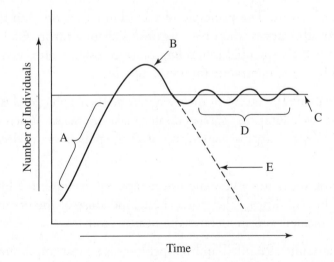

22. Which letter shows a mature, well-established population in favorable conditions?

    (A) A
    (B) B
    (C) C
    (D) D
    (E) E

23. Which letter shows the carrying capacity of the environment?

    (A) A
    (B) B
    (C) C
    (D) D
    (E) E

24. Which letter shows a population in unfavorable conditions?

    (A) A
    (B) B
    (C) C
    (D) D
    (E) E

---

25. Many poisonous animals are brightly colored (red or orange) as a warning to predators. This special coloration is called

    (A) Müllerian mimicry
    (B) Batesian mimicry
    (C) aposematic coloration
    (D) mutualistic coloration
    (E) commensal coloration

Answers to Multiple-Choice Questions

1. **(D)** Organisms at higher trophic levels have a greater concentration of accumulated toxins. The principle of exclusion has to do with the competition that arises when two organisms share a niche. Ecological succession is the sequential rebuilding of an ecosystem after it has been destroyed by some natural or human-made disaster.

2. **(D)** Only about 10 percent of the energy from one trophic level is transferred to the next level. The other choices do not make any sense. When referring to choice E, the more organic matter, the larger the population of decomposing bacteria.

3. **(C)** Because water has a very high heat capacity, it requires a lot of time to heat up and cool down. Therefore, the temperature of the oceans remains relatively constant and moderates the nearby land.

4. **(A)** The amount of rainfall and temperature are most important in distribution of biomes. The length of day changes everywhere on earth, so choice E must be eliminated. Choice B must be eliminated because it is irrelevant. Altitude does not, in itself, determine climate.

5. **(E)** Decomposers are bacteria, living things. All the others are abiotic factors, not living.

6. **(A)** From the most specific to the most general: individual, population, species, community, and ecosystem.

7. **(E)** As you go toward the North Pole or up in elevation from the equator, the temperatures decrease.

8. **(A)** There is high survival rate of the young and death occurs in old age. The curve is flat in the beginning, and then drops at old age.

9. **(C)** Fertilization in the sea star (starfish) is external, so survival of the young is poor. That is why the curve dips steeply initially.

10. **(C)** Fertilization is external with no parenting, so survival of the young is poor. This is why the curve dips steeply initially.

11. **(A)** This is a K-strategist. K-strategists are characterized by intensive parenting of a few young whose maturation is slow.

12. **(B)** Population pattern is not tied to r-strategists or K-strategists. Think of *r* as standing for "risky." There is little or no parenting and high mortality of the young in the r-strategists.

13. **(E)** Fires are a natural occurrence and are independent of population.

14. **(B)** Plants may secrete toxins that keep other plants from growing nearby. This minimizes competition for limited resources.

15. **(B)** This is a restatement of Gause's principle of exclusion.

16. **(C)** An example of commensalism is the barnacle that attaches to the bottom of the whale. The whale is unaware of the barnacle, which gains a varied food source as the whale swims to different areas.

17. **(A)** Species A is the producer because both B and D feed on it.

18. **(C)** Species C would accumulate the most toxins because it is at the top of the food chain. The longest chain runs from A to D to E to C.

19. **(A)** The producer always has the greatest biomass.

20. **(C)** The top consumer always has the least biomass.

21. **(E)** Eutrophication means "true feeding" and results from runoff of sewage and manure from pastures; phosphates, nitrates, and sulfates are the major contaminants.

22. **(D)** The population at D fluctuates around the carrying capacity (C) for that environment.

23. **(C)** The carrying capacity is the maximum population size that can be supported by the available resources. It is symbolized as K.

24. **(E)** This population is crashing.

25. **(C)** This is the definition of aposematic coloration.

Free-Response Questions

Directions: Answer all questions. You must answer the question in essay—
not outline—form. You may use labeled diagrams to supplement your essay,
but diagrams alone are *not* sufficient. Before you start to write, read each
question carefully so that you understand what the question is asking.

1. Describe the process of ecological succession.
2. Name five properties of populations and explain what they tell us about a population.
3. Explain the water, nitrogen, and carbon cycles.
4. Discuss four environmental issues that affect humans today.
5. Discuss the flow of energy through a food web. Include the recycling of energy.

Typical Free-Response Answers

Note: The essays in this section are simple and straightforward. Answering these essays is good practice to help you learn the vocabulary and the concept in the study of ecology. One essay below has the answer written out. Some other questions are listed below but with no answers, only key words to guide you as you write your essay. It would be good practice for you to answer those on your own after reviewing the material in this chapter. Once again, terms that you should include are written in bold.

1. **Ecological succession** is a sequential building or rebuilding of an entire ecosystem. The process is called **primary succession** if it begins in a virtually lifeless area where soil has yet to form. The first organisms to inhabit a barren area are **pioneer organisms** like **lichens** (a symbiont consisting of algae and fungi) and mosses, which are introduced into the area as spores by the wind. Soil develops gradually as rocks weather and organic matter accumulates from the decomposed remains of the pioneer organisms. Once soil is present, pioneer organisms are overrun by other larger organisms: grasses, bushes, and then trees, the prevalent form of vegetation for that community. Primary succession can take hundreds or thousand of years to reach the **climax community**, the final stable community that develops. **Secondary succession** occurs when an existing community has been severely damaged by some disturbance that leaves the soil intact. Examples are the fire in Yellowstone Park in 1988 or the massive volcanic explosion of Mount Saint Helens in 1980 or the **clear-cutting** of forests that is ongoing in the Pacific Northwest, where all trees are removed. This process does not take as long as primary succession. Often, though, the same community does not return because the new environment is very different from the way it was when the climax community formed years before.

2. Key words/topics:
   Size
   Density
   Dispersion
   Survivorship curves
   Age structure diagrams

3. Key words/topics:
   •Interdependence of organisms on earth
   Water—evaporation, condensation
   Photosynthesis—requires CO_2 and releases O_2
   Cellular respiration—requires O_2 and releases CO_2
   Bacteria of decay
   Nitrogen-fixing bacteria
   Nitrifying bacteria
   Denitrifying bacteria

4. Key words/topics:
   Interdependence of organisms on earth
   Eutrophication of lakes
   Acid rain
   Toxins in the food chain
   Global warming
   Depleting the ozone layer
   Introducing new species

5. Key words/topics:
   Interdependence of organisms in an ecosystem and on earth
   Sunlight/energy
   Producers
   Primary consumers
   Secondary consumers
   Tertiary consumers
   Food webs
   Food chains
   Biological magnification or amplification of toxins
   Decomposers and the cycling of nutrients

Animal Behavior

CHAPTER 17

- Fixed action pattern
- Learning
- Social behavior

INTRODUCTION

An organism's behavior is critical for its survival and for the successful production of offspring. The study of behavior and its relationship to its evolutionary origins is called **ethology**. Foremost in the field of ethology are three scientists who shared the Nobel Prize in 1973: Karl von Frisch, Konrad Lorenz, and Niko Tinbergen. **Karl Von Frisch** is known for his extensive studies of honeybee communication and his famous description of the **waggle dance** in bees. **Niko Tinbergen** is known for his elucidation of the **fixed action pattern**. **Konrad Lorenz** is famous for his work with **imprinting**. Here are some basics concepts in the field of animal behavior.

FIXED ACTION PATTERN

A **fixed action pattern (FAP)** is an innate, *highly stereotypic behavior,* that once begun is continued to completion, no matter how useless. FAPs are initiated by external stimuli called **sign stimuli**. When these stimuli are exchanged between members of the same species, they are known as **releasers**. An example of an FAP studied by **Tinbergen** involves the **stickleback fish**, which attacks other males that invade its territory. The **releaser** for the attack is the red belly of the intruder. The stickleback will not attack an invading male stickleback lacking a red underbelly, but it will readily attack a nonfishlike wooden model as long it has a splash of red visible.

LEARNING

Learning is a sophisticated process in which the responses of the organism are modified as a result of experience. The capacity to learn can be tied to length of life span and complexity of the brain. If the animal has a very short life span, like a fruit fly, it has no time to learn, even if it has the ability. It must therefore rely on fixed action patterns. In contrast, if the animal lives for a long time and has a complex brain, then a large part of its behavior depends on prior experience and learning.

Habituation

Habituation is one of the simplest forms of learning. An animal comes to ignore a persistent stimulus so it can go about its business. If you tap the dish containing a hydra, it will quickly shrink and become immobile. If you keep tapping, after a while the hydra will begin to ignore the tapping, elongate, and continue moving about. It has becomes **habituated** to the stimulus.

Associative Learning

Associative learning is one type of learning in which one stimulus becomes linked to another through experience. Examples of associative learning are classical conditioning and operant conditioning.

- **Classical conditioning**, a type of **associative learning**, is widely accepted because of the ingenious work of **Ivan Pavlov** in the 1920s. Normally, dogs salivate when exposed to food. Pavlov trained dogs to associate the sound of a bell with food. The result of this conditioning was that dogs would salivate upon merely hearing the sound of the bell, even though no food was present.
- **Operant conditioning**, also called **trial and error** learning, is another type of **associative learning**. An animal learns to associate one of its own behaviors with a reward or punishment and then repeats or avoids that behavior. The best-known studies involving operant conditioning were done by **B. F. Skinner** in the 1930s. In one study, a rat was placed into a cage containing a lever that released a pellet of food. At first, the rat would depress the lever only by accident and would receive food as a reward. The rat soon learned to associate the lever with the food and would depress the lever at will. Similarly, an animal can learn to carry out a behavior to avoid punishment. Such systems of rewards and punishment are the basis of most animal training.

Imprinting

Imprinting is learning that occurs during a **sensitive** or **critical period** in the early life of an individual and is **irreversible** for the length of that period. When you see ducklings following closely behind their mother, you are seeing the result of successful imprinting. Mother-offspring bonding in animals that depend on parental care is critical to the safety and development of the offspring. If the pair does not bond, the parent will not care for the offspring and the offspring will die. At the end of the juvenile period, when the offspring can survive without the parent, the response disappears.

STUDY TIP
The most commonly asked questions on this topic concern: • Imprinting • Fixed action pattern • Sign stimulus • Releasers

Classic imprinting experiments were carried out by **Konrad Lorenz** with geese. Geese hatchlings will follow the first thing they see that moves. Although the object is usually the mother goose, it can be a box tied to a string or, in the case of the classic experiment, it was Konrad Lorenz himself. Lorenz was the first thing the hatchlings saw, and they became **imprinted** on the scientist. Wherever he went, they followed.

SOCIAL BEHAVIOR

Social behavior is any kind of interaction among two or more animals, usually of the same species. It is a relatively new field of study, developed in the 1960s. Types of social behaviors are **cooperation**, **agonistic**, **dominance hierarchies**, **territoriality**, and **altruism**.

Cooperation

Cooperation enables the individuals to carry out a behavior, such as hunting, that they can do as a group more successfully than they can do separately. Lions or wild dogs will hunt in a pack, enabling them to bring down a larger animal than an individual could ever bring down alone.

Agonistic Behavior

Agonistic behavior is aggressive behavior. It involves a variety of threats or actual combat to settle disputes among individuals. These disputes are commonly held over access to food, mating, or shelter. This behavior involves both real aggressive behavior as well as ritualistic or symbolic behavior. One combatant does not have to kill the other. The use of symbolic behavior often prevents serious harm. A dog shows aggression by baring its teeth and erecting its ears and hair. It stands upright to appear taller and looks directly at its opponent. If the aggressor succeeds in scaring the opponent, the loser engages in submissive behavior that says, "You win, I give up." Examples of submissive behaviors are looking down or away from the winner. Dogs or wolves put their tail between their legs and run off. Once two individuals have settled a dispute by agonistic behavior, future encounters between them usually do not involve combat or posturing.

Dominance Hierarchies

Dominance hierarchies are pecking order behaviors that dictate the social position an animal has in a culture. This is commonly seen in hens where the alpha animal (top-ranked) controls the behaviors of all the others. The next in line, the beta animal, controls all others except the alpha animal. Each animal threatens all animals beneath it in the hierarchy. The top-ranked animal is assured of first choice of any resource, including food after a kill, the best territory, or the most fit mate.

Territoriality

A **territory** is an area an organism defends and from which other members of the community are excluded. Territories are established and defended by *agonistic behaviors* and are used for capturing food, mating, and rearing young. The size of the territory varies with its function and the amount of resources available.

Altruism

Altruism is a behavior that reduces an individual's reproductive fitness (the animal may die) while increasing the fitness of the group or family. When a worker honeybee stings an intruder in defense of the hive, the worker usually dies. However, this action increases the fitness of the queen bee, who lays all the eggs. How can altruism evolve if the altruistic individual dies? The answer is called **kin selection**. When an individual sacrifices itself for the family, it is sacrificing itself for relatives (the kin) that share similar genes. The *kin* are *selected* as the recipients of the altruistic behavior. They are saved and can pass on their genes. Altruism evolves because it increases the number of copies of a gene common to a related group.

Multiple-Choice Questions

Questions 1–5

For questions 1–5, choose from the list of scientists below.

   (A) B. F. Skinner
   (B) Karl von Frisch
   (C) Niko Tinbergen
   (D) Ivan Pavlov
   (E) Konrad Lorenz

   1. Studied communication in bees

   2. Classical conditioning

   3. Imprinting

   4. Fixed action pattern

   5. Operant conditioning

Questions 6–11

For questions 6–11, choose categories of animal behavior from the list. You may use each one more than once or not at all.

   (A) Fixed action pattern
   (B) Associative learning
   (C) Classical conditioning
   (D) Imprinting
   (E) Operant conditioning

   6. Pavlov's dogs

   7. One stimulus becomes linked to another

   8. Ducklings follow their mother

   9. Trial and error learning

10. Innate, highly stereotypic behavior that must continue until it is completed

11. Involves releasers and sign stimuli

---

12. Pavlov's dogs learned to associate hearing a bell with food. Simply hearing a bell caused them to salivate. This is an example of

    (A) habituation
    (B) operant conditioning
    (C) classical conditioning
    (D) imprinting
    (E) a fixed action pattern

13. Ethology is the study of

    (A) endocrinology
    (B) animal behavior and its relationship to its evolutionary history
    (C) the brain and nervous system
    (D) operant conditioning
    (E) biology and ethics

14. "Mary had a little lamb; its fleece was white as snow. And everywhere that Mary went, the lamb was sure to go." The behavior of the lamb is best described as

    (A) habituation
    (B) imprinting
    (C) operant conditioning
    (D) classical conditioning
    (E) fixed action pattern

15. You want to train your puppy to wait at the curb until you tell him to cross the road. Your friend advises you to give your dog a treat every time he does as you ask. Your friend is advising that you train the dog using

    (A) operant conditioning
    (B) classical conditioning
    (C) imprinting
    (D) fixed action pattern
    (E) habituation

16. A sign stimulus that functions as a signal to trigger a certain behavior in another member of the same species is called

    (A) a ritual
    (B) a fixed action stimulus
    (C) an inducer
    (D) a precursor
    (E) a releaser

17. To begin the mating dance, the male ostrich moves his head in a particular bobbing fashion. This initiates a specific response from the female, and the ritualized mating dance can begin. The male head bobbing is

    (A) an imprinting stimulus
    (B) a habituation
    (C) a fixed action stimulus
    (D) an inducer
    (E) a releaser

18. Animals that help other animals are expected to be

    (A) stronger than other animals
    (B) related to the animals they help
    (C) male
    (D) female
    (E) disabled in some way

19. Which of the following is related to altruistic behavior?

    (A) kin selection
    (B) fixed action pattern
    (C) a search image
    (D) imprinting
    (E) classical conditioning

Answers to Multiple-Choice Questions

1. **(B)** Karl von Frisch studied and named the waggle dance in bees.

2. **(D)** Classical conditioning involves learning to associate an arbitrary stimulus with a reward or punishment. Ivan Pavlov "trained" dogs to salivate at the sound of a bell.

3. **(E)** Imprinting is responsible for bonding between mother animals and their babies. It occurs during a sensitive period and disappears later in the animals' development. Konrad Lorenz is famous for imprinting ducks onto himself. They followed Lorenz as if he were their mother.

4. **(C)** A fixed action pattern enables an animal to engage in complex behavior automatically without having to "think" about it. An FAP is a sequence of behaviors that is unchangeable and usually carried out to completion once initiated. Niko Tinbergen is most associated with FAPs.

5. **(A)** Operant conditioning is also called trial and error learning. An animal learns to associate one of its own behaviors with a reward or punishment and then tends to avoid that behavior. The best known lab studies in operant conditioning were done by B. F. Skinner in the 1930s.

6. **(C)** An example of classical conditioning is how Pavlov trained his dogs to salivate at the sound of a bell.

7. **(B)** When one stimulus becomes associated with one response, it is associative learning.

8. **(D)** Konrad Lorenz imprinted his ducks onto himself. They thought he was their mother and followed him everywhere.

9. **(E)** Operant conditioning is trial and error learning.

10. **(A)** This is the definition of fixed action pattern.

11. **(A)** Sign stimuli initiate a fixed action pattern. Releasers are sign stimuli that are exchanged among members of the same species.

12. **(C)** Pavlov is known for his work in classical conditioning by inducing dogs to salivate at the sound of a bell.

13. **(B)** Ethology originated in the 1930s with naturalists who were studying animals in their natural habitat. They studied animal behavior and how it connected to evolution and ecology.

14. **(B)** The reason the lamb followed Mary everywhere is because the lamb was imprinted on Mary. The lamb will continue to follow Mary until the animal is mature enough to live on its own and the sensitive period has passed.

15. **(A)** Operant conditioning is a type of learning that is the basis for most animal training. The trainer encourages a behavior by rewarding the animal. Eventually, the animal will perform the behavior without necessarily receiving a reward.

16. **(E)** A releaser is a sign stimulus that triggers a fixed action pattern among members of the same species.

17. **(E)** The releaser is a signal between two members of the same species that initiates a fixed action pattern. In this case, the releaser is the bobbing of the head. In the stickleback, it is the red color on the underbelly.

18. **(B)** Animals that help other animals are engaging in altruistic behavior. Altruistic behavior is seemingly selfless behavior that may save kin carrying genes similar to the individual that sacrificed itself.

19. **(A)** When an individual sacrifices itself for the family or group, it is sacrificing itself for relatives (the kin) that share similar genes. The kin are selected as the recipients of the altruistic behavior. They are saved and can pass on their genes. Altruism evolves because it increases the number of copies of a gene common to a related group.

Free-Response Question

Explain Darwin's theory of evolution by natural selection. Each of the following refers to one aspect of evolution. Discuss each term and explain it in terms of natural selection. For a. and b., see the chapter "Evolution."

a. Convergent evolution
b. Insecticide resistance
c. **Fixed action pattern**
d. **Imprinting**

Typical Free-Response Answer

Note: Most likely, a free-response question based on the material would appear in a question about evolution. The question below is a modification of a free-response question that is answered in the chapter "Evolution." Terms you should include in your essay are in bold letters.

Natural selection favors behavioral patterns that enhance survival and reproductive success. If a behavior does not increase reproductive success, it will be selected against and disappear from the gene pool. Fixed action pattern and imprinting are two behavioral patterns that exist because they favor reproductive success.

A **fixed action pattern (FAP)** is an innate, highly stereotypic behavior that once begun, continues to completion. FAPs are initiated by external stimuli called **sign stimuli**. When these stimuli are exchanged between members of the same species, they are known as **releasers**. An example of an FAP studied by **Tinbergen** involves the stickleback fish, which attacks other males that invade its territory. The releaser for the attack is the red belly of the intruder. The stickleback will not attack an invading male stickleback lacking a red underbelly, but it will readily attack a non-fishlike wooden model as long as a splash of red is visible. If a fixed action pattern enables an animal to survive long enough to reproduce, that animal will have a selective advantage over animals that do not automatically carry out the behavior.

Imprinting is learning that occurs during a sensitive or critical period in the early life of an individual and is irreversible for the length of that period. Konrad Lorenz studied this behavior pattern in geese. The hatchling responds to the first thing it sees that moves, becoming imprinted on the mother and following her everywhere. In species with parental care, mother-offspring bonding is critical to the survival of the offspring. If bonding does not occur, the parent will not initiate care and the offspring will die. If an animal hatches offspring that do not become imprinted on her, she has not reproduced successfully. Her genes will then be selected against and will be lost from the gene pool.

LABORATORY
SECTION

Laboratory Review

GRAPHING

The purpose of showing data on a graph is to make it clear and easily understood. If you are directed to draw a graph on this exam, you will not be given line-by-line instructions and will have to rely on your own expertise. Here are some guidelines, reminders, and perhaps something new.

Label Your Graph

a. Title it.
b. Use the x-axis for the **independent variable**, the value that you control, such as time.
c. Use the y-axis for the **dependent variable**, the value that changes as a result of changes in the independent variable. For example, if in the course of an experiment, you take a measurement every minute and the chunk of potato gets heavier and heavier, then time is independent and the mass of the potato is dependent.

Make Sure Your Intervals on Any One Axis Are Equal

Table 18.1 shows some sample data.

Since the intervals of time are NOT equal, you cannot place them directly on the x-axis. You must create equal intervals, such as 3:00, 3:15, 3:30, 3:45, 4:00, and so on, and then plot the data.

TABLE 18.1

Time vs. Mass	
Time	**Mass**
3:00	5.30 g
3:15	5.80 g
3:45	6.24 g
4:15	7.10 g
4:30	8.25 g
5:00	9.01 g

Plot the Data Points, Then Draw a Best-Fit Line

If you are instructed to connect the dots, do so. Chances are, you will **not** be so instructed. You should draw a **best-fit line** (or curve). This is probably not something you ever encountered in math class because the math teacher always provided you with values to graph that formed a perfectly straight line. In science, when you collect data, the numbers rarely fall into a straight line. However, if the line is not straight, you cannot make predictions by simply extending the line. Somehow, you must translate your rough data into a straight line. Here is how.

STUDY TIP

Can you draw a best fit line? Learn how.

Plots your points as usual. Then, using a ruler (preferably a transparent one), draw a straight line that best shows the trend (slope) and that takes into account the location of all the data points. If one data point really differs from all the others, you may ignore it. Also, when drawing a best-fit line, your line does not have to pass through any data points and it does not have to pass through zero. Figure 18.1 shows a graph with data points plotted and best-fit line drawn.

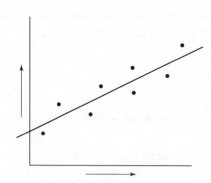

Figure 18.1   Best-Fit Line

DESIGNING AN EXPERIMENT

The laboratory-based essay question may direct you to design an experiment. This makes many students nervous. However, if you follow these guidelines, you will have a good basis for devising satisfactory experiment and writing a fine essay.

1. **State a clear hypothesis**, what you expect to happen.
2. The **experiment must be reasonable to carry out and must work**. For example, having to set up 100 fish tanks is not reasonable.
3. A "controlled" experiment must **have a control**. The control must be exactly like the experimental except for the single factor you are testing. The control can be an organism left untreated.
4. Any experiment must have **only one variable**. For example, in an experi-ment where you are testing the effect of various light intensities on the growth of several plants, light intensity must be constant for each plant. A heat sink must be placed between the light and the plant to absorb the heat from the light. Without something to absorb the heat from the light source, temperature will become a second variable.
5. **Have a large enough sample** to draw a reasonable conclusion. A sample of one organism is not acceptable.
6. **Experimental organisms must be as similar as possible.** They must all be of the same variety, size, and/or mass, whichever is appropriate. State that fact.
7. **State that the experiment must be repeated.** This reduces the possibility that a change occurred by chance, some random factor, or individual variations in the experimental organism.

LAB 1: DIFFUSION AND OSMOSIS

Introduction

Water potential (ψ)

Is measured in units of pressure called megapascals (MPa)
Measures the relative tendency for water to move from one place to another
Is the result of the combined effects of solute concentration and pressure

$$\begin{array}{ccccc} \text{water potential} & = & \text{pressure potential} & + & \text{solute potential} \\ \psi & = & \psi_p & + & \psi_s \end{array}$$

Water moves from high water potential to low water potential.

Objective

To investigate the processes of diffusion, osmosis, and water potential in a model membrane system and in living cells. This lab has four parts.

Part 1a

To observe diffusion across a semipermeable membrane

Fashion a length of semipermeable dialysis tubing into a bag, and fill it with two solutions: starch and glucose. Place the bag into a Lugol's iodine solution, and allow the system to stand for 30 minutes. The contents of the bag turn blue-black because iodine molecules are small enough to diffuse into the bag and react with the starch. However, the iodine solution in the beaker remains unchanged because the starch molecules are too large to diffuse out of the bag and mix with the iodine; see Figure 18.2.

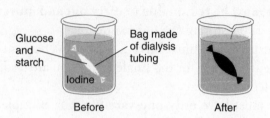

Before   After

Figure 18.2

Part 1b

To learn how to calculate the molarity of an unknown solution by observing osmosis

Fashion 6 lengths of semipermeable dialysis tubing into 6 bags, and fill them with sucrose solutions of varying molarities, 0.2 M, 0.4 M, 0.6 M, 0.8 M, and 1.0 M, and one with distilled water. (The molecules of sucrose are too large to diffuse through the dialysis membrane.) Remove as much air from each bag as possible to allow room for expansion. Blot and weigh each bag, and place into a beaker with enough distilled water to cover the bag. Allow them to sit for 30 minutes. Then remove them from the solutions, and blot and weigh them as before. Calculate the percent change in mass for each bag, record and graph the data with the dependent variable (percent change in mass) on the *y*-axis and the various molarities (independent variables) on the *x*-axis. The results show that the mass of the bag containing distilled water did not change because it is the control. *The mass of the other bags increased. The bag with the lowest molarity increased the least, and the bag with the highest molarity increased the most*; see Figure 18.3.

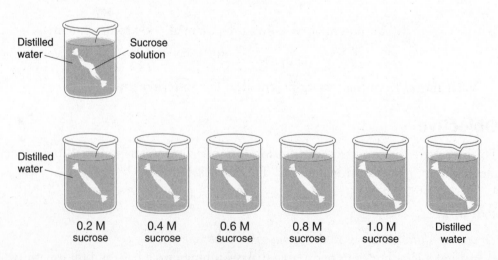

Figure 18.3

Part 1c

To learn how to determine the molarity of a living (potato) cell

Cut identical-size small squares of potato and weigh them. Place each into a beaker covered with 300 mL of the following sucrose solutions: 0.2 M, 0.4 M, 0.6 M, 0.8 M, and 1.0 M and one with distilled water. Allow them to sit overnight, then weigh them again. Some pieces of potato will gain mass and some will lose mass. Calculate the percent change in mass for each. Plot a best-fit line on a graph, with the molarities of the solutions, on the *x*-axis and the percent change in mass on the *y*-axis. The point of the best-fit line, which crosses the *x*-axis, is the molarity where there was no change in mass of the potato and represents the molarity inside the potato cell; see Figures 18.4 and 18.5.

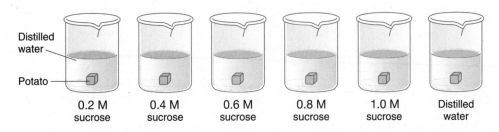

Figure 18.4

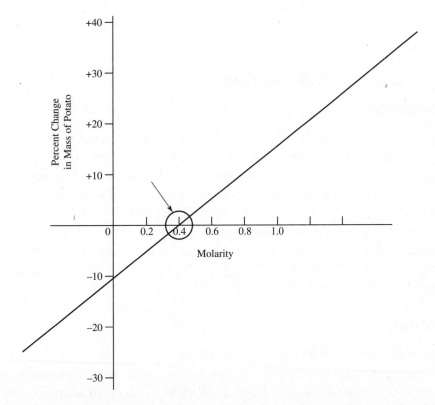

Figure 18.5   Determining the Molarity of the Potato Cell

Part 1d

To observe plasmolysis in a living plant (elodea) cell

Tear a thin piece of healthy elodea tissue, and prepare a wet mount. Locate a single layer of cells under the light microscope at 40× with clearly visible chloroplasts. While observing the slide, expose the tissue to 5% (hypertonic) sodium chloride solution. Observe changes in the cytoplasm of the cell as water suddenly diffuses out of the cell from high water potential to low water potential. The cytoplasm shrinks (plasmolysis), and the chloroplasts condense into a small circle. If you wash away the salt water solution and rehydrate the cells with distilled water, you can see the cell fill up with water (turgor) and appear similar to the way it was initially. However, the cell is no longer alive; see Figure 18.6.

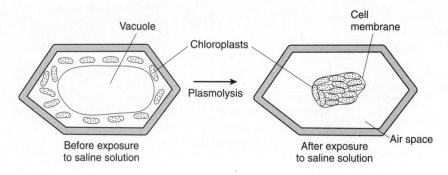

Figure 18.6

LAB 2: ENZYME CATALYSIS

Introduction

In general, the rate at which enzymes catalyze reactions is altered by environmental factors such as pH, temperature, inhibitors, and salt concentration. In this lab, the enzyme used is one that is found in all aerobic cells, **catalase**. Its function in the cell is to decompose hydrogen peroxide, a by-product of cell respiration, into a less-harmful substance. Here is the reaction: $2\ H_2O_2 \rightarrow 2\ H_2O + O_2\uparrow$. In this lab, the enzyme catalase is added to peroxide of known concentration and bubbling is observed. The bubbles are oxygen gas. (The test for the presence of oxygen gas is a glowing splint. It will burst into flames in the presence of O_2.) The quantity of bubbling is proportional to the activity of the catalase.

Objective

To observe the effects that acid and high heat have on enzyme function.

For this experiment, extract fresh catalase from liver tissue by homogenizing it in a blender and filtering the liquid. Keep in on ice at all times. In addition, you must establish how much active peroxide is in each peroxide sample because it decomposes readily on its own. This is done by titrating the hydrogen peroxide to which H_2SO_4 has been added with $KMnO_4$. When a persistent pink color remains, the titration is complete and the concentration of H_2O_2 is known because *the concentration of KMnO_4 is proportional to the concentration of H_2O_2.* The less H_2O_2 is left in the beaker, the less $KMnO_4$ will be needed to turn the liquid in the beaker pink or light brown.

Here is the equation that is the basis for the experiment.

$$5 \ H_2O_2 + 2 \ KMnO_4 + 3 \ H_2SO_4 \rightarrow K_2SO_4 + 2 \ MnSO_4 + 8 \ H_2O + 5 \ O_2$$

To observe the reaction to be studied, take three samples of fresh catalase. Leave one unchanged as a control, boil one, and acidify the other. Compare the effectiveness of decomposing H_2O_2 by observing the bubbling. To quantify how much H_2O_2 actually remains in each sample, run a titration. See Figure 18.7.

The results will demonstrate that high heat and strong acid denature the enzyme catalase. There will be no bubbling in those beakers.

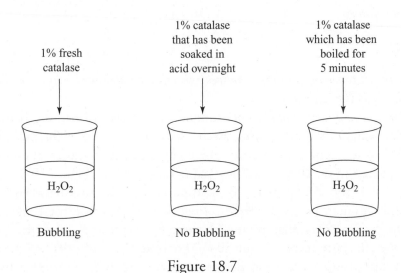

Figure 18.7

LAB 3: MITOSIS AND MEIOSIS
Introduction

For a review of mitosis and meiosis, see the chapter entitled "Cell Division."

Objective

In this lab, you investigate the processes of meiosis and mitosis.

Part 3a

Observe the stages of mitosis in the apical meristem of an onion root tip or a whitefish blastula. Determine how long a cell undergoing mitosis remains in each stage.

Study 200 cells undergoing interphase, prophase, metaphase, anaphase, and telophase under the light microscope at 40×, and make sketches of each phase. Tally how many cells are in each stage of mitosis. Consider that it takes 24 hours (1,440 minutes) for onion root tip cells to complete the cell cycle. Use the following formula to calculate the amount of time spent in each phase of the cell cycle from the percentage of cells at that stage.

Percentage of cells in a stage × 1,440 minutes = ___ minutes of cell cycle spent in a stage

Cells spend the greatest amount of time in interphase. Once cell division begins, cells spend the longest time in prophase and the least time in anaphase.

Part 3b

Prepare your own stained slide of an onion root tip.

Part 3c

Use different-colored beads, rubber tubing, and string to simulate meiosis, including crossing over and recombination. Use the beads to demonstrate the differences between meiosis and mitosis.

Part 3d

Calculate the distance in map units between the gene for spore color and the centromere in the fungus Sordaria fimicola.

Sordaria fimicola is an **ascomycete** fungus that consists of fruiting bodies called **perithecia**, which hold many **asci**. Each ascus encapsulates eight **ascospores** or **spores**. When ascospores are mature, the ascus ruptures, releasing the ascospores. Each ascospore can develop into a new haploid fungus; see Figure 18.8.

By observing the arrangements of ascospores from a cross between wild-type (black) and a mutant tan type *Sordaria fimicola* fungus, you can estimate the percentage of crossing-over that occurs between the centromere and the gene that controls the tan spore color. Count 50–100 hybrid asci, and calculate the distance in map units between the gene for spore color and the centromere using this formula.

Divide the number of crossover asci (2:2:2:2 or 2:4:2)
by the total number of asci × 100 and divide by 2

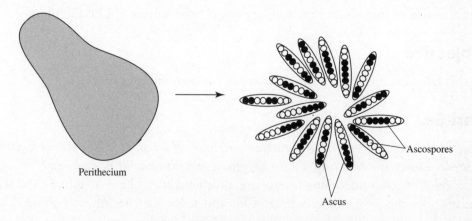

Figure 18.8

LAB 4: PLANT PIGMENTS AND PHOTOSYNTHESIS

Introduction

For a review of the light reactions of photosynthesis, review the chapter entitled "Photosynthesis."

Paper chromatography is a technique developed for separating two or more substances from a mixture. A solvent moves up the paper by capillary action, which occurs as a result of the attraction between the solvent molecules and the fibers of the paper. As solvent moves up the paper, it carries along any substances dissolved in it. The pigments are carried along at different rates because they are not equally soluble in the solvent and not equally attracted to the paper.

Objective

Separate plant pigments using chromatography.
Measure the rate of photosynthesis in isolated plant pigments or chloroplasts.

Part 4a

Separate plant pigments using chromatography.

Extract the photosynthetic pigments from some spinach leaves in either of two ways that work equally well. One technique is to mash spinach leaves with a cold sucrose solution in a blender and filter the mixture. Another is to use a dime placed over a spinach leaf to press some green pigment directly onto the chromatography paper.

Place the mixture of photosynthetic pigments extracted from a green leaf, onto the chromatography paper, and submerge the bottom edge of the paper into solvent. Allow the mixture to run up the paper. You should see four separate bands. **Beta-carotene** is carried along near the front because it is very soluble in the solvent and because it forms no hydrogen bonds with the cellulose in the paper. Another pigment, **xanthophyll**, runs farther from the solvent front because it is less soluble and forms more hydrogen bonds with the cellulose. **Chlorophylls *a* and *b*** bond even more tightly to the paper and run behind the other two pigments. See Figure 18.9.

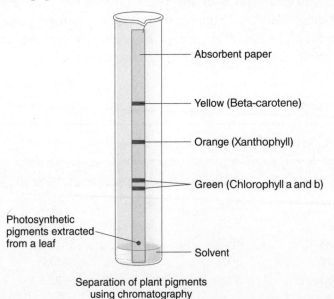

Separation of plant pigments
using chromatography

Figure 18.9   Separation of Plant Pigments Using Chromatography

Determining R_f Values

The R_f value is the relationship of the distance moved by a pigment to the distance moved by the solvent. This constant can be calculated for each of the four pigments using this formula:

$$R_f = \frac{\text{Distance pigment migrated (mm)}}{\text{Distance solvent migrated (mm)}}$$

Part 4b

*Measure the rate of photosynthesis in isolated plant pigments or chloroplasts using **DPIP**,* a chemical dye.

During the light reactions of photosynthesis, light energy boosts electrons from chlorophyll *a* to higher energy levels and into an electron transport chain. Missing electrons are replaced from molecules of water. The protons that remain reduce NADP, which carries them to the light-independent reactions for carbon fixation. The greater the light energy, the faster the rate of photosynthesis and the faster the reduction of NADP. In this experiment, a dye, DPIP (2,6-dichlophenol-indophenol) substitutes for NADP and is *used to measure the rate of photosynthesis*. Blue DPIP turns colorless as it is reduced. A **spectrophotometer** is used to measure the color change by measuring the percent transmittance of light of the DPIP solutions.

The chloroplasts or photosynthetic pigments that were extracted from spinach leaves are used in this experiment. Two beakers of pigment extract are prepared; one contains boiled extract and one contains fresh extract. Both are kept on ice at all times. Place fresh extract into two test tubes and an equal amount of boiled extract into another two test tubes. Place one tube of each pair in the dark and the other tubes in the light. To control the amount of light the plants are exposed to, set up a light source aimed at the two tubes. Also use a **heat sink** between the light source and the tube of plant extract to ensure that the energy from the light does not transform into heat and alter the conditions of the experiment. Every five minutes, take samples of each extract and measure the transmittance of light in the spectrophotometer. The lighter the color, the greater the transmittance of light.

Table 18.2 shows the transmittance of the five samples.

TABLE 18.2

Transmittance Values

Cuvette	0 min.	5 min.	10 min.	15 min.
1. Fresh/dark	30.5	31.6	32.2	32.9
2. Fresh/light	30.5	55.8	65.3	66.3
3. Boiled/dark	30.9	31.2	31.8	32.0
4. Boiled/light	30.9	31.6	32.0	32.1
5. No chloroplasts	30.5	30.5	30.5	30.5

Cuvette 2 is the only one to change. Transmittance of light increases because DPIP turns colorless as photosynthesis proceeds. Photosynthesis does not occur in cuvette 1 because it is placed in the dark. In cuvettes 3 and 4, the chlorophyll was destroyed by boiling. There are no chloroplasts in cuvette 5, therefore no photosynthesis occurred.

LAB 5: CELL RESPIRATION

Introduction

Cell respiration is the process by which living cells produce energy by the oxidation of glucose with the formation of the waste product CO_2. The rate of cell respiration can be measured by the volume of oxygen used. The cells used to study respiration in this experiment are sprouted sweet pea seeds, and the control consists of sweet pea seeds that are dormant and not carrying out respiration.

Objective

Construct a **respirometer** that will measure oxygen consumption of sprouted or unsprouted seeds. Observe the effects that different temperatures have on the rate of respiration, see Figure 18.10.

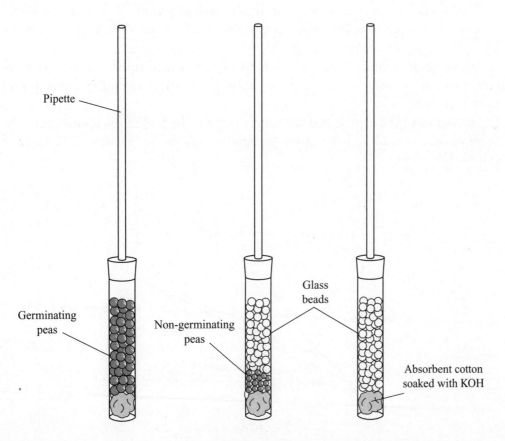

Pipette

Glass beads

Germinating peas

Non-germinating peas

Absorbent cotton soaked with KOH

Figure 18.10

Since sprouted seeds have absorbed water and are larger than unsprouted seeds, glass beads are added to the respirometer holding the unsprouted seeds to equalize the volume of empty space in both vessels. Set up six respirometers as shown in Table 18.3.

TABLE 18.3

Effects of Temperature on Respiration Rate		
Respirometer	Temperature of Water	Contents
1	25°C	Germinating seeds
2	25°C	Dry seeds + beads
3	25°C	Glass beads alone
4	10°C	Germinating seeds
5	10°C	Dry seeds + beads
6	10°C	Glass beads alone

Place a thin layer of cotton at the bottom of each respirometer soaked with KOH to absorb CO_2 that is given off by respiration. This allows the change in volume inside the respirometer to measure *only* oxygen consumption.

Submerge the respirometers in water. As oxygen is used up inside the respirometer, water flows into the respirometer through the pipette and is a measure of oxygen consumption.

The results of the experiment are simply stated. The higher the temperature, the more oxygen is used and the faster the rate of respiration. Figure 18.11 shows a graph of the data.

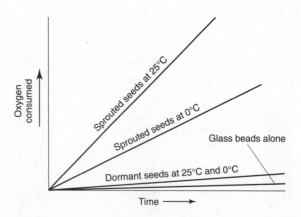

Figure 18.11

LAB 6: MOLECULAR BIOLOGY

Introduction

In this lab, you will explore the tools of biotechnology. Restriction enzymes have been extracted from bacteria which use them to fend off attacks by invading bacteriophages. Restriction enzymes cut DNA at specific recognition sequences. The DNA used in this experiment is **Lambda DNA**, which is extracted from the most widely studied bacteriophage. **Gel electrophoresis** is a technique that separates large molecules of DNA on the basis of their rate of movement through the agarose gel in an electric field.

Objective

Cut lambda DNA with different restriction enzymes, and transform bacteria into an antibiotic resistant form by inserting a plasmid into it.

Part 6a

Bacterial transformation

Make bacteria **competent** to uptake a plasmid by using **heat shock**, a combination of alternating hot and cold, in the presence of calcium ions that help disrupt the cell membrane. Once competent, the bacteria are incubated with plasmids that carry the resistance to a particular antibiotic. A control sample is run along with the experimental one, treated in exactly the same way except it does not get exposed to a plasmid. After the cells are allowed to rest, they are poured onto four petri dishes that contain Luria broth. Two petri dishes also contain antibiotic, two do not. Table 18.4 shows the setup.

TABLE 18.4

Experimental Data			
Test Tube	Plasmid	Petri Dish	Growth
1	+	+ Antibiotic	+
2	+	No Antibiotic	+
3 (Control)	–	+ Antibiotic	–
4 (Control)	–	No Antibiotic	+

Figure 18.12 is a sketch of the petri dishes showing the results of the experiment.

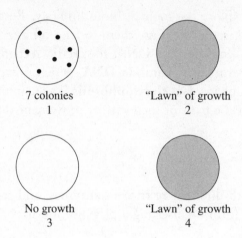

Figure 18.12

- Plate 1. These bacteria were incubated with plasmid. There are 7 colonies growing here. Each colony consists of bacteria that have been **transformed** and are resistant to the antibiotic because they absorbed the plasmid.
- Plate 2. This plate contains a lawn of bacteria. Any bacteria would grow on this plate; there is no antibiotic to prevent it. However, this culture did receive plasmid. We can assume that this plate contains bacteria resistant to the antibiotic as well as those susceptible to it.
- Plate 3. There is no growth on this dish. These bacteria were not incubated with plasmid and the plate contains no growth because the dish contains antibiotic that killed the bacteria.
- Plate 4. The bacteria were not incubated with plasmid and the plate also contains a lawn of growth because there is no antibiotic to kill the bacteria.

Part 6b

DNA Restriction Digest

In this experiment, lambda DNA is incubated with two different restriction enzymes and then run through an electrophoresis gel. Place the lambda DNA into three test tubes. Incubate one sample with *Eco*R1, one with *Hind*III, and leave one untreated as a control. The two restriction enzymes digest the DNA of two samples, cutting them into characteristic pieces called **restriction fragments**. The three samples are placed into the wells on an agarose gel and drawn from the **cathode** to the **anode** by electric current. After staining, a pattern results on the gel that reveals the **restriction fragments**. The control sample of DNA remains uncut and can be seen as a large block of DNA near the well in lane 4. The two other samples were cut with restriction enzymes, and the restriction fragment or banding pattern is visible. By measuring the distance of each fragment from the well (similar to determining the R_f value in paper chromatography) and comparing this distance against a standard, the size of the fragment of DNA can be determined; see Figure 18.13.

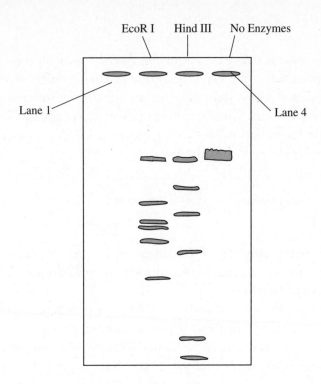

Figure 18.13

LAB 7: GENETICS—THE FRUIT FLY

Introduction

The life cycle of the fruit fly is about 12–14 days long from egg to adult. Hundreds of eggs are laid by each female, which then undergo three **instar** (larval) stages. They **molt** (shed) their skin after the first two instar stages and then **pupate**, forming a hard cocoon from which they hatch as an adult. The females remain virgins for about 8–12 hours after hatching. Therefore, if you wish to mate a female with a certain type of male, you must isolate the females soon after they hatch in order to ensure that they are virgins when you place them into a vial with the males. Common mutations in fruit flies are white eyed, which is **sex-linked recessive**; bar eyed, which is **sex-linked dominant**; and vestigial, which is **autosomal recessive**.

Objective

To determine the pattern of inheritance of a particular trait in fruit flies
To analyze your data using a **chi-square test**.

Part 7a

You must learn to distinguish males from females, how to manipulate the animals, and how to put them to sleep. Once the chosen animals mate, you must analyze 100 F_1 flies. Allow those F_1 individuals to interbreed and then analyze 100 F_2 flies. From this data, and the data from the results of the reciprocal cross, you can determine the pattern of inheritance of the cross you carried out.

Part 7b

Use a chi-square test on the results from the F_2 cross to determine if the variation in your results from what you expected is due to chance or to experimental error. The **null hypothesis** states that there is no significant difference between the observed and expected data. Good results mean you accept the null hypothesis. Poor results mean that the variation in your experiment was due to experimental error and you must reject the null hypothesis.

Apply the chi square test:

$$\text{chi-square} = \chi^2 = \frac{\Sigma\,(\text{observed} - \text{expected})^2}{\text{expected}}$$

Determine how many **degrees of freedom** are in your experiment. Degrees of freedom are the number of phenotypes minus 1. If you have four phenotypes, you have three degrees of freedom.

Refer to a critical values chart, such as Table 18.5, to determine if your error was due to chance or experimental error. If your calculated chi-square value is less than the critical value from the table, then the null hypothesis is accepted. If the value is equal to or greater than the critical value, you have experimental error and must find reasons for it.

TABLE 18.5

Critical Values of the Chi-Square Distribution

Probability	Degrees of Freedom (df)				
	1	2	3	4	5
0.05	3.84	5.99	7.82	9.49	11.1
0.01	6.64	9.21	11.3	13.2	15.1
0.001	10.8	13.8	16.3	18.5	20.5

LAB 8: POPULATION GENETICS AND EVOLUTION

Introduction

Hardy-Weinberg theory describes the characteristics of a population that is stable or nonevolving. In a stable population, the allelic frequencies do not change from generation to generation. The Hardy-Weinberg equation enables us to calculate the frequencies of alleles in the population. The equation is:

$$p^2 + 2pq + q^2 = 1 \qquad \text{or} \qquad p + q = 1.$$

Sample problems can be found in the chapter entitled "Evolution" earlier in this book or at the end of this chapter.

In this lab, your class will represent an entire breeding population, and the allele in question is the ability to taste the chemical PTC. A bitter taste indicates the dominant allele. A person who is homozygous recessive cannot taste PTC.

LAB 9: TRANSPIRATION

Introduction

Transpiration is the loss of water from a leaf. The **theory of transpirational pull–cohesion tension** explains that as one molecule of water evaporates from a leaf, one molecule of water is pulled into the plant through the roots. An increase in sunlight increases the rate of transpiration by causing more water to evaporate from the leaves and also increases the rate of photosynthesis. Environmental conditions that increase photosynthesis and transpiration are low humidity, wind, and increased sunlight.

Objective

In this lab, you will measure the rate of transpiration in a plant under various conditions in a controlled experiment.

Insert two-week old bean seedlings (*Phaseolus vulgaris*) into a **potometer**, and measure the amount of water used by the plants under varying conditions of humidity (misting the leaves), light (a strong lightbulb shining on the plant), and wind (a fan). Set up several potometers: one for each variable and one with only ambient conditions as the control; see Figure 18.14.

Here is what you can expect to happen.

- Increasing humidity, by placing a bag over or misting the leaves, will decrease transpiration.
- Increasing light will increase transpiration.
- Increasing wind with a fan will lower the humidity around the leaves and thereby increase transpiration.

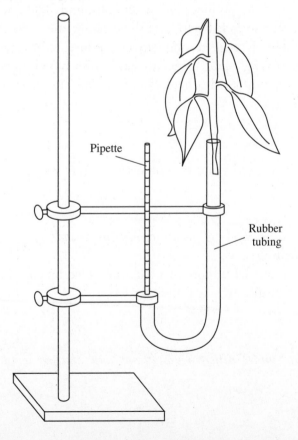

Figure 18.14   Potometer

When you increase the light intensity, make sure that you also do not increase the temperature surrounding the plant, thus, introducing another variable. To accomplish this, place a **heat sink**, a large bowl or beaker of water, between the light and the plant to absorb the heat while allowing the light to pass through.

Since each plant does not have the same leaf surface area, your results are meaningless unless you can demonstrate consistency. You must measure the area of the leaves on the plants and calculate the water loss in terms of square meters. To measure the surface area of the leaf, trace all the leaves from each plant onto graph paper with squares of known size. Construct a graph with "water loss (mL/m^2)" on the y-axis and "time (min)" on the x-axis.

LAB 10: PHYSIOLOGY OF THE CIRCULATORY SYSTEM

Introduction

Blood pressure is an important measurable aspect of the circulatory system. The arterial blood is measured by a **sphygmomanometer**. This device consists of an inflatable cuff connected to rubber hoses, a hand pump, and a pressure gauge graduated in millimeters of mercury. The examiner listens to the sounds of blood flow in the brachial artery by placing the stethoscope inside the elbow, below the biceps.

At rest, blood normally flows through arteries without impediment and there is no sound when one listens. When the sphygmomanometer cuff is inflated to a pressure above the systolic pressure, the flow of blood is stopped and the artery is also silent. As the pressure of the cuff is released, blood is pushed through compressed vessels and **heart sounds**, known as **Korotkoff sounds**, can be heard. There are five phases of sounds based on loudness and quality of sound. The systolic measurement is taken when the first sounds begin, and the diastolic measurement is taken when the sounds end. The average resting blood pressure for every adult is **120/80 (systole/diastole)**. Young people generally have lower blood pressure, while elderly people who have accumulated atherosclerotic plaque in their blood vessels have higher blood pressure.

Objective

Explore blood pressure and pulse in humans and heart rate in an arthropod.

Part 10a

Measure blood pressure under different conditions: standing, sitting, reclining, and during and after exercise.

Compare your results with those in a chart provided by your instructor to determine your level of cardiovascular fitness.

Part 10b

Measure pulse rate under different conditions and analyze it in terms of relative fitness.

Work in pairs and take each other's pulse lying down (supine), sitting, standing, and during exercise. Explore the **baroreceptor reflex**. The baroreceptor reflex is one of the body's homeostatic mechanisms for maintaining blood pressure.

Take your partner's pulse while lying down and then again *immediately* upon standing. There is an increase in pulse initiated by baroreceptors in the carotid artery and aortic arch.

Normal resting pulse is around 70. However, very active people who have good cardiac fitness have lower pulse rates because they pump more blood with each contraction than do people who are in poor physical condition. Blood pressure is lowest when lying down, higher when standing, and highest when exercising.

Part 10c

Measure heart rate in the water flea, Daphnia magna.

In **ectotherms**, there is a direct relationship between the rate of many physiological activities and environmental temperature. For every 10°C increase in environmental temperature, metabolic rate doubles. This is known as the Q_{10} **effect**, with Q_{10} being the multiple by which a particular enzymatic reaction or overall metabolic process increases with a 10°C increase in body temperature. For example, if the rate of glycogen hydrolysis is 1.75 times greater in frogs at 30°C than at 20°C, then the Q_{10} for that reaction is 1.75.

Place the **daphnia** into a depression microscope slide with water at 10°C and observe it under a light microscope. Measure the heart rate as you count the beats of the heart, which is clear through the animal's transparent "skin." Add warm water and measure the change in heart rate. Continue to collect data and record it in a chart. Graph the data with "temperature" on the *x*-axis and "heart rate beats per minute" on the *y*-axis. This is not a controlled experiment. It is merely observation of a phenomenon.

LAB 11: ANIMAL BEHAVIOR

Introduction

Ethology is the study of animal behavior. See the chapter entitled "Animal Behavior" for more information. **Orientation behaviors** place an animal in a more favorable environment, toward or away from a stimulus. A **taxis** is a response seen in simple organisms like Protista, where the organism responds to changes in light, heat, sound, and so on. **Kinesis** is a general term for random movement which does not result in orientation with respect to a stimulus. **Agonistic behavior** is exhibited when animals respond to each other in an aggressive or submissive way. **Mating behaviors** involve a complex series of activities.

Objective

In this lab, you will observe and describe some aspects of animal behavior mentioned above. There are two parts to this lab.

Part 11a

Explore habitat preference.

To working with pill bugs or vestigial-winged fruit flies, fashion a **choice chamber** by attaching two petri dishes together and cutting an opening between them. Line one chamber with moist filter paper and one with dry filter paper. While using a soft brush, place 5 bugs onto each side of the chamber and cover the chamber. Do not interfere with the animals in any way, merely observe and describe their behavior, using the terms in the introduction as a guide. Record the numbers of pill bugs in each side of the chamber in 30-second intervals. Graph this data.

Part 11b

Study mating behavior.

Isolate 4 newly hatched female flies (*Drosophila melanogaster*) in a culture tube with food for 2–3 days. Then place them into a culture tube with 4 male flies. Place the vial on its side and observe them under a dissecting microscope.

The animals follow a strict behavioral pattern. Five phases can be distinguished: orientation, male song (wing vibration), licking the female genitalia, attempted copulation, copulation, and rejection (extrusion of the ovipositor).

Observe and record your observations. Include quantitative analysis; you may include pictures.

LAB 12: DISSOLVED OXYGEN AND AQUATIC PRIMARY PRODUCTIVITY

Introduction

Primary productivity of an ecosystem is defined as the amount of light energy converted to chemical bond (glucose) energy by an ecosystem's autotrophs during a given period of time. The rate of oxygen production can be used as a basis for measuring primary productivity. Since some of the oxygen produced is used for cell respiration, **net primary production (NPP)** equals **gross primary production (GPP)** minus the energy used by the **primary producers for respiration** (R).

$$NPP = GPP - R$$

Objective

In this experiment, you will measure primary productivity based on changes in dissolved oxygen in a controlled experiment and investigate the effects of changing light intensity on primary productivity in a controlled experiment. Kits to test dissolved oxygen are commercially available.

Fill bottles with lake water, stream water, or water containing algae. Mark the bottles I (initial), D (dark), 100%, 65%, 25%, 10%, and 2%. Determine the oxygen content of the bottle labeled I. This is the amount of oxygen the water had initially. Cover the bottle labeled D with aluminum foil. If no light enters this bottle, then the change in dissolved oxygen (DO) will be due only to respiration. Cover the remaining bottles with varying amounts of screening to control the light entering those bottles. Place the bottles on their sides and leave for 24 hours under room light, then determine the DO in all bottles. Use the information from the bottle labeled D to help calculate net primary productivity.

Multiple-Choice Questions

1. If a piece of potato is allowed to sit out on the counter and dry out, the water potential of the potato cells would

   (A) increase
   (B) decrease
   (C) remain the same
   (D) increase, then decrease
   (E) It cannot be determined.

2. Which of the following correctly indicates the gradient of water potential from the highest to the lowest?

   (A) root, soil stem, leaf, air
   (B) soil, root, stem, leaf, air
   (C) air, leaf, stem, root, soil
   (D) root, leaf, stem soil, air
   (E) air, leaf, root, soil, stem

3. In a diffusion lab with 6 bags of dialysis tubing, why do you use the percent change in mass rather than simply change in mass?

   (A) Doing so was arbitrary and applies to these instructions only; it really does not matter.
   (B) Doing so is the convention.
   (C) The percent change in mass is not as accurate.
   (D) The bags did not all weigh exactly the same mass at the start.
   (E) These were the controls in the experiment.

4. Which is the test for the presence of oxygen?

   (A) Limewater turns from clear to cloudy.
   (B) A glowing splint bursts into flames.
   (C) Phenolphthalein turns from clear to pink.
   (D) Cobalt chloride turns from blue to white.
   (E) A lighted splint will make a popping noise.

5. During the titration of catalase with $KMnO_4$, what is the role of H_2SO_4?

   (A) It denatures the catalase, thus stopping the reaction.
   (B) It denatures the H_2O_2.
   (C) It decomposes the H_2O_2, thus there is none left to react with the catalase.
   (D) It is a strong base that denatures the enzyme.
   (E) It acts as a dye that makes the product visible.

6. Which is not an important event that occurs during or as a result of meiosis?

   (A) independent assortment of alleles
   (B) crossing-over
   (C) dividing the chromosome number by two
   (D) production of gametes
   (E) cloning of cells

7. Refer to the data table in Lab 4, the experiment that measures the rate of photosynthesis using DPIP. Which cuvette(s) in the experiment represents the control?

   (A) 1 only
   (B) 1 and 3
   (C) 1 and 2
   (D) 1, 3, and 5
   (E) 5 only

Question 8

The list below contains choices for question 8.

   (A) KOH
   (B) the vial containing glass beads
   (C) unsprouted (dormant seeds)

8. Which component of the experiment that uses respirometers is the control?

   (A) A and B
   (B) both B and C
   (C) C only
   (D) all of the above
   (E) A only

9. If two bands on an electrophoresis gel have nearly the same base pair size, how might you best separate them?

   (A) Run them in a gel that is more concentrated.
   (B) Run them in a gel that is less concentrated.
   (C) Turn up the current and run them through the gel faster.
   (D) Stain them.
   (E) Leave them unstained.

10. Which is true of the genetics of fruit flies?

   (A) The correct order of development is egg → pupa → larva.
   (B) There are four instar stages in fruit flies.
   (C) The males are larger than the females.
   (D) The females remain virgins for only 4 hours after hatching.
   (E) White-eyed trait is sex-linked recessive.

11. The allele for the hair pattern called widow's peak is dominant over the allele for no widow's peak. In a population of 1,000 individuals, 841 show the dominant phenotype. How many individuals would you expect to be hybrid for widow's peak?

    (A) 16%
    (B) 24%
    (C) 48%
    (D) 100%
    (E) none

12. In a certain population, the dominant phenotype of a certain condition occurs 91% of the time. What is the frequency of the dominant allele?

    (A) 0.3
    (B) 3%
    (C) 0.7
    (D) 7%
    (E) 49%

13. If a population is in Hardy-Weinberg equilibrium and the frequency of the recessive phenotype for a particular trait appears in 16% of the population, what will the frequency for the **hybrid condition** be in 500 years?

    (A) 16%
    (B) 24%
    (C) 48%
    (D) 64%
    (E) It is impossible to say.

14. All of the following are correct about the circulatory system EXCEPT

    (A) High blood pressure is dangerous because it damages delicate blood vessels in the kidneys and brain.
    (B) In general, athletes have a slower pulse rate than the population average.
    (C) The baroreceptor reflex is responsible for raising blood pressure if you quickly stand up from a lying down position.
    (D) Systole is a measurement of pressure when the ventricles of the heart relax.
    (E) Korotkoff sounds are sounds heard when taking blood pressure.

15. The Q_{10} effect refers to which of the following?

    (A) gel electrophoresis
    (B) habitat selection
    (C) paper chromatography
    (D) enzyme catalysis
    (E) body temperature

Answers to Multiple-Choice Questions

1. **(B)** As tissue drives out, its water potential decreases. Water flows from higher water potential and osmotic potential to lower. Water would tend to flow into the dehydrated cells.

2. **(B)** Water moves from the highest water potential to the lowest. In a plant, water moves from the soil to the roots to the stems and leaves and into the air spaces in the spongy mesophyll and out to the environment.

3. **(D)** Since the bags did not all weigh the same mass, the only way to control for the variation is to show change in mass as a percentage.

4. **(B)** A lighted splint in the presence of oxygen is explosive. Limewater is the test for CO_2. Phenolphthalein is the test for a base. Cobalt chloride tests for water or moisture. A lighted splint pops in the presence of hydrogen.

5. **(A)** The acid H_2SO_4 denatures the enzyme catalase, which stops the reaction.

6. **(E)** A clone is an identical copy. The cloning of cells is characteristic of mitosis, not meiosis. The purpose of meiosis is to produce gametes that are different from the parent cells. All the other events are important parts of meiosis.

7. **(D)** There are two sets of controls. Cuvette 1 is a control for cuvette 2. Cuvette 3 is a control for cuvette 4. Cuvette 5 is a control for cuvettes 1 and 3. Cuvettes 1, 3, and 5 should get similar results, and they do.

8. **(B)** The unsprouted seeds are used as a comparison for the sprouted seeds. However, the glass beads that were introduced to equalize the volume in the various containers, must be included to demonstrate that they have no effect on the results in vials #2 and #5.

9. **(A)** If the gel is more concentrated, it will present more of an impediment and separate the two bands more effectively. Another method would be to lower the current and allow them to run very slowly and overnight. One must simply try both techniques to see which works better.

10. **(E)** White-eyed trait is sex-linked recessive. The correct order of development is egg → larva or instar → pupa. There are 3 instar stages in the fruit fly's development. The female is larger than the male fruit fly. Females remain virgin for about 10 hours.

11. **(C)** If approximately 84% of the population shows the dominant trait, then 16% show the recessive phenotype and are homozygous recessive. Therefore, the frequency of the recessive allele is 0.4 and of the dominant allele is 0.6. The frequency of the hybrid in the Hardy-Weinberg equation is $2pq$: $2(0.4 \times 0.6) = .48 = 48\%$.

12. **(C)** The frequency of a dominant allele is q; the frequency of a recessive allele is p. If the dominant phenotype appears 91% of the time, then the recessive phenotype appears 9% of the time. If $q^2 = 0.9$, then $q = 0.3$ and $p = 0.7$.

13. **(C)** The population is in equilibrium, it will not change. If $q^2 = .16$, then $q = .4$ and $p = .6$. Therefore, the frequency of the hybrid is 48%.

14. **(D)** Systole is a measure of blood pressure when the ventricles of the heart contract, not relax.

15. **(E)** The Q_{10} effect refers to body temperature and its relationship to metabolic rate in ectotherms.

EXTRA PREPARATION
FOR THE AP EXAM

Five Themes to Help You Write a Great Essay

CHAPTER 19

- Energy transfer
- Relationship of structure to function
- Regulation
- Interdependence of nature
- Evolution

In preparing for the AP Exam, you should be aware that there are major concepts that unify the study of biology and that appear over and over, particularly in Part II questions. You should be prepared to discuss these ideas and to give examples of each at the **molecular**, **cellular**, **organ**, **organism**, and **population** level. Below is a list of the major themes and a review of the ideas in each.

> Energy transfer
> Relationship of structure to function
> Regulation
> Interdependence of nature
> Evolution

> **STUDY TIP**
>
> Studying the material in this chapter will really help you with the essay part of the test.

ENERGY TRANSFER

At the Molecular/Cellular Level

- During the process of photosynthesis, plants convert solar energy to chemical bond energy. This is an energy conversion upon which all life depends.
- During the electron transport chain of cellular respiration, the exergonic flow of electrons is coupled with the energy-absorbing formation of a proton gradient.
- A proton gradient across membranes in mitochondria and chloroplasts powers the synthesis of ATP through a process called oxidative phosphorylation.
- Energy is transferred from ATP to energy-absorbing reactions. For example, ATP provides the energy to pump sodium and potassium ions across the cell membrane of the axon of a neuron.

At the Organ/Organism/Population Level

- Plants transfer light energy into chemical bond energy.
- Energy is transferred from producers to animals through the food chain. Food chains are never more than four or five trophic levels because energy is lost along the chain.
- Energy transfer is about 10% from one trophic level to the next.

RELATIONSHIP OF STRUCTURE TO FUNCTION

At the Molecular/Cellular Level

- The structure of the cell membrane (fluid-mosaic model) explains how various substances can cross the membrane, into or out of the cell.
- The structure of DNA, two strands connected by hydrogen bonds, enables scientists to understand how DNA replicates itself.
- The winding internal membrane (cristae) of the mitochondria allows for vastly increased ATP production.
- The many layers of thylakoid membranes that make up the grana of chloroplasts are responsible for increased absorption of light by thousands of photosystems.
- A nerve cell has a special shape to accommodate its function of transmitting impulses long distances. Some nerve cells in humans are more than 1 foot (3 m) long.
- Sclerenchyma cells in plants have almost no cytoplasm. They are made of very thick, primary and secondary cell walls impregnated with lignin and provide support for a plant.
- Lipid-storing cells consist almost entirely of a vacuole that stores fat. The nucleus is pushed to the edge, and there is very little cytoplasm.
- The particular conformation of an enzyme determines what substrate it will act upon. If the enzyme is denatured (changes its shape), it will not function at all.
- When hemoglobin combines with one substrate molecule (oxygen), the binding triggers a slight conformational change in the hemoglobin that amplifies hemoglobin's ability to bind with more oxygen.

At the Organ/Organism/Population Level

- The long small intestine in humans provides increased surface area for enhanced digestion and absorption.
- The loop of Henle in the mammalian nephron is a long, thin tube involved in osmoregulation and the production of urine. The longer the loop, the more water is reabsorbed back into the body. Mammals that evolved in very dry environments have longer loops than their cousins who evolved in wetter environments.
- The type of teeth an animal has reflects its dietary habits. Rodents that are vegetarians have sharp, chisellike incisors that facilitate gnawing and no canines. Predatory carnivores, like the large cats, have large canines that can kill prey and rip an animal apart. Humans are omnivores and have three different types of teeth suited to eating a variety of foods.

REGULATION

At the Molecular/Cellular Level

- The operon is a major regulatory mechanism of gene transcription in bacteria.
- Examples of cells communicating are gap junctions, tight junctions, and desmosomes.
- The plasma membrane, by virtue of its structure, controls what enters and leaves a cell.
- Hox genes control the expression of other genes and the development of a growing embryo.
- Enzymes control the rate of reactions by speeding them up or slowing them down. They speed up reactions by lowering the activation energy. They slow down or inhibit reactions through competitive, noncompetitive, and allosteric inhibitions.
- The cell cycle and cell division are under strict control by a built-in system of checkpoints or restriction points in G_1, G_2, and M. This timing is regulated by cyclins and cyclin-dependent kinases (CDKs).
- The glycocalyx of the cell membrane helps regulate cell and population growth. If cells become too crowded, they stop dividing. This is called contact inhibition.

At the Organ/Organism/Population Level

- The nervous and endocrine systems regulate the body overall.
- Negative feedback mechanisms in the endocrine system maintain homeostasis.
- Body temperature, breathing rate, and heart rate are under the control of the brain and nervous system.
- Hormones like ecdysone and juvenile hormone control the development of a metamorphosing insect.
- Plant hormones, such as auxins, ethylene, gibberellins, abscisic acid, and cytokinins, help to maintain homoeostasis in plants.
- The role of the kidney and the nephron is one of osmoregulation.
- Countercurrent heat exchange maintains stable body temperatures in polar animals.
- Apoptosis is programmed cell death. This built-in suicide mechanism that is controlled by genes is essential to the normal development of the nervous and immune systems in animals.

INTERDEPENDENCE OF NATURE

At the Molecular/Cellular Level

- Plants and animals are mutually dependent. Photosynthesis requires CO_2 and produces O_2; while respiration requires O_2 and releases CO_2.

At the Organ/Organism/Population Level

- All life depends on the ability of plants as producers. The solar energy that plants convert to chemical energy is passed along the food chain from consumer to consumer.
- Organisms depend on other organisms through various symbiotic relationships: mutualism, commensalism and parasitism.
- Decomposers, such as bacteria and fungi, recycle nutrients within an ecosystem. Without them, life would not exist.

EVOLUTION

At the Molecular/Cellular Level

- The theory of endosymbiosis states that millions of years ago, small prokaryotic cells were engulfed by larger prokaryotic cells. Those small cells became the chloroplasts and mitochondria that are critical for life as it is known.
- Structural and functional modifications for dry environments include:
    C4 and CAM photosynthesis
    Kranz anatomy and the Hatch-Slack pathway
    Seeds
- Countercurrent exchange is a mechanism that evolved to enhance diffusion rates where necessary.

At the Organ/Organism/Population Level

- Mutations and genetic recombination generate heritable variation that is subject to natural selection.
- The peppered moth population changed drastically from white to dark in less than 50 years in response to the industrialization in England. The phenomenon is referred to as *industrial melanism.*
- Bacteria become resistant to antibiotics quickly after exposure to antibiotics. After a course of antibiotics, bacteria that are not resistant to the antibiotic are killed, while those that happen to be resistant to the antibiotic reproduce rapidly with no competition. This results in a new population that is resistant to the antibiotic.
- The current system of taxonomy reflects the current understanding of phylogenetic relationships among organisms.
- Plants and animals have evolved many modifications to move from sea to land. Plants have evolved a waxy cuticle, supporting tissue like collenchyma and sclerenchyma, roots, and vascular vessels. Greatly reduced leaf size in conifers reduces water loss.
- Animals have evolved limbs for support and movement. Scales minimize water loss, and lungs allow breathing. Nephrons or nephridia produce hypertonic urine to remove nitrogenous waste and to conserve water.
- Social behavior is any kind of interaction among two or more animals, usually of the same species. Types of social behaviors include cooperation, agonistic, dominance hierarchies, territoriality, and altruism.

Learn How to Grade an Essay

CHAPTER 20

- Sample essay A
- Analysis of essay A
- Sample essay B
- Analysis of essay B

INTRODUCTION

If you want to write a good essay, you have to know how it is graded.

Here are two sample essays similar to those written by students. The essays have already been graded. Each time a student gives the correct terminology or a correct explanation of a concept, a point is given in the margin. Below each answer is an explanation of what the student did or did not receive credit for. For sample essay B, cover the numbers in the margin on the right, read the essay, then try to grade it yourself.

PRACTICE

How many points would you give this essay?

Question: *Explain, in detail, how enzymes speed up and slow down reactions.*

SAMPLE ESSAY A

Enzymes, which are proteins, help a reaction speed up. It does this by using   **1**
a catalyst. The function of enzymes relates to the structure of the enzyme. Certain enzymes can help only those that are right for it. They work by "lock and key." Enzymes speed up reactions by lowering the reaction barrier. They slow down reactions by raising the reaction barrier. Enzymes attach to a specific substrate. They can also inhibit reactions through competitive   **1** and noncompetitive inhibition. There is also denaturization. Enzymes are affected badly by heat and strong pH. In the wrong environment, they will
be denatured, their shape will change, and they will not work in any   <u>**1**</u>
reaction.   **Total**
   3 pts.

Analysis of Essay A

The numbers on the right side of the essay represent points the student earned. Here is what the student received credit for:

1 pt.   Enzymes are proteins.

1 pt.   Enzymes attach to a specific substrate.

1 pt.   Enzymes can be denatured (their shape altered) by strong heat or pH. Because they are denatured, they no longer function.

---

3 pts.   Total

Here is what the student seemed to be trying to say but for which he/she received no credit.

In the first sentence, the student states that enzymes speed up reactions. That is true. However, that information was given in the question.

Enzymes *are* catalysts; they do not work by *using* a catalyst.

Enzymes do not work by "lock and key." They work by "induced fit."

Enzymes speed up reactions by lowering the energy of activation and facilitating more effective collisions.

Although it is true that enzymes slow reactions by competitive and non-competitive inhibition, these terms were only mentioned, not explained.

SAMPLE ESSAY B

Enzymes are organic catalysts. They can speed up a reaction by lowering the $\quad$ 1
energy of activation, but they cannot make a reaction happen that wouldn't $\quad$ 1
normally happen. Enzymes are globular proteins that exhibit tertiary struc- $\quad$ 1
ture due to various intramolecular attractions, including Van der Waals $\quad$ 1
forces, hydrophobic and hydrophilic interactions, and disulfide bridges.
They have a particular conformation or 3-D shape that determines how they
function. Since enzymes are not used up in a reaction, an organism has a $\quad$ 1
limited number of enzymes. Enzymes have *specificity*, meaning that one
enzyme catalyzes one specific chemical reaction. The molecule an enzyme $\quad$ 1
bonds to is called a substrate, and when the two connect an enzyme-substrate $\quad$ 1
complex is formed. An enzyme bonds to a substrate at an active site. The
process by which enzymes actually catalyze reactions can be explained using
the induced-fit model. Since all reactions need effective collisions for them
to occur, the angle at which particles collide with each other is important.
Enzymes, by binding with the substrate, increase the angle of collisions.
Because the collisions between particles become more effective, less energy
is needed to activate the reaction. An example of an enzyme that catalyzes $\quad$ 1
a reaction is catalase. Catalase assists H_2O_2 into breaking down into H_2O
and O_2 gas. The catalase binds with the substrate H_2O_2 at the active site. $\quad$ 1
Enzymes are named after the substrate they work on plus the suffix "ase." $\quad$ 1
For example, maltase is the name of the enzyme that breaks apart maltose
into glucose molecules. $\quad$ 1

In addition to speeding up reactions, enzymes can also slow them down or inhibit them competitively, noncompetitively, or allosterically. In competitive inhibition, the enzyme has one active site but two substrates. If there is a higher concentration of substrate *A* than substrate *B*, the enzyme will catalyze a reaction involving substrate *A*. In noncompetitive inhibition, the enzyme has two active sites and two substrates. When one substrate goes into its active site, it blocks the other active site, preventing the other substrate from fitting into it. In allosteric regulation, a modifier is needed to bind with the enzyme at the allosteric site, to change its conformation so it can bind with its substrate. If the modifier is not there, the substrate's reaction will be impeded.

1

1

1

I have explained the structure of enzymes and how they can speed up and slow down reactions.

Total 14 pts.

Analysis of Essay B

The numbers on the right side of the essay represent points the student earned. Although this is not a beautifully written essay, there is a substantial amount of correct information given, and it would earn full credit, 10 points.

> **REMEMBER**
>
> Explain or define all scientific terms you use.

1 pt.	Enzymes are organic catalysts.
1 pt.	They speed up reactions by lowering the energy of activation.
1 pt.	They cannot make a reaction happen that would not normally happen.
1 pt.	Enzymes are globular proteins that exhibit tertiary structure due to—
1 pt.	Various intramolecular attractions, as listed.
1 pt.	Form/conformation dictates function.
1 pt.	Enzymes are not used up and are needed in limited quantity.
1 pt.	Enzymes are specific.
2 pts.	Enzymes bond to a substrate at an active site, forming an enzyme-substrate complex.
1 pt.	More effective collisions result.
1 pt.	Example: catalase
1 pt.	Competitive inhibition is correctly explained
1 pt.	Noncompetitive inhibition is correctly explained.

What the Student Did Not Get Credit For

The student did not receive credit for mentioning "induced fit," because the concept was not explained.

Allosteric inhibition is not clearly explained. The following should be added. "There are two active sites, one for the substrate and one for the allosteric inhibitor."

The student states, "Since enzymes are not used up . . . an organism has a limited number of enzymes." This would have been better stated as "Since enzymes are not used up, very small amounts are needed by the body."

Students can receive points for discussing something in depth. The mention of "catalase" as an example of an enzyme would give the student a point for "depth of understanding." The instructions did not state to include an example; but doing so was a good idea.

Notice, there is no mention of denaturing as in the first essay. **You do not have to discuss every single point to get full credit**. You gain points as you state correct information using related terminology. **Points are never taken away on an essay if information is incorrect.**

PRACTICE TESTS

Answer Sheet
AP BIOLOGY MODEL TEST 1

1 Ⓐ Ⓑ Ⓒ Ⓓ Ⓔ   26 Ⓐ Ⓑ Ⓒ Ⓓ Ⓔ   51 Ⓐ Ⓑ Ⓒ Ⓓ Ⓔ   76 Ⓐ Ⓑ Ⓒ Ⓓ Ⓔ
2 Ⓐ Ⓑ Ⓒ Ⓓ Ⓔ   27 Ⓐ Ⓑ Ⓒ Ⓓ Ⓔ   52 Ⓐ Ⓑ Ⓒ Ⓓ Ⓔ   77 Ⓐ Ⓑ Ⓒ Ⓓ Ⓔ
3 Ⓐ Ⓑ Ⓒ Ⓓ Ⓔ   28 Ⓐ Ⓑ Ⓒ Ⓓ Ⓔ   53 Ⓐ Ⓑ Ⓒ Ⓓ Ⓔ   78 Ⓐ Ⓑ Ⓒ Ⓓ Ⓔ
4 Ⓐ Ⓑ Ⓒ Ⓓ Ⓔ   29 Ⓐ Ⓑ Ⓒ Ⓓ Ⓔ   54 Ⓐ Ⓑ Ⓒ Ⓓ Ⓔ   79 Ⓐ Ⓑ Ⓒ Ⓓ Ⓔ
5 Ⓐ Ⓑ Ⓒ Ⓓ Ⓔ   30 Ⓐ Ⓑ Ⓒ Ⓓ Ⓔ   55 Ⓐ Ⓑ Ⓒ Ⓓ Ⓔ   80 Ⓐ Ⓑ Ⓒ Ⓓ Ⓔ
6 Ⓐ Ⓑ Ⓒ Ⓓ Ⓔ   31 Ⓐ Ⓑ Ⓒ Ⓓ Ⓔ   56 Ⓐ Ⓑ Ⓒ Ⓓ Ⓔ   81 Ⓐ Ⓑ Ⓒ Ⓓ Ⓔ
7 Ⓐ Ⓑ Ⓒ Ⓓ Ⓔ   32 Ⓐ Ⓑ Ⓒ Ⓓ Ⓔ   57 Ⓐ Ⓑ Ⓒ Ⓓ Ⓔ   82 Ⓐ Ⓑ Ⓒ Ⓓ Ⓔ
8 Ⓐ Ⓑ Ⓒ Ⓓ Ⓔ   33 Ⓐ Ⓑ Ⓒ Ⓓ Ⓔ   58 Ⓐ Ⓑ Ⓒ Ⓓ Ⓔ   83 Ⓐ Ⓑ Ⓒ Ⓓ Ⓔ
9 Ⓐ Ⓑ Ⓒ Ⓓ Ⓔ   34 Ⓐ Ⓑ Ⓒ Ⓓ Ⓔ   59 Ⓐ Ⓑ Ⓒ Ⓓ Ⓔ   84 Ⓐ Ⓑ Ⓒ Ⓓ Ⓔ
10 Ⓐ Ⓑ Ⓒ Ⓓ Ⓔ   35 Ⓐ Ⓑ Ⓒ Ⓓ Ⓔ   60 Ⓐ Ⓑ Ⓒ Ⓓ Ⓔ   85 Ⓐ Ⓑ Ⓒ Ⓓ Ⓔ
11 Ⓐ Ⓑ Ⓒ Ⓓ Ⓔ   36 Ⓐ Ⓑ Ⓒ Ⓓ Ⓔ   61 Ⓐ Ⓑ Ⓒ Ⓓ Ⓔ   86 Ⓐ Ⓑ Ⓒ Ⓓ Ⓔ
12 Ⓐ Ⓑ Ⓒ Ⓓ Ⓔ   37 Ⓐ Ⓑ Ⓒ Ⓓ Ⓔ   62 Ⓐ Ⓑ Ⓒ Ⓓ Ⓔ   87 Ⓐ Ⓑ Ⓒ Ⓓ Ⓔ
13 Ⓐ Ⓑ Ⓒ Ⓓ Ⓔ   38 Ⓐ Ⓑ Ⓒ Ⓓ Ⓔ   63 Ⓐ Ⓑ Ⓒ Ⓓ Ⓔ   88 Ⓐ Ⓑ Ⓒ Ⓓ Ⓔ
14 Ⓐ Ⓑ Ⓒ Ⓓ Ⓔ   39 Ⓐ Ⓑ Ⓒ Ⓓ Ⓔ   64 Ⓐ Ⓑ Ⓒ Ⓓ Ⓔ   89 Ⓐ Ⓑ Ⓒ Ⓓ Ⓔ
15 Ⓐ Ⓑ Ⓒ Ⓓ Ⓔ   40 Ⓐ Ⓑ Ⓒ Ⓓ Ⓔ   65 Ⓐ Ⓑ Ⓒ Ⓓ Ⓔ   90 Ⓐ Ⓑ Ⓒ Ⓓ Ⓔ
16 Ⓐ Ⓑ Ⓒ Ⓓ Ⓔ   41 Ⓐ Ⓑ Ⓒ Ⓓ Ⓔ   66 Ⓐ Ⓑ Ⓒ Ⓓ Ⓔ   91 Ⓐ Ⓑ Ⓒ Ⓓ Ⓔ
17 Ⓐ Ⓑ Ⓒ Ⓓ Ⓔ   42 Ⓐ Ⓑ Ⓒ Ⓓ Ⓔ   67 Ⓐ Ⓑ Ⓒ Ⓓ Ⓔ   92 Ⓐ Ⓑ Ⓒ Ⓓ Ⓔ
18 Ⓐ Ⓑ Ⓒ Ⓓ Ⓔ   43 Ⓐ Ⓑ Ⓒ Ⓓ Ⓔ   68 Ⓐ Ⓑ Ⓒ Ⓓ Ⓔ   93 Ⓐ Ⓑ Ⓒ Ⓓ Ⓔ
19 Ⓐ Ⓑ Ⓒ Ⓓ Ⓔ   44 Ⓐ Ⓑ Ⓒ Ⓓ Ⓔ   69 Ⓐ Ⓑ Ⓒ Ⓓ Ⓔ   94 Ⓐ Ⓑ Ⓒ Ⓓ Ⓔ
20 Ⓐ Ⓑ Ⓒ Ⓓ Ⓔ   45 Ⓐ Ⓑ Ⓒ Ⓓ Ⓔ   70 Ⓐ Ⓑ Ⓒ Ⓓ Ⓔ   95 Ⓐ Ⓑ Ⓒ Ⓓ Ⓔ
21 Ⓐ Ⓑ Ⓒ Ⓓ Ⓔ   46 Ⓐ Ⓑ Ⓒ Ⓓ Ⓔ   71 Ⓐ Ⓑ Ⓒ Ⓓ Ⓔ   96 Ⓐ Ⓑ Ⓒ Ⓓ Ⓔ
22 Ⓐ Ⓑ Ⓒ Ⓓ Ⓔ   47 Ⓐ Ⓑ Ⓒ Ⓓ Ⓔ   72 Ⓐ Ⓑ Ⓒ Ⓓ Ⓔ   97 Ⓐ Ⓑ Ⓒ Ⓓ Ⓔ
23 Ⓐ Ⓑ Ⓒ Ⓓ Ⓔ   48 Ⓐ Ⓑ Ⓒ Ⓓ Ⓔ   73 Ⓐ Ⓑ Ⓒ Ⓓ Ⓔ   98 Ⓐ Ⓑ Ⓒ Ⓓ Ⓔ
24 Ⓐ Ⓑ Ⓒ Ⓓ Ⓔ   49 Ⓐ Ⓑ Ⓒ Ⓓ Ⓔ   74 Ⓐ Ⓑ Ⓒ Ⓓ Ⓔ   99 Ⓐ Ⓑ Ⓒ Ⓓ Ⓔ
25 Ⓐ Ⓑ Ⓒ Ⓓ Ⓔ   50 Ⓐ Ⓑ Ⓒ Ⓓ Ⓔ   75 Ⓐ Ⓑ Ⓒ Ⓓ Ⓔ   100 Ⓐ Ⓑ Ⓒ Ⓓ Ⓔ

AP Biology
Model Test 1

MULTIPLE-CHOICE QUESTIONS

80 minutes
100 questions
60% of total grade

Directions: Select the best answer in each case.

1. Which of the following is the dominant stage of the life cycle in moss?

    (A) diploid
    (B) sporophyte
    (C) flowering stage
    (D) haploid
    (E) none of the above is correct

2. What is the chance of a couple having 4 daughters in a row?

    (A) $\frac{1}{8}$
    (B) $\frac{1}{4}$
    (C) $\frac{1}{2}$
    (D) $\frac{1}{16}$
    (E) $\frac{1}{32}$

3. The nitrogenous base adenine is found in which three of the following?

    (A) ATP, DNA, and NAD
    (B) proteins, chlorophyll, and vitamin A
    (C) carbohydrates, NAD, and proteins
    (D) cellulose, cytochromes, and hemoglobin
    (E) ATP, DNA, and proteins

4. The process by which some bacteria convert nitrate to free nitrogen is called

    (A) ammonification
    (B) nitrogen fixing
    (C) denitrifying
    (D) nitrifying
    (E) denitrogenation

5. Which of the following is the frozen desert?

    (A) taiga
    (B) tundra
    (C) chaparral
    (D) grasslands
    (E) temperate plains

6. Which is true of aerobic respiration but not true of anaerobic respiration?

    (A) CO_2 is required.
    (B) ATP is produced.
    (C) Water is produced.
    (D) Alcohol is produced.
    (E) Pyruvate is produced.

7. Which of the following is most effective in lowering the body temperature?

    (A) increased flow of blood to the skin
    (B) shivering
    (C) erection of fur
    (D) increased thyroid activity
    (E) formation of goose bumps

8. The most common type of population dispersion in nature is

    (A) random
    (B) dispersive
    (C) uniform
    (D) clumped
    (E) conforming

9. Species that utilize the same source of nutrition within a food web in one area can best be described as

    (A) autotrophs
    (B) occupying the same niche
    (C) secondary consumers
    (D) producers
    (E) symbionts

10. Which of the following requires ATP?

    (A) the movement of water up the xylem of a tree
    (B) the facilitated diffusion of glucose into a cell
    (C) countercurrent exchange
    (D) the diffusion of oxygen into a fish's gills
    (E) the uptake of cholesterol by a cell

11. In fruit flies, the mutation for bar eyes is a sex-linked dominant trait. If a pure bar-eyed female is crossed with a red-eyed male, what proportion of the male offspring will have bar eyes?

    (A) 0%
    (B) 25%
    (C) 50%
    (D) 75%
    (E) 100%

12. You want to train your puppy to wait at the curb until you tell him to cross the road. Your friend advises you to give your dog a treat every time he does as you ask. Your friend is advising that you train the dog using

    (A) operant conditioning
    (B) classical conditioning
    (C) imprinting
    (D) fixed action pattern
    (E) habituation

13. Which is true about water potential?

    (A) The water potential of pure water is negative.
    (B) The water potential of pure water is positive.
    (C) The water potential of pure water is zero.
    (D) The symbol for water potential is S.
    (E) Water moves from a low water potential to a high water potential.

14. Phototropisms in plants are controlled by

    (A) gibberellins
    (B) guard cells
    (C) ethylene
    (D) auxins
    (E) cytokinins

15. The working of the Lac operon is important because

   (A) it represents how mammals utilize lactose
   (B) it illustrates how RNA is processed after it is transcribed
   (C) it illustrates possible control on the cell cycle and may lead to an understanding about the nature of a malignancy
   (D) it is proof of semiconservative replication of DNA
   (E) it represents a principle means by which genes are regulated in prokaryotes

16. What is the substrate of salivary amylase?

   (A) protein
   (B) starch
   (C) amino acid
   (D) glucose
   (E) vitamins

17. Which pair is matched correctly?

   (A) grasshopper-external respiratory surface
   (B) arthropods-hemocyanin
   (C) earthworms-spiracles
   (D) humans-positive breathing mechanism
   (E) earthworms-hemocoels

18. A diploid cell has four pairs of homologous chromosomes: *AaBbCcDd*. How many different gametes can this cell produce?

   (A) 4
   (B) 8
   (C) 16
   (D) 32
   (E) 64

19. Which of the following has an open circulatory system?

   (A) hydra
   (B) earthworm
   (C) human
   (D) sponge
   (E) lobster

20. The population of peppered moths (*Biston betularia*) in England changed from white to black in less than 100 years. This is an example of

   (A) directional selection
   (B) genetic drift
   (C) disruptive selection
   (D) sexual selection
   (E) artificial selection

21. In order to move from the right side of the heart to the left side of the heart, blood must pass through the

    (A) mitral valve
    (B) vena cava
    (C) right atrioventricular valve
    (D) lungs
    (E) aorta

22. Which is true about photosynthesis?

    (A) Water provides replacement electrons for the light-dependent reactions.
    (B) The light-dependent reactions occur in the stroma.
    (C) Oxygen is the final hydrogen acceptor in the light reactions.
    (D) The product of photophosphorylation is PGAL.
    (E) The photosystems on which photosynthesis depends are located in the stroma membranes.

23. Which is NOT a cause of sympatric speciation?

    (A) geographic isolation
    (B) polyploidy
    (C) behavioral isolation
    (D) mechanical isolation
    (E) temporal isolation

24. Which of the following is an endocrine gland?

    (A) salivary gland
    (B) gallbladder
    (C) sebaceous gland
    (D) lachrymal gland
    (E) pineal gland

25. Drinking alcoholic beverages has the effect of

    (A) decreasing production of antidiuretic hormone
    (B) increasing blood pressure
    (C) decreasing the amount of urine released
    (D) increasing the production of concentrated urine
    (E) increasing the filtration rate in the nephron

26. If a stimulus on the dendrite of a sensory neuron is above the threshold level, a further increase in the intensity of this stimulus will most likely cause

    (A) the impulse to move faster
    (B) the neuron to become more polarized (hyperpolarized)
    (C) the neuron to stop firing entirely
    (D) the strength of the impulse to increase
    (E) the frequency of the impulse to increase

27. A calico cat is mated with a male black cat. What is the chance that they will produce any calico cats?

    (A) 0%
    (B) 25%
    (C) 50%
    (D) 75%
    (E) 100%

28. Which best explains why the incidence of polydactyly is so much higher in the Amish population than in the rest of the population in the United States?

    (A) gene flow
    (B) disruptive selection
    (C) mutation
    (D) bottleneck effect
    (E) founder effect

29. In humans, most nutrients are absorbed in the

    (A) stomach
    (B) small intestine
    (C) large intestine
    (D) both small and large intestine
    (E) mouth

30. Which of the following hormones controls metamorphosis in insects and crustaceans?

    (A) cortisol
    (B) adrenaline
    (C) ecdysone
    (D) insulin
    (E) thyroxin

31. The threshold potential of a membrane of a neuron

    (A) is always 70 mV
    (B) is always 90 mV
    (C) is always 50 mV
    (D) can never change
    (E) is a measure of the polarization of the membrane

32. A Barr body is normally found in the nucleus of which kinds of cells?

    (A) somatic female cells
    (B) sperm cells
    (C) both male and female somatic cells
    (D) somatic male cells
    (E) ova

33. Most energy during cell respiration is harvested during

    (A) the Krebs cycle
    (B) oxidative phosphorylation
    (C) glycolysis
    (D) anaerobic respiration
    (E) fermentation

34. The most immediate danger if the temperature on earth rises due to global warming is that

    (A) oxygen levels in the air will decline
    (B) carbon dioxide levels in the air will rise
    (C) the rate at which the continents drift will increase
    (D) the polar ice caps will melt and some major coastal cities could be in danger
    (E) biological magnification will increase to dangerous levels

35. A sign stimulus that functions as a signal that triggers a certain behavior in another member of the same species is called

    (A) a ritual
    (B) a fixed action stimulus
    (C) an inducer
    (D) a precursor
    (E) a releaser

36. The harmless viceroy butterfly resembles the poisonous monarch butterfly. This is an example of

    (A) Müllerian mimicry
    (B) commensal coloration
    (C) camouflage coloration
    (D) mutualistic coloration
    (E) Batesian mimicry

37. When a fish is taken out of water, it soon dies. The reason for this is that

    (A) oxygen gas is poisonous to fish
    (B) there is inadequate CO_2 in the air
    (C) there is inadequate O_2 in the air
    (D) the gills dry out and oxygen cannot diffuse across dry membranes
    (E) fish gills dissolve in air

38. A man who is color-blind marries a woman with normal vision but whose father was color-blind. What is the chance they will have a color-blind child?

(A) 0%
(B) 25%
(C) 50%
(D) 75%
(E) 100%

39. Segregation does NOT occur in which of the following groups of organisms?

(A) insects
(B) bacteria
(C) roses
(D) ferns
(E) mammals

40. Which of the following processes by the kidney is the LEAST selective?

(A) filtration
(B) secretion
(C) reabsorption
(D) the target of ADH
(E) the pumping of sodium ions out of the tubule

41. The relative location of four genes on a chromosome can be mapped from the following data on crossover frequencies.

Genes	Frequency of Crossover
A and *D*	60%
B and *A*	20%
C and *B*	10%
D and *C*	30%

Which is the correct order of genes on the chromosomes?

(A) *ABCD*
(B) *ABDC*
(C) *ACBD*
(D) *ADBC*
(E) *ADCB*

42. Which is a component of a molecule of chlorophyll?

    (A) iron
    (B) copper
    (C) magnesium
    (D) calcium
    (E) phosphorous

43. How many chromosomes are in this cell?

    (A) 4
    (B) 8
    (C) 16
    (D) 2
    (E) 18

44. In guinea pigs, black coat color (B) is dominant. What is the most likely genotype of the parents if they have 10 offspring, 5 of which are white?

    (A) *Bb* × *bb*
    (B) *Bb* × *Bb*
    (C) *BB* × *bb*
    (D) *bb* × *bb*
    (E) *BB* × *BB*

45. The ATP produced during fermentation is generated by which of the following?

    (A) electron transport chain
    (B) substrate level phosphorylation
    (C) the Krebs cycle
    (D) chemiosmosis
    (E) citric acid cycle

46. Which of the following best characterizes the structure of the plasma membrane?

    (A) rigid and unchanging
    (B) rigid but varying from cell to cell
    (C) fluid but unorganized
    (D) very active
    (E) rigid and organized

47. In pea plants, the trait for tall plants (T) is dominant to the trait for dwarf plants (t). The trait for yellow seed color (Y) is dominant to the trait for green seed color (y). In a particular cross of pea plants, the probability of the offspring being tall is $^1/_2$ and the seed being green is $^1/_4$. Which of the following most probably represents the parental genotypes?

   (A) *TTYy* × *TtYY*
   (B) *TTYy* × *ttYy*
   (C) *TtYy* × *ttyy*
   (D) *TtYy* × *ttYy*
   (E) *TtYy* × *TtYy*

48. Which one of the following would *not* normally diffuse through the lipid bilayer of a plasma membrane?

   (A) CO_2
   (B) amino acid
   (C) starch
   (D) water
   (E) O_2

49. What is the approximate size of a human red blood cell?

   (A) .01 micrometer
   (B) 8 micrometers
   (C) 80 micrometers
   (D) 8 nanometers
   (E) 1,000 micrometers

50. Which of the following is *present* in a prokaryote cell?

   (A) ribosomes
   (B) mitochondria
   (C) endoplasmic reticulum
   (D) chloroplasts
   (E) nuclear membrane

51. If a cell has 24 chromosomes at the beginning of meiosis, how many chromosomes will it have at the end of meiosis?

   (A) 12
   (B) 24
   (C) 48
   (D) 64
   (E) It varies with the species.

52. Which of the following principles is NOT part of Darwin's theory of evolution by natural selection?

    (A) Evolution is a gradual process that occurs over a long period of time.
    (B) Variation occurs among individuals in a population.
    (C) Individuals possessing the most favorable traits have the best chance to survive and to pass their traits to offspring.
    (D) Organisms tend to overpopulate.
    (E) Mutations are the main source of all variation in any population.

53. Which of the following occurs during the light-independent reactions of photosynthesis?

    (A) photolysis
    (B) ATP is produced
    (C) oxygen is released
    (D) carbon is reduced
    (E) CO_2 is released

54. SnRPs are most closely related to

    (A) cell division
    (B) the electron transport chain
    (C) how transposons jump
    (D) RNA processing
    (E) replication of DNA

55. Which environmental conditions would have the highest rate of transpiration?

    (A) hot and humid
    (B) hot and dry
    (C) cool and humid
    (D) cool and dry
    (E) conditions fluctuating between hot and humid, and cool and humid

56. If blond hair, green eyes, and freckles are often inherited together, it is most likely that the genes for these traits

    (A) are located on separate chromosomes
    (B) are located far apart on the same chromosome
    (C) are located close together on the same chromosome
    (D) are sex-linked
    (E) assort independently

57. Which of the following are lacking in plant cells?

    (A) centrioles
    (B) mitochondria
    (C) ribosomes
    (D) Golgi bodies
    (E) nucleoli

58. Bryophytes have all of the following characteristics EXCEPT

    (A) a dependent sporophyte
    (B) specialized cells and tissue
    (C) multicellularity
    (D) chlorophyll
    (E) roots

Directions: Each group of questions below relates to one topic. When there is a list of terms to choose from, choose the best answer. Each term may be used once, more than once, or not at all.

Questions 59–61

    (A) Golgi apparatus
    (B) Microtubules
    (C) Rough endoplasmic reticulum
    (D) Mitochondria
    (E) Nucleoli

59. Produces proteins

60. Synthesizes ribosomes

61. Directly assists with cell division

Questions 62–64

    (A) B cells
    (B) Macrophages
    (C) Helper T cells
    (D) Cytotoxic T cells
    (E) Natural killer cells

62. Form from monocytes and turn into antigen-presenting cells

63. Responsible for humoral immunity

64. Secrete cytokines to stimulate other lymphocytes

The following questions refer to Hardy-Weinberg equilibrium

65. The frequency of a particular allele in a population in Hardy-Weinberg equilibrium is 0.4. After 500 years, if the population is still in Hardy-Weinberg equilibrium, what would the frequency of the allele be?

   (A) It cannot be determined.
   (B) The frequency would increase because it is well adapted to the environment.
   (C) 0.4
   (D) It would probably die out because 500 years is a long time for an allele to survive in a population.
   (E) It cannot be determined, but it would definitely change because the environment would change.

66. In a school population, the percentage of students with blue eyes is 16%. What is the percentage of students with pure brown eyes?

   (A) .4%
   (B) 4%
   (C) 16%
   (D) 36%
   (E) 48%

Questions 67–70
Refer to the drawing of the circulatory system below.

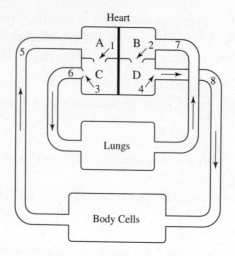

67. Left ventricle

   (A) A
   (B) B
   (C) C
   (D) D
   (E) none of the above

68. Mitral valve

   (A) 1
   (B) 2
   (C) 3
   (D) 4
   (E) none of the above

69. Aorta

   (A) 5
   (B) 6
   (C) 7
   (D) 8
   (E) none of the above

70. Pulmonary vein

   (A) 5
   (B) 6
   (C) 7
   (D) 8
   (E) none of the above

Model Test 1

Questions 71–72

Questions 71–72 refer to the graph below of the dissociation curve of hemoglobin.

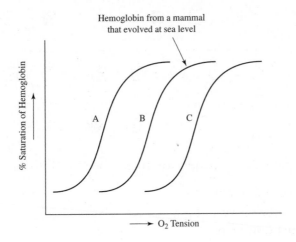

71. Identify the curve that is characteristic of hemoglobin of a mammal that evolved at high elevations.

    (A) A
    (B) B
    (C) C
    (D) both A and C but at different times
    (E) cannot be determined

72. Which hemoglobin has the greatest affinity for oxygen?

    (A) A
    (B) B
    (C) C
    (D) both A and C but at different times
    (E) cannot be determined

Questions 73–75

Questions *73–75* refer to the pedigree below for one form of albinism. Shading indicates a person who is albino.

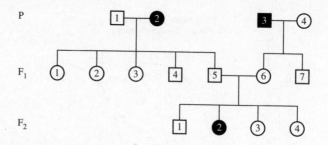

73. What is the most likely pattern of inheritance for this trait?

   (A) sex-linked dominant
   (B) sex-linked recessive
   (C) autosomal dominant
   (D) autosomal recessive
   (E) holandric

74. What is the most likely genotype for person #5 in the F_1 generation?

   (A) *A/A*
   (B) *A/a*
   (C) *a/a*
   (D) *X–Y*
   (E) *X–X–*

75. What is/are the possible genotype(s) for F_2 person #4?

   (A) *A/A* and *a/a*
   (B) *A/A* and *A/a*
   (C) *a/a* only
   (D) *A/a* only
   (E) XX

Questions 76–79
Questions 76–79 refer to DNA.

76. Which is directly responsible for the removal of introns during RNA processing in the eukaryotic nucleus?

    (A) GTP
    (B) the poly(A) tail
    (C) exons
    (D) the promoter
    (E) SNRPs

77. What are the regions of DNA that direct the production of polypeptides?

    (A) introns
    (B) codons
    (C) anticodons
    (D) exons
    (E) transposons

78. If AUU is the codon, what is the anticodon?

    (A) AUU
    (B) TAA
    (C) UUA
    (D) UAA
    (E) TUU

79. To which part of the DNA strand does RNA polymerase attach in the synthesis of proteins?

    (A) the primer
    (B) the operator
    (C) the promoter
    (D) the regulator
    (E) the cAMP binding site

Questions 80–81

Questions 80–81 refer to this sketch of a gel with four samples of DNA.

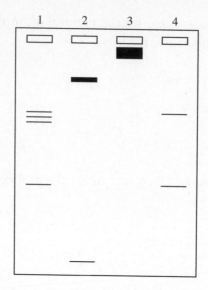

80. Which well/s contained DNA that was not first cut by restriction enzymes?

   (A) 1
   (B) 2
   (C) 3
   (D) 4
   (E) both 3 and 4

81. Which lane contains the smallest piece of DNA?

   (A) 1
   (B) 2
   (C) 3
   (D) 4
   (E) cannot be determined

Questions 82–85

   (A) Cytoplasm
   (B) Cristae membrane
   (C) Inner matrix
   (D) Outer compartment

82. Electron transport chain

83. Where ATP synthetase is located

84. Glycolysis

85. Krebs cycle

Questions 86–87

You are carrying out an experiment to study the rate of transpiration in plants using two-week-old bean plants. You set up a potometer by cutting the stem of the plant at its base and securing it into a calibrated pipette that is set into some clear, flexible tubing that is filled with water. You prepare five such potometers and expose each one to a different environment. You measure the amount of water lost by the plant by taking readings from the pipettes every 10 minutes. Below is a graph of the data.

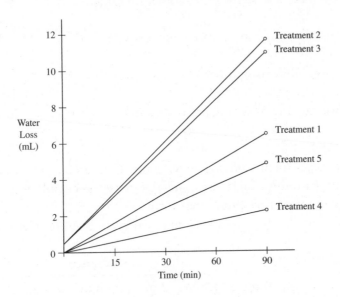

86. According to the data, which condition provides the greatest rate of transpiration?

    (A) treatment 1
    (B) treatment 2
    (C) treatment 3
    (D) treatment 4
    (E) treatment 5

87. Which of the following might cause the results seen in treatment 4?

    (A) increasing the light intensity
    (B) placing a heat sink between the light and the plant
    (C) placing a plastic bag over the leaves of the plant
    (D) increasing the green wavelengths of light
    (E) using a fan to circulate the air around the plant

Questions 88–91

    (A) Annelida
    (B) Chordata
    (C) Echinodermata
    (D) Arthropoda
    (E) Platyhelminthes

88. Closed circulatory system, protostome, no skeleton

89. Open circulatory system, protostome, exoskeleton

90. Acoelomate, bilateral symmetry

91. Deuterostome, radial symmetry as an adult

Questions 92–94

In a laboratory experiment studying enzyme function, an enzyme is combined with a substrate and a product forms as a precipitate. A spectrophotometer is used to measure the amount of precipitate that forms. An increase in absorption indicates more product formed. Measurements are taken every minute. Here are the data.

Enzyme Preparation	Absorbance						
	0 min.	1 min.	2 min.	3 min.	4 min.	5 min.	6 min.
I. 1 mL of fresh enzyme 5 mL of substrate pH 7	0.0	0.13	0.19	0.24	0.29	0.29	0.29
II. 1 mL of boiled enzyme 5 mL of substrate pH 7	0.0	0.01	0.02	0.02	0.03	0.03	0.03
III. 1 mL of fresh enzyme 5 mL of substrate pH 4	0.0	0.13	0.18	0.23	0.29	0.29	0.29
IV. 1 mL of boiled enzyme 5 mL of substrate pH 2	0.0	0.0	0.01	0.02	0.03	0.03	0.03

92. Regarding setup I, what is the most likely reason that the absorption does not change after three minutes?

    (A) The pH is not a favorable environment for this enzyme.
    (B) The enzyme itself breaks down after four minutes.
    (C) The data was not accurately collected.
    (D) All of the substrate was broken down after three minutes.
    (E) The enzyme was used up after three minutes.

93. Which of the following statements is best supported by the data?

    (A) A pH of 7 is optimal for this enzyme.
    (B) A pH of 4 is optimal for this enzyme.
    (C) The enzyme has no activity at pH 3.
    (D) This is most likely a human enzyme.
    (E) This enzyme can function at more than one pH.

94. If you were asked to carry out this experiment again, what would you change?

    (A) I would use more enzyme and substrate to get better results.
    (B) I would take measurements every minute up to ten minutes to get a better idea of what is going on.
    (C) I would use another instrument to measure absorbance because the spectrophotometer is not accurate when a precipitate is involved.
    (D) I would use an enzyme that was not affected by temperature or pH.
    (E) I would control the variables better. The enzyme in setup IV should be at the same pH as the pH of setup III.

Questions 95–97

Questions 95–97 refer to the graph below, which shows the changes in electric potential across the membrane of a neuron during the transmission of an electrical impulse.

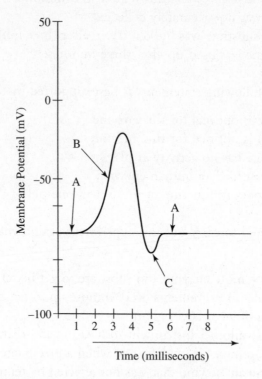

95. Identify where the membrane is pumping sodium and potassium ions to in order to return the membrane to the resting potential.

 (A) A
 (B) B
 (C) C
 (D) A and B
 (E) B and C

96. Identify the action potential

 (A) A
 (B) B
 (C) C
 (D) A and B
 (E) B and C

97. What is the measurement of the membrane potential at rest?

 (A) −100 mV
 (B) −25 mV
 (C) −50 mV
 (D) −70 mV
 (E) 0 mV

Questions 98–100

Questions 98–100 refer to this drawing of an embryo.

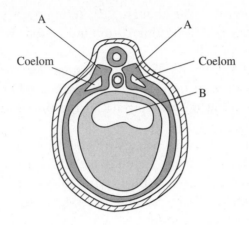

98. This embryo could NOT be from a

   (A) flatworm
   (B) frog
   (C) mollusk
   (D) spider
   (E) earthworm

99. The area labeled B will become the

   (A) brain and nerve cord
   (B) body cavity
   (C) primitive gut
   (D) muscles and bones
   (E) all of the above

100. The part labeled A will become the

   (A) nervous system
   (B) blood and bones
   (C) viscera
   (D) body cavity
   (E) skin

If there is still time remaining, you may review your answers.

READING PERIOD

10 minutes

Read all four essays, organize your thoughts, make any notes you need to assist you on your question sheet. Following this 10-minute period, you will be directed to answer the four essays formally.

You must answer all four essays. Because each question will be weighted equally, you are advised to divide your time equally among them. Do not spend too much time on any one essay. You must keep track of the time yourself; no one will do that for you.

You are to write your answers in the answer booklet only. Use black or blue ink. Be sure to write clearly and legibly. If you make a mistake, you may save time by crossing out with a single line rather than trying to erase it. Number each answer as the question is numbered in the examination. Begin each answer on a new page. *You may answer the essays in any order.*

Each answer should be organized, well-balanced, and as comprehensive as time permits. You must write in prose form; outline form is not acceptable. Do not spend time restating the questions; simply answer the questions. If a specific number of examples are asked for, no credit will be given for extra examples. Diagrams may be used to help explain your answer, but in no case will a diagram alone suffice.

WRITING PERIOD

1 hour and 30 minutes
Suggested writing time per question—22 minutes.

Question 1

Discuss the adaptations that have enabled animals and plants to overcome the problems associated with moving to land. Use examples from the animal *and* plant kingdoms to discuss each of the problems below.

a. Dehydration or water conservation
b. The absence of support for the animal or plant body
c. The absence of an aquatic environment for reproduction

Question 2

Many scientists have contributed to the knowledge of genes, the structure of DNA, and how genes function. People such as

a. Hershey and Chase
b. Griffith
c. Rosalind Franklin
d. Meselsohn and Stahl
e. Barbara McClintock
f. Watson and Crick
g. Avery, MacLeod, and McCarty

Choose *four* from the list above and discuss their contributions to science.

Question 3

Describe how energy is produced in animal cells by aerobic respiration. Include the

a. structure of the mitochondrion
b. electron transport chain

Question 4

Enzymes control every reaction in the body.

a. Relate the structure of an enzyme to its function
b. Design a controlled experiment to demonstrate that enzymes are adversely affected by heat *or* pH. The experiment must be reasonable to set up and must work.

If there is still time remaining, you may review your answers.

Answer Key
MODEL TEST 1

Multiple-Choice Questions

1.	D	26.	E	51.	A	76.	E
2.	D	27.	B	52.	E	77.	D
3.	A	28.	E	53.	D	78.	D
4.	C	29.	B	54.	D	79.	C
5.	B	30.	C	55.	B	80.	C
6.	C	31.	E	56.	C	81.	B
7.	A	32.	A	57.	A	82.	B
8.	D	33.	B	58.	E	83.	B
9.	B	34.	D	59.	C	84.	A
10.	E	35.	E	60.	E	85.	C
11.	E	36.	E	61.	B	86.	B
12.	A	37.	D	62.	B	87.	C
13.	C	38.	C	63.	A	88.	A
14.	D	39.	B	64.	C	89.	D
15.	E	40.	A	65.	C	90.	E
16.	B	41.	A	66.	D	91.	C
17.	B	42.	C	67.	D	92.	D
18.	C	43.	A	68.	B	93.	E
19.	E	44.	A	69.	D	94.	E
20.	A	45.	B	70.	C	95.	C
21.	D	46.	D	71.	A	96.	B
22.	A	47.	D	72.	A	97.	D
23.	A	48.	C	73.	D	98.	A
24.	E	49.	B	74.	B	99.	C
25.	A	50.	A	75.	B	100.	B

ANSWER EXPLANATIONS TO MODEL TEST 1

Multiple-Choice Questions

1. **(D)** The dominant life cycle of mosses is the gametophyte, which is haploid. The diploid sporophyte grows out of and is dependent on the gametophyte. In ferns, the two generations are separate from each other, but the sporophyte is larger and longer lived and is therefore dominant. In flowering plants, the sporophyte is dominant, while the gametophyte generation is located inside the ovary and is dependent on it.

2. **(D)** The chance of having a daughter at each birth is $\frac{1}{2}$. Each birth is an independent event, so the rule of multiplication applies. You must multiply the chance of each event occurring by the number of events: $\frac{1}{2} \times \frac{1}{2} \times \frac{1}{2} \times \frac{1}{2} = \frac{1}{16}$.

3. **(A)** Adenine is present in all nucleotides (DNA and RNA), in nicotinamide adenine dinucleotide (NAD), and in flavin adenine dinucleotide (FAD). Proteins are polymers of amino acids and contain the elements sulfur, phosphorous, carbon, oxygen, hydrogen, and nitrogen. Cytochromes and hemoglobin are proteins. Cellulose is a carbohydrate.

4. **(C)** Denitrifying bacteria convert NO_3 to free nitrogen. Nitrifying bacteria convert the ammonium ion into nitrites and then into nitrates. Nitrogen-fixing bacteria convert free nitrogen into the ammonium ion.

5. **(B)** The tundra is bitterly cold with high winds that are responsible for the absence of trees. It is a frozen desert with a permafrost, ground that never thaws. It is considered a desert because of the lack of water; ice and snow are not available sources of water. Tundra is characteristic of central Alaska, the arctic, and high mountaintops at all latitudes.

6. **(C)** During aerobic respiration when oxygen combines with hydrogen as the final hydrogen acceptor from the electron transport chain, water is formed and released. CO_2 is produced, not required, as a by-product of the Krebs cycle of aerobic respiration and in alcoholic fermentation. Alcohol is produced during alcoholic (anaerobic) fermentation. Pyruvate is the product of glycolysis and is the raw material for the Krebs cycle (aerobic respiration).

7. **(A)** Increased blood flow to the skin causes a lowering of body temperature as the heat from the warm blood radiates away from the body. All the others are ways the body keeps itself warm. The thyroid controls metabolism, and an increase in metabolism increases body temperature.

8. **(D)** Clumped dispersion of organisms is the most common in nature because there is safety in numbers. Uniform dispersion results from complex interactions among individuals such as aggressive territorial behavior. Random spacing occurs in the absence of strong attractions or repulsions among individuals of a population. It is least common in nature.

9. **(B)** A niche is determined by what an organism feeds on. Species that utilize the same source of nutrition in one area occupy the same niche and are in competition. Autotrophs are producers. Secondary consumers feed on primary consumers. Only when two secondary consumers feed on the same primary consumer are they sharing a niche and competing with each other. Symbionts are organisms that live closely together. Examples of symbiotic relationships are parasitism, commensalism, and mutualism.

10. **(E)** The movement of water up a tree is passive and occurs by a combination of transpirational pull and cohesion tension. Diffusion is always passive transport requiring no energy. Facilitated diffusion is a type of diffusion and, as such, does not require energy. Countercurrent exchange is a mechanism that maximizes the rate of diffusion. By the process of elimination, the correct answer is the uptake of cholesterol, which occurs by receptor-mediated endocytosis.

11. **(E)** The genotype of the bar-eyed female is X–X–, the red-eyed (normal) male is XY, and the bar-eyed male is X–Y. Since this trait is sex-linked, if the male has one affected gene, he has the condition.

    Here is the cross.

	X	Y (males)
X–	X–X	X–Y
X–	X–X	X–Y

    (females)

    All the males have one affected allele and are bar-eyed.

12. **(A)** Operant conditioning is also called trial-and-error learning. An animal learns to associate one of its behaviors with a reward or punishment and then repeats or avoids that behavior. This is usually how humans train their pets.

13. **(C)** If water has anything dissolved in it, the water potential is negative (less than one).

14. **(D)** Phototropisms are controlled by an unequal distribution of auxins. Gibberellins cause excessive stem elongation. Guard cells control the opening and closing of the stomates and, to some extent, are under hormonal control. Ethylene promotes fruit ripening. Cytokinins stimulate cell division and delay aging.

15. **(E)** The operon exists solely in prokaryotic cells and was discovered by Jacob and Monod. It is important as a model for gene regulation in bacteria. The operon consists of a cluster of related genes and their switches, the promoter and operator.

16. **(B)** Amylase is an enzyme that hydrolyzes starch. It is released from salivary glands in the mouth, where starch digestion begins.

17. (**B**) The molecule in arthropods that functions as hemoglobin does in humans is hemocyanin, which has copper as its active central atom.

18. (**C**) This type of problem can be solved by following this rule: 2 raised to the power of the number of pairs of alleles. There are four pairs of alleles, so $2^4 = 16$.

19. (**E**) All mollusks and arthropods, including lobsters, have open circulatory systems. The sponge and the hydra have no circulatory system. Earthworms and humans have a closed circulatory system where blood is transported within vessels.

20. (**A**) Directional selection describes evolution in which one phenotype is replaced by another phenotype over time. In the case of industrial melanism (the peppered moths), evolution occurred because of strong pressure from an environment that changed drastically in a short time. Genetic drift is a change in a gene pool due to chance. Disruptive selection increases the extreme phenotypes in a population. Sexual selection usually relates to the selection of mates, the plumage on the male peacock, for example. Artificial selection is the process by which humans control matings. The existence of different breeds of dogs is a good example of artificial selection.

21. (**D**) From the right ventricle, blood passes into the pulmonary arteries, lungs, and pulmonary veins, then back to the heart by entering the left atrium.

22. (**A**) Oxygen plays the role of the final hydrogen acceptor in aerobic respiration, not in photosynthesis. The product of photophosphorylation is ATP, not PGAL, which is the product of the Calvin cycle. The light reactions occur in thylakoid membranes of the grana. There are no membranes in the stroma.

23. (**A**) Geographic isolation is an example of allopatric speciation. Allopatric speciation is caused by separation of one or more populations by geography, such as mountains, canyons, rivers, or even wide highways. Sympatric speciation is speciation that occurs within one area. Choices B–E are all examples of sympatric speciation.

24. (**E**) Endocrine glands release hormones. Hormones are substances that are produced in one part of the body and that exert their influence in another part of the body. The gallbladder stores bile (an emulsifier) that is produced in the liver and is not a gland. The other glands are exocrine glands; they release substances out of the body. Lachrymal glands produce tears, sebaceous glands produce oils to protect facial skin, and salivary glands release digestive enzymes into the mouth.

25. (**A**) Drinking alcohol blocks production of antidiuretic hormone, which causes the collecting tube of the nephron to become less permeable to water. This causes an increase in urination because less water is absorbed out of the nephron and back into the bloodstream. Drinking alcohol actually lowers blood pressure because it dilates blood vessels. It has no effect on the filtration rate in the nephron, which occurs at Bowman's capsule.

26. (**E**) Either a neuron fires or it does not. The way in which the body distinguishes a strong stimulus from a weak one is by the frequency of the impulse. The stronger the stimulus, the more frequent the impulses. Once a nerve is at its threshold, a further increase in the intensity of a stimulus will cause an increase in only the frequency of nerve impulses, not an increase in the strength of the impulse.

27. (**B**) Here is the cross.

	X^B	Y
X^B	$X^B X^B$	$X^B Y$
X^Y	$X^B X^Y$	$X^Y Y$

$X^B X^B$ is a black female. $X^B Y$ is a black male. $X^B X^Y$ is a calico female. $X^Y Y$ is a yellow male. The chance of getting a calico cat is 25%.

28. (**E**) One or more Amish settlers to this country carried the rare but dominant trait for polydactyly. Due to the extreme isolation and intermarriage in this close community, this population now has a high incidence of this trait. This is an example of the founder effect; it is one type of genetic drift. Remember, if a trait is dominant, you know it will show if it is present in an individual. That says nothing about the frequency of the allele in the population. For example, a dominant gene that is disadvantageous will decline in frequency in a population; whereas, an advantageous gene will increase in frequency.

29. (**B**) All digestion and absorption is completed in the small intestine. The first part of the small intestine digests food, while the later section of the small intestine absorbs nutrients with the assistance of the villi. The large intestine is responsible for removal of undigested waste, vitamin production, and removing any excess water from the material that passes through it. The stomach is the site of protein digestion. The mouth is where starch digestion (hydrolysis) begins.

30. (**C**) Ecdysone and juvenile hormone control metamorphosis in insects. Cortisol, adrenaline, and insulin all control blood sugar levels in mammals. Thyroxine controls the rate of metabolism and metamorphosis from a tadpole to a frog.

31. (**E**) The membrane potential of an axon of a neuron is commonly measured at −70 mV. The measurement is always negative because the inside of the cell is more negative than the outside of the cell. However, the threshold potential of a neuron varies. If a neuron becomes hyperpolarized (less likely to fire) the membrane potential becomes lower, for example, −80 mV.

32. (**A**) Barr bodies are condensed, inactive X chromosomes. When there are two X chromosomes in a cell, one will always deactivate. Thus, normal body cells in a human female, which have two X chromosomes, will always have one Barr body. Therefore, the presence of a Barr body indicates the presence of two X chromosomes.

33. **(B)** Very little energy (4 ATP) is released from anaerobic respiration of one molecule of glucose, which includes glycolysis and fermentation. Most energy comes from the aerobic phase of respiration, which includes the Krebs cycle and the electron transport chain. Oxidative phosphorylation is the process by which energy is produced during the electron transport chain. ADP gets phosphorylated into ATP as protons flow through the ATP synthetase channel in the cristae membrane.

34. **(D)** Increased carbon dioxide levels in the atmosphere are responsible for the greenhouse effect, which is causing the average air temperature to rise worldwide. (Choice B states the reverse.) The consequence can be dire as stated in the question. Continental drift is always occurring but is not affected by global warming. Biological magnification is not related to global warming. It is the amplification of toxins which have entered and accumulate at the top of the food chain.

35. **(E)** This is the definition of a releaser. A male stickleback fish attacks a male who wanders into his territory. The attack is an example of a fixed action pattern and the releaser is the red abdomen of the invading fish.

36. **(E)** Batesian mimicry is copycat coloring, where a harmless individual mimics a poisonous individual to protect itself.

37. **(D)** Diffusion occurs only through moist membranes. When the gills dry out, the fish cannot get any oxygen and it dies. The same is true for the earthworm, which exudes a mucus that keeps its skin moist. The earthworm relies on its moist skin to breathe.

38. **(C)** Here is the cross

	X–	Y (Husband)
X–	X–X–	X–Y
X	X–X	XY

(Wife)

The wife is a carrier (X–X) because her father was color-blind. X–X– is a color-blind female; X–Y is a color-blind male. There is a 50 percent chance that a child will be color-blind; this includes half the female offspring and half the male offspring.

39. **(B)** Segregation occurs only during meiosis for the production of gametes. Bacteria reproduce asexually by binary fission, and they have one circular chromosome. Segregation does not occur in asexually reproducing organisms.

40. **(A)** Filtration is the nonselective diffusion of all substances small enough to pass through the glomerulus membrane and into Bowman's capsule.

41. **(A)** Begin with the two genes that are the farthest apart, *A* and *D*. Then place the other genes between those two genes. *A* to *B* are 20 units apart, so *B* must be 40 units from *D*. Since *D* and *C* are 30 units apart and *B* and *C* are 10 units apart, *C* must be between *B* and *D*. Therefore, the order is *ABCD*.

42. **(C)** The central atom in the head of a molecule of chlorophyll is magnesium and a major component of any commercial fertilizer. The central and active atom in a molecule of hemoglobin is iron, and the active atom in hemocyanin is copper.

43. **(A)** The number of centromeres is equal to the number of chromosomes. There are four replicated chromosomes in this cell and eight chromatids.

44. **(A)** Out of 10 offspring, if 5 are white, then 5 are black, and the ratio of black to white is 1:1. From that ratio, you should know that the parents are *B/b* and *b/b*. Here is the cross. Since half of the offspring are white, each parent must contribute a recessive (*b*) allele. In addition, one parent must have two recessive (*b*) alleles.

	b	*b*
B	*Bb*	*Bb*
b	*bb*	*bb*

Bb appears black and *bb* appears white.

45. **(B)** The production of ATP in anaerobic respiration (fermentation) is by substrate level phosphorylation, where a kinase enzyme moves a phosphate from 1 molecule to ADP, forming ATP.

46. **(D)** The plasma membrane as described as a fluid mosaic by S. J. Singer is very active, is highly organized, and consists of phospholipids and proteins that are very fluid. The composition and characteristics of any particular membrane varies with the type of cell. For example, the plasma membrane of a mitochondrion contains more protein than the plasma membrane of a red blood cell.

47. **(D)** When you are given two traits to analyze, approach each trait separately at first. If the probability of the offspring being tall is $\frac{1}{2}$, then the probability is also $\frac{1}{2}$ that it will be short. Therefore, the ratio of tall to short is 1:1. Whenever the ratio of dominant to recessive is 1:1, the parents are *T/t* (hybrid) and *t/t* (homozygous recessive). Now, address seed color. If the probability of the offspring being green is $\frac{1}{4}$, then the chance it will be yellow is $\frac{3}{4}$, and the ratio is 3:1, dominant to recessive trait. Therefore, we must be dealing with a monohybrid cross, and the parents are *Y/y* and *Y/y*. Now, put the two traits together. The parents must be *T/t Y/y* and *t/t Y/y*.

48. **(C)** Starch is too large a molecule to diffuse through a plasma membrane. All the others are small enough to diffuse through the membrane.

49. **(B)** Cells vary in size, a plant parenchymal cell can be about 200 μm. A red blood cell is a small cell, about 8 μm or 8,000 nm. Bacteria are even smaller, about 2 μm, about the length of a mitochondrion.

50. **(A)** Prokaryotic cells have no internal membranes, no E.R., Golgi, nucleus, vacuoles, chloroplasts, or mitochondria. They contain small ribosomes for protein production.

51. **(A)** The process of meiosis cuts the chromosome number in half. The diploid number before meiosis becomes the haploid number after meiosis. Human diploid cells contain 46 chromosomes, while human gametes contain 23 chromosomes.

52. **(E)** Mutations were not discovered until around 1900 when DeVries, while working with polyploidy plants, coined the term. Darwin could not explain the origin of all the variation he observed among organisms. It was the weak part of his theory of natural selection.

53. **(D)** During the light-independent reactions, carbon is reduced (carbon fixation) by the Calvin cycle and PGAL is formed. CO_2 is not released during photosynthesis at all; it is the source of carbon that is reduced in carbon fixation. Oxygen is released during the light reactions as a result of the photolysis of water. ATP is produced during the light reactions and utilized in huge quantities during the light-independent reactions.

54. **(D)** SnRPs stands for small nuclear ribonucleic proteins. Along with certain other proteins, they assist in the processing of RNA after transcription. They edit newly transcribed RNA and splice out the introns. The other choices do not relate to RNA processing.

55. **(B)** Transpiration is loss of water from the leaf by evaporation. The highest rate of evaporation of water would be in an environment that is hot and dry. The lowest rate would be in an environment that is humid and cool.

56. **(C)** If two traits are usually inherited together, they must be on the same chromosome and close together because they never (or rarely) get separated by crossing over.

57. **(A)** Plant cells lack centrioles. Instead, they have microtubule-organizing regions that function as centrioles to pull chromosomes apart during anaphase. In addition, plant cells usually lack lysosomes, although that is not a choice in this question.

58. **(E)** Bryophytes do not have transport vessels (xylem and phloem). As a result, they do not have roots.

59. **(C)** Rough E.R. has ribosomes attached to it that produce proteins.

60. **(E)** The nucleoli produce ribosomal RNA, which is a main component of the ribosome.

Answer Explanations

61. **(B)** Microtubules organized in a 9 triplet configuration make up the spindle fibers that pull apart the chromosomes during anaphase of cell division. Microtubules with the same 9 triplet configuration also make up the centrioles that anchor the spindle fibers.

62. **(B)** Macrophages are very active phagocytosing cells that gobble up huge numbers of microbes and present pieces of them on their cell surface. This presentation of antigens mobilizes the immune system and selects the correct B and T cells (clonal selection) to fight a specific infection.

63. **(A)** Humoral immunity refer to antibodies, which are produced only by B lymphocytes.

64. **(C)** Cytokines are released by helper T cells. Cytokines are a chemical alarm that stimulates lymphocytes into action.

65. **(C)** A population in Hardy-Weinberg equilibrium, as the question states, means a population is stable and nonevolving. By definition, if the population is not evolving, the frequency of the allele does not change; it remains at 0.4.

66. **(D)** The percentage of students with blue eyes (homozygous recessive) is 16%, so that means $q^2 = .16$ and $q = 0.4$. Since $p + q = 1$, $p = 0.6$. Since the percent of students with pure brown eyes is p^2, then $.6 \times .6 = 36\%$.

67. **(D)** The left ventricle connects to the aorta and has the thickest wall of all the chambers because it has to pump the blood out to the entire body.

68. **(B)** The mitral valve keeps blood flowing in one direction from the left atrium to the left ventricle.

69. **(D)** The aorta sends blood from the left ventricle out to the body. Although not shown in this sketch, it divides into an ascending aorta which sends blood to the head and a descending aorta that sends blood downward to the major organs and the rest of the body.

70. **(C)** The pulmonary vein brings blood from the lungs to the left atrium. It is the only vein that carries oxygenated blood.

71. **(A)** The more to the left on the graph, the greater the affinity hemoglobin has for oxygen. At high elevations less oxygen is available, so the hemoglobin needs to have a greater affinity for oxygen.

72. **(A)** The more to the left on the graph, the greater the affinity hemoglobin has for oxygen. Make up values for the *x*- and *y*-axis for yourself to see that this is true.

73. **(D)** Approach this type of problem by trying to eliminate the wrong choices first. Can you eliminate autosomal or sex-linked dominant? Yes. Daughter #2 in the F2 generation has the condition but neither parent has it. You can also eliminate sex-linked recessive as a possibility for the same reason as above. In order for daughter #2 in the F2 generation to have the condition, she must have received a gene from both parents. Her dad cannot carry a sex-linked trait; if he has the gene, he must have the condition. Holandric

means carried on the Y chromosome; such traits have to do with maleness. This must be eliminated because two females have the condition. The trait must therefore be transmitted as autosomal recessive.

74. **(B)** The trait is inherited as an autosomal recessive. Person #5 does not have the condition and neither does his wife, #6, and yet they have a child with the condition. Therefore, person #5 is hybrid, *A/a*.

75. **(B)** Person #4 in the F_2 generation does not have the condition, so we know she has at least one dominant (good) trait. She has a sister with the condition; therefore the parents must be hybrid. However, we do not know if either parent passed the recessive trait onto #4. Person #4 could be either *A/A* or *A/a*.

76. **(E)** SnRPs, which is short for small nuclear ribonucleoproteins, along with other proteins are responsible for removing the introns from a new RNA strand. This is an important part of RNA processing.

77. **(D)** Exons are coding or expressed sequences. They code for polypeptides. Introns are intervening sequences and do not get translated into proteins.

78. **(D)** The codon is the triplet associate with mRNA. The anticodon is the triplet associated with tRNA. They are complementary to each other. Remember, there is no thymine (T) in RNA; it is replaced with uracil (U).

79. **(C)** When RNA polymerase attaches to the promoter, DNA replication can proceed.

80. **(C)** When DNA is uncut, it is too large to run through the gel, so it remains near the well as one large piece.

81. **(B)** The smaller the piece of DNA, the faster and farther it runs through the gel.

82. **(B)** The electron transport chain is a collection of molecules embedded within the cristae membrane of the mitochondrion. It pumps protons from the matrix into the outer compartment to create a steep proton gradient.

83. **(B)** The ATP synthase, or ATP synthetase, is a special channel located within the cristae membrane where protons flow down a steep gradient from the outer compartment into the matrix. As protons flow through the ATP synthase channel, they generate energy that is used to phosphorylate ADP into the high-energy ATP. This is the way in which the greatest amount of energy is generated during all of aerobic respiration.

84. **(A)** Glycolysis, the breakdown of glucose into pyruvic acid, occurs in the cytoplasm. The pyruvic acid then moves into the matrix of the mitochondrion where the Krebs cycle occurs.

85. **(C)** Pyruvate formed during glycolysis in the cytoplasm moves into the matrix of the mitochondrion where the Krebs cycle occurs.

86. **(B)** Treatment 2, increased light, shows the most transpiration. On the graph, it is the line with the steepest slope.

87. **(C)** The plastic bag would increase the humidity around the leaves and decrease the rate of transpiration producing the results seen in treatment 4. The fan would increase transpiration. Increasing the light increases the rate of transpiration. Green light does not affect the rate of photosynthesis because green light is refected by a green plant.

88. **(A)** The earthworm is an annelid. It has red blood containing hemoglobin carried within vessels in a closed system.

89. **(D)** The arthropods include all insects, crustaceans, and spiders. They are protostomes with an open circulatory system and an exoskeleton.

90. **(E)** The platyhelminthes, flatworms, includes the planaria. They have three cell layers: endoderm, ectoderm, and mesoderm. Like all triploblastic animals, they show bilateral symmetry. However, they are the most primitive of the triploblastic animals and have no coelom.

91. **(C)** The echinoderms, including sea stars, are advanced animals. They are deuterostomes, as are humans, and they show bilateral symmetry as larvae. However, as adults they revert to the primitive radial symmetry.

92. **(D)** After three minutes, there was no more substrate to break down. If the pH were not favorable, there would not have been any reaction from the beginning. There is no reason that the enzyme would break down or was used up. Enzymes do not get used up; they are reused and unchanged in any reaction. The data makes sense and there is no evidence that it was not collected correctly.

93. **(E)** From the data, you can see that the enzyme functions at pH 7 and pH 4. There is no data about what is optimum for this enzyme as choices A and B state. There is also no evidence that this is a human enzyme.

94. **(E)** There are too many variables in this experiment as performed. One variable is fresh or boiled enzyme. Another variable is pH. The spectrophotometer measures the clarity of a mixture. It is the perfect machine to use for this experiment.

95. **(C)** This is called the refractory period and no impulse can pass.

96. **(B)** This is where the impulse is passing. The membrane is depolarizing.

97. **(D)** The horizontal line represents the membrane at rest or resting potential. It measures −70 mV.

98. **(A)** This drawing of an embryo shows a coelom arising from within the mesoderm. All the animals listed have a coelom except the flatworm, which is acoelomate.

99. **(C)** This is the archenteron, which will become the primitive gut. It replaced the blastocoel.

100. **(B)** This area surrounding the coelom is the mesoderm and will become the blood, bones, and muscles.

What Topics Do You Need to Work On?

Table 21.1 shows an analysis by topic for each question on Model Test 1.

TABLE 21.1

Topic Analysis			
Biochemistry and Enzymes (Ch. 3)	**Cells and Cell Division (Ch. 4, 7)**	**Cell Respiration and Photosynthesis (Ch. 5, 6)**	**Heredity and Molecular Genetics (Ch. 8, 9)**
13, 16, 92, 93, 94	10, 18, 37, 43, 46, 48, 49, 50, 51, 57, 59, 60, 61	6, 22, 33, 42, 45, 53, 82, 83, 84, 85	2, 3, 11, 15, 27, 32, 38, 39, 41, 44, 47, 54, 56, 73, 74, 75, 76, 77, 78, 79, 80, 81
Classification and Evolution (Ch. 10, 11)	**Animals (Ch. 13, 14, 15, 17)**	**Plants (Ch. 12)**	**Ecology (Ch. 16)**
12, 20, 23, 28, 35, 36, 52, 65, 66	7, 17, 19, 21, 24, 25 26, 29, 30, 31, 40, 62, 63, 64, 67, 68, 69, 70, 71, 72, 88, 89, 90, 91, 95, 96, 97, 98, 99, 100	1, 14, 55, 57, 58, 86, 87	4, 5, 8, 9, 34

How to Score Your Essay

After you have written the best essay you can write in about 20 minutes, you are ready to grade it. First, though, take a short break to clear your head. When you are ready, reread your essay. Put a 1 to the left or right of any line where you explain any point that is listed in the following **scoring standard**. If you simply list examples, like "induced fit" or "sodium-potassium pump," you get NO points. *You must explain each term.* Also, if you try to explain something without using the scientific term, you get NO credit. Here is an example. If you say, "There are protein channels in the plasma membrane" without saying what they do or how they function, you get no credit.

After reading and analyzing each part of the question, add up all the points you placed at the end of the line. If a question has three parts, the maximum number of points you can earn in any one part is 3–4. If, in your essay, you happen to include every point listed, you will get a maximum of 10 points. Be honest. You must use and explain scientific terminology and explain all concepts clearly so that anyone would understand what you are trying to say.

Scoring Standard for Free-Response Questions

Question 1
Total—10 points

Dehydration or water conservation

	1 pt.	Kidney/nephrons can produce a hypertonic urine > 300 mosm.
	1 pt.	Collecting tube of the nephron is under hormonal control of ADH; this regulates how much urine the body excretes
Animals	1 pt.	Large intestine in humans for reabsorption of water used in digestion
	1 pt.	Birds and insects excrete uric acid as a dry paste to conserve water
	1 pt.	Waxy cuticle/cutin
Plants	1 pt.	Stomates
	1 pt.	Stomatal crypts
	1 pt.	roots/root hairs
	1 pt.	CAM plants
	1 pt.	C4 photosynthesis/Kranz anatomy/Hatch-Slack pathway
	4 pts.	Maximum

The absence of support for the animal or plant body

	1 pt.	Internal skeleton consists of bones that grow
Animals	1 pt.	External skeleton provides support and protection—must be shed periodically
	1 pt.	Muscles/tendons/ligaments for support and mobility
	1 pt.	Cell walls
Plants	1 pt.	Collenchyma and sclerenchyma cells
	1 pt.	Lignin in some plant cell walls
	1 pt.	Root pressure to assist moving water up the plant passively
	1 pt.	Roots anchor plants
	1 pt.	Prop roots, example: corn
	1 pt.	Secondary growth—lateral cambium—girth lends support

4 pts. Maximum

The absence of aquatic environment for reproduction

	1 pt.	Internal fertilization
	1 pt.	Internal development
Animals	1 pt.	Placenta
	1 pt.	Fallopian tubes and uterus
	1 pt.	Pollen/pollination
	1 pt.	Gametangia
Plants	1 pt.	Sporopollenin
	1 pt.	Seeds/seed coat/seed dormancy
	1 pt.	Fruit
	1 pt.	Reduced gametophyte/Increased sporophyte
	1 pt.	Internal fertilization

4 pts. Maximum

Question 2
Total—10 points

Three points maximum are granted for each part of the answer.
You cannot get full credit unless you discuss all four scientists or sets of scientists.

a. Hershey and Chase—Published their classic findings that Griffith's transformation factor is, in fact, DNA. This provided direct experimental evidence that DNA was the inherited material.

b. Griffith performed experiments with several different strains of the bacterium *Diplococcus pneumoniae*. Some strains were virulent and caused pneumonia in humans and mice, and some strains were harmless. Virulent strains are encapsulated in a polysaccharide capsule and are free to multiply and cause disease. Harmless strains, in contrast, are not encapsulated and are readily engulfed and destroyed by phagocytic cells in the host's immune system. Griffith discovered that bacteria have the ability to transform harmless cells into virulent ones by transferring some genetic factor from one bacterium cell to another. This process is known as bacterial transformation, and the experiment is known as the transformation experiment.

c. Rosalind Franklin, while working in the lab of Maurice Wilkins, carried out the X-ray crystallography analysis of DNA that showed DNA to be a helix. Her work was critical to Watson and Crick in developing their now-famous model of DNA. Although Maurice Wilkins shared the Nobel Prize with Watson and Crick, Rosalind Franklin did not. She had died by the time the prize was awarded, and the prize is not awarded posthumously.

d. Meselsohn and Stahl—Proved that DNA replicates in a semiconservative fashion. They cultured bacteria in a medium containing heavy nitrogen (^{15}N), allowing the bacteria to incorporate this heavy nitrogen into their DNA as they replicated and divided. These bacteria were then transferred to a medium containing light nitrogen (^{14}N) and allowed to replicate and divide only once. The bacteria that resulted from this final replication were spun in a centrifuge and found to be midway in density between the bacteria grown in heavy nitrogen and those grown in light nitrogen. This demonstrated that the new bacteria contained DNA consisting of one heavy strand and one light strand, thus confirming Crick's hypothesis that replication of DNA is semiconservative.

e. Barbara McClintock—Studied the genetics of corn in the 1940s and 1950s. She discovered transposons. Some transposons jump, in a cut-and-paste fashion, from one part of the genome to another. Others make copies of themselves, which move to another region of the genome, leaving the original behind. There are two classes of jumping genes, insertion sequences and complex transposons.

f. Watson and Crick, while working at Cambridge University, proposed the double helix structure of DNA. They received the Nobel Prize in 1962 for their work. The DNA molecule is a double helix, shaped like a twisted ladder, consisting of two strands running in opposite directions. One runs 5' to 3' (right-side up), the other 3' to 5' (upside-down). DNA is a polymer of repeating units of nucleotides, which in DNA, consists of a 5-carbon sugar (deoxyribose), a phosphate, and a nitrogen base that are connected by phosphodiester linkages. Understanding the structure of DNA gives a foundation to understand how DNA can replicate itself.

g. Avery, MacLeod, and McCarty published their classic findings that Griffith's transformation factor is, in fact, DNA. This provided direct experimental evidence that DNA was the genetic material. In the laboratory, they separated carbohydrates, lipids, proteins, DNA, and RNA from (Griffith's) heat-killed virulent bacteria and added each of these fractions to a separate suspension of living, nonvirulent pneumococcus bacteria. Next, they cultured the bacteria from each suspension and found that some bacteria that had been exposed to DNA from the dead bacteria had become virulent. In contrast, bacteria treated with other chemicals (carbohydrates, lipids, proteins, and RNA) from the dead bacteria remained unchanged. This research proved that DNA was the agent that carried the genetic characteristics from the virulent dead bacteria to the living, nonvirulent bacteria. *This provided direct experimental evidence that DNA was the genetic material.*

Question 3
Total—10 points

Structure of the mitochondrion

1 pt.	Double outer membrane
1 pt.	Cristae membrane
1 pt.	Outer compartment
1 pt.	Inner matrix
1 pt.	Size = 1–10 µm long
1 pt.	ATP synthetase channel in the cristae membrane
4 pts.	Maximum

Electron transport chain

1 pt.	Structure of ATP/currency of energy for the cell
1 pt.	NAD/FAD are coenzymes/vitamin derivatives/proton carriers
1 pt.	Oxidative phosphorylation is the name for how the energy is produced during the electron transport chain
1 pt.	Mitchell hypothesis/explanation of how ATP is formed by chemiosmosis
1 pt.	Proton motive force/proton gradient between the outer and inner compartments
1 pt.	Cytochromes/proteins, parts of the electron transport chain/ within the cristae membrane
1 pt.	Oxygen is the final proton acceptor in the electron transport chain/pulls the protons down the electron transport chain into the inner matrix to form water, which is exhaled
1 pt.	Theoretical yields: 1 NAD $\rightarrow$ 3 ATP; 1 FAD $\rightarrow$ 2 ATP
7 pts.	Maximum

Answer Explanations

Question 4
Total—10 points

Relate the structure of an enzyme to its function

1 pt.	Proteins—polypeptides
1 pt.	3-D shape/conformation/tertiary structure
1 pt.	Shape explained by intramolecular bonding/Van der Waals, disulfide bridges
1 pt.	Active site defined, not merely mentioned
1 pt.	Substrate defined, not merely mentioned
1 pt.	Induced fit (not lock and key)
1 pt.	Denatured/alteration of shape due to environmental conditions
1 pt.	Lower activation energy by facilitating effective collisions
1 pt.	Modifier/allosteric enzymes

5 pts.   Maximum

Design a controlled experiment to demonstrate that enzymes are adversely affected by heat *or* pH. The experiment must be reasonable to set up and must work.

If the student sets up two experiments, one to test pH and one to test temperature, read the first one only.

1 pt.	Describe which variable is being studied/details
1 pt.	Eliminate all variables except the one being studied
1 pt.	There must be a negative control, with no enzyme or substrate
1 pt.	Measure product/disappearance or appearance of
1 pt.	Record data
1 pt.	Draw correct conclusions about the results
1 pt.	For depth of understanding/independent vs. dependent variable/importance of replicating experiment

4 pts.   Maximum

Answer Sheet

AP BIOLOGY MODEL TEST 2

1 Ⓐ Ⓑ Ⓒ Ⓓ Ⓔ 26 Ⓐ Ⓑ Ⓒ Ⓓ Ⓔ 51 Ⓐ Ⓑ Ⓒ Ⓓ Ⓔ 76 Ⓐ Ⓑ Ⓒ Ⓓ Ⓔ

2 Ⓐ Ⓑ Ⓒ Ⓓ Ⓔ 27 Ⓐ Ⓑ Ⓒ Ⓓ Ⓔ 52 Ⓐ Ⓑ Ⓒ Ⓓ Ⓔ 77 Ⓐ Ⓑ Ⓒ Ⓓ Ⓔ

3 Ⓐ Ⓑ Ⓒ Ⓓ Ⓔ 28 Ⓐ Ⓑ Ⓒ Ⓓ Ⓔ 53 Ⓐ Ⓑ Ⓒ Ⓓ Ⓔ 78 Ⓐ Ⓑ Ⓒ Ⓓ Ⓔ

4 Ⓐ Ⓑ Ⓒ Ⓓ Ⓔ 29 Ⓐ Ⓑ Ⓒ Ⓓ Ⓔ 54 Ⓐ Ⓑ Ⓒ Ⓓ Ⓔ 79 Ⓐ Ⓑ Ⓒ Ⓓ Ⓔ

5 Ⓐ Ⓑ Ⓒ Ⓓ Ⓔ 30 Ⓐ Ⓑ Ⓒ Ⓓ Ⓔ 55 Ⓐ Ⓑ Ⓒ Ⓓ Ⓔ 80 Ⓐ Ⓑ Ⓒ Ⓓ Ⓔ

6 Ⓐ Ⓑ Ⓒ Ⓓ Ⓔ 31 Ⓐ Ⓑ Ⓒ Ⓓ Ⓔ 56 Ⓐ Ⓑ Ⓒ Ⓓ Ⓔ 81 Ⓐ Ⓑ Ⓒ Ⓓ Ⓔ

7 Ⓐ Ⓑ Ⓒ Ⓓ Ⓔ 32 Ⓐ Ⓑ Ⓒ Ⓓ Ⓔ 57 Ⓐ Ⓑ Ⓒ Ⓓ Ⓔ 82 Ⓐ Ⓑ Ⓒ Ⓓ Ⓔ

8 Ⓐ Ⓑ Ⓒ Ⓓ Ⓔ 33 Ⓐ Ⓑ Ⓒ Ⓓ Ⓔ 58 Ⓐ Ⓑ Ⓒ Ⓓ Ⓔ 83 Ⓐ Ⓑ Ⓒ Ⓓ Ⓔ

9 Ⓐ Ⓑ Ⓒ Ⓓ Ⓔ 34 Ⓐ Ⓑ Ⓒ Ⓓ Ⓔ 59 Ⓐ Ⓑ Ⓒ Ⓓ Ⓔ 84 Ⓐ Ⓑ Ⓒ Ⓓ Ⓔ

10 Ⓐ Ⓑ Ⓒ Ⓓ Ⓔ 35 Ⓐ Ⓑ Ⓒ Ⓓ Ⓔ 60 Ⓐ Ⓑ Ⓒ Ⓓ Ⓔ 85 Ⓐ Ⓑ Ⓒ Ⓓ Ⓔ

11 Ⓐ Ⓑ Ⓒ Ⓓ Ⓔ 36 Ⓐ Ⓑ Ⓒ Ⓓ Ⓔ 61 Ⓐ Ⓑ Ⓒ Ⓓ Ⓔ 86 Ⓐ Ⓑ Ⓒ Ⓓ Ⓔ

12 Ⓐ Ⓑ Ⓒ Ⓓ Ⓔ 37 Ⓐ Ⓑ Ⓒ Ⓓ Ⓔ 62 Ⓐ Ⓑ Ⓒ Ⓓ Ⓔ 87 Ⓐ Ⓑ Ⓒ Ⓓ Ⓔ

13 Ⓐ Ⓑ Ⓒ Ⓓ Ⓔ 38 Ⓐ Ⓑ Ⓒ Ⓓ Ⓔ 63 Ⓐ Ⓑ Ⓒ Ⓓ Ⓔ 88 Ⓐ Ⓑ Ⓒ Ⓓ Ⓔ

14 Ⓐ Ⓑ Ⓒ Ⓓ Ⓔ 39 Ⓐ Ⓑ Ⓒ Ⓓ Ⓔ 64 Ⓐ Ⓑ Ⓒ Ⓓ Ⓔ 89 Ⓐ Ⓑ Ⓒ Ⓓ Ⓔ

15 Ⓐ Ⓑ Ⓒ Ⓓ Ⓔ 40 Ⓐ Ⓑ Ⓒ Ⓓ Ⓔ 65 Ⓐ Ⓑ Ⓒ Ⓓ Ⓔ 90 Ⓐ Ⓑ Ⓒ Ⓓ Ⓔ

16 Ⓐ Ⓑ Ⓒ Ⓓ Ⓔ 41 Ⓐ Ⓑ Ⓒ Ⓓ Ⓔ 66 Ⓐ Ⓑ Ⓒ Ⓓ Ⓔ 91 Ⓐ Ⓑ Ⓒ Ⓓ Ⓔ

17 Ⓐ Ⓑ Ⓒ Ⓓ Ⓔ 42 Ⓐ Ⓑ Ⓒ Ⓓ Ⓔ 67 Ⓐ Ⓑ Ⓒ Ⓓ Ⓔ 92 Ⓐ Ⓑ Ⓒ Ⓓ Ⓔ

18 Ⓐ Ⓑ Ⓒ Ⓓ Ⓔ 43 Ⓐ Ⓑ Ⓒ Ⓓ Ⓔ 68 Ⓐ Ⓑ Ⓒ Ⓓ Ⓔ 93 Ⓐ Ⓑ Ⓒ Ⓓ Ⓔ

19 Ⓐ Ⓑ Ⓒ Ⓓ Ⓔ 44 Ⓐ Ⓑ Ⓒ Ⓓ Ⓔ 69 Ⓐ Ⓑ Ⓒ Ⓓ Ⓔ 94 Ⓐ Ⓑ Ⓒ Ⓓ Ⓔ

20 Ⓐ Ⓑ Ⓒ Ⓓ Ⓔ 45 Ⓐ Ⓑ Ⓒ Ⓓ Ⓔ 70 Ⓐ Ⓑ Ⓒ Ⓓ Ⓔ 95 Ⓐ Ⓑ Ⓒ Ⓓ Ⓔ

21 Ⓐ Ⓑ Ⓒ Ⓓ Ⓔ 46 Ⓐ Ⓑ Ⓒ Ⓓ Ⓔ 71 Ⓐ Ⓑ Ⓒ Ⓓ Ⓔ 96 Ⓐ Ⓑ Ⓒ Ⓓ Ⓔ

22 Ⓐ Ⓑ Ⓒ Ⓓ Ⓔ 47 Ⓐ Ⓑ Ⓒ Ⓓ Ⓔ 72 Ⓐ Ⓑ Ⓒ Ⓓ Ⓔ 97 Ⓐ Ⓑ Ⓒ Ⓓ Ⓔ

23 Ⓐ Ⓑ Ⓒ Ⓓ Ⓔ 48 Ⓐ Ⓑ Ⓒ Ⓓ Ⓔ 73 Ⓐ Ⓑ Ⓒ Ⓓ Ⓔ 98 Ⓐ Ⓑ Ⓒ Ⓓ Ⓔ

24 Ⓐ Ⓑ Ⓒ Ⓓ Ⓔ 49 Ⓐ Ⓑ Ⓒ Ⓓ Ⓔ 74 Ⓐ Ⓑ Ⓒ Ⓓ Ⓔ 99 Ⓐ Ⓑ Ⓒ Ⓓ Ⓔ

25 Ⓐ Ⓑ Ⓒ Ⓓ Ⓔ 50 Ⓐ Ⓑ Ⓒ Ⓓ Ⓔ 75 Ⓐ Ⓑ Ⓒ Ⓓ Ⓔ 100 Ⓐ Ⓑ Ⓒ Ⓓ Ⓔ

AP Biology
Model Test 2

MULTIPLE-CHOICE QUESTIONS

80 minutes
100 questions
60% of total grade

Directions: Select the best answer in each case.

1. Selection acts directly on

    (A) the phenotype
    (B) the genotype
    (C) an allele
    (D) the entire genome
    (E) a community

2. Which of these ecosystems accounts for the largest amount of earth's nutritional resources?

    (A) oceans
    (B) tropical rain forest
    (C) taiga
    (D) grasslands
    (E) temperate deciduous forest

3. The cells of which of the following structures have a triploid chromosome number?

    (A) pollen
    (B) fruit
    (C) cotyledon
    (D) embryo
    (E) all cells have the diploid chromosome number

4. All of the following are true about enzymes EXCEPT

   (A) enzymes accelerate reactions by lowering the energy of activation
   (B) some enzymes are allosteric
   (C) they are all proteins
   (D) all enzymes in the human are most active at 37°C and a pH around 7
   (E) they catalyze reactions in BOTH directions

5. In plants, phytochromes are most closely associated with control of

   (A) sexual reproduction
   (B) circadian rhythm
   (C) asexual reproduction
   (D) tropisms
   (E) photosynthesis

6. Which is not correctly matched?

   (A) bile; emulsifier
   (B) lacteal; absorption of fatty acid and glycerol
   (C) pepsinogen; stomach
   (D) secretin; small intestine
   (E) small intestine; water absorption

7. In traveling from a forest ecosystem to grasslands, trees gradually give way to grasses. The critical factor responsible for this shift is due to

   (A) amount of sunlight
   (B) availability of oxygen
   (C) altitude
   (D) length of growing season
   (E) availability of water

8. The final stage in eutrophication of a lake is that

   (A) it is not safe to eat the fish because they are not healthy
   (B) the temperature near the lake rises
   (C) the lake becomes too acidic to support life
   (D) the lake dries up
   (E) runoff from the land causes nutrients to pour into the lake

9. In flowering plants, a mature male gametophyte contains

   (A) one haploid nucleus and a diploid coat
   (B) two sperm nuclei and one tube nucleus
   (C) three diploid nuclei
   (D) one haploid gamete
   (E) four haploid gametes

10. A couple has two children who have cystic fibrosis. Neither parent has the condition. What is the probability that their next child will have cystic fibrosis?

    (A) 0%
    (B) 25%
    (C) 50%
    (D) 75%
    (E) 100%

11. Which was not present in the atmosphere of the earth 4 billion years ago?

    (A) CH_3
    (B) H_2O
    (C) CO_2
    (D) N_2
    (E) O_2

12. Use the list of the kingdoms below to answer the following question.

    (A) Fungi
    (B) Plantae
    (C) Animalia
    (D) Protista

    Which kingdoms include autotrophs?

    (A) A and B
    (B) B and D
    (C) A, B, and D
    (D) B only
    (E) all of the above

13. What is the main means of seed dispersal in conifers?

    (A) carried by wind
    (B) carried by animals
    (C) as fruit eaten by animals
    (D) all of the above are correct
    (E) none of the above is correct

14. Homeotic or Hox genes

    (A) are controlled by the dorsal lip
    (B) are master genes that control the expression responsible for anatomical structures
    (C) are turned off immediately after fertilization and prior to cleavage
    (D) are found only in the most primitive organisms
    (E) all of the above are correct

15. Which is not necessary for blood to clot normally?

    (A) fibrin
    (B) thrombin
    (C) thromboplastin
    (D) macrophages
    (E) calcium ions

16. Which tenet of the theory of evolution was the weakest because Darwin was unable to explain it fully?

    (A) Evolution occurs as advantageous traits accumulate in a population.
    (B) In any population, there is variation and an unequal ability of individuals to survive and reproduce.
    (C) Only the best-fit individuals survive and get to pass their traits on to offspring.
    (D) Populations tend to grow exponentially, overpopulate, and exceed their resources.
    (E) Overpopulation results in competition and a struggle for existence.

17. What surface molecules on a plasma membrane are most important for cell-to-cell recognition?

    (A) phospholipids
    (B) glycoproteins
    (C) cholesterol
    (D) protein channel
    (E) single ion channels

18. In a population that is in Hardy-Weinberg equilibrium, the frequency of the allele for blue eyes is 0.3. What is the percent of the population that is homozygous for brown eyes?

    (A) 3%
    (B) 9%
    (C) 21%
    (D) 42%
    (E) 49%

19. Which of the following is a density-dependent factor limiting human population growth?

    (A) fires
    (B) earthquakes
    (C) floods
    (D) famine
    (E) storms

20. Which term best describes a change in allelic frequencies due to chance?

    (A) gene flow
    (B) genetic drift
    (C) divergent evolution
    (D) balanced polymorphism
    (E) adaptive radiation

21. Which of the following is NOT made of microtubules?

    (A) Golgi apparatus
    (B) basal bodies
    (C) spindle fibers
    (D) centrioles
    (E) cilia

22. A common method of speciation that is important in the evolution of flowers but not in the evolution of animals is

    (A) genetic drift
    (B) geographic isolation
    (C) polyploidy
    (D) mutation
    (E) behavioral isolation

23. Muscle cells in oxygen deprivation convert pyruvate to _____ and during this process gain _____ .

    (A) lactate; ATP
    (B) alcohol; CO_2
    (C) lactate; NAD^+
    (D) ATP; NAD^+
    (E) alcohol; ATP

24. The smallest biological unit that can evolve over time is

    (A) a cell
    (B) an individual
    (C) a species
    (D) a population
    (E) a community

25. In a famous experiment, the dorsal lip from one frog embryo was grafted to the ventral side of a second embryo, and a notochord and neural tube developed in the recipient embryo at the location of the graft. Subsequently, other organs and structures developed, creating a nearly complete second embryo attached to the recipient embryo. This process of one tissue instructing and causing adjacent tissue to transform is referred to as

    (A) polarity development
    (B) morphogenesis
    (C) instructional cleavage
    (D) embryonic induction
    (E) epibolic movement

26. The high level of pesticides in birds of prey is an example of

    (A) the principle of exclusion
    (B) biological magnification
    (C) exponential growth
    (D) cycling of nutrients by decomposers
    (E) ecological succession

27. Which of the following is an example of polygenic inheritance?

    (A) pink flowers in Japanese four o'clocks
    (B) the ABO blood type in humans
    (C) Marfan's syndrome
    (D) epistasis
    (E) skin color in humans

28. Which of the following occurs first in an inflammation?

    (A) fever
    (B) mobilization of cytotoxic T cells
    (C) release of histamine
    (D) the proliferation of B cells
    (E) the production of antibodies

29. Which of the following is NOT a requirement for maintaining a stable population according to the Hardy-Weinberg theory?

    (A) random mating
    (B) no migration, in or out
    (C) no mutation
    (D) a small population
    (E) no natural selection

30. Facilitated diffusion differs from simple diffusion in that facilitated diffusion

    (A) requires small amounts of energy
    (B) moves materials against a gradient
    (C) requires membrane transport proteins
    (D) A and C only
    (E) A, B, and C

31. Here are the names of three organisms.

    (A) *Pseudotriton rubrum*
    (B) *Acer rubrum*
    (C) *Acer sucre*

    Which of the following statements is true about these three names?

    (A) A and B are most closely related.
    (B) A and C are most closely related.
    (C) B and C are most closely related.
    (D) All three are closely related.
    (E) None of the organisms are closely related.

32. Which of the following cell types would provide the best opportunity to study lysosomes?

    (A) neurons
    (B) macrophages
    (C) T lymphocytes
    (D) liver cells
    (E) plant cells

33. The earth is estimated to be 4.6 billion years old. How old is the oldest pro-karyotic cell?

    (A) 1.6 billion years
    (B) 2.6 billion years
    (C) 3.6 billion years
    (D) 4.6 billion years
    (E) 4.6 million years

34. Which is mismatched?

    (A) smooth endoplasmic reticulum; detoxification
    (B) centrioles; cell division
    (C) nucleolus; ribosome synthesis
    (D) plasma membrane; transport
    (E) desmosome; cell division

35. The pioneer organisms of the tundra are the

    (A) bacteria
    (B) lichens
    (C) hawks
    (D) polar bears
    (E) grasses

36. Which of the following does not contain nucleic acid?

    (A) a chloroplast in a plant cell
    (B) an animal mitochondrion
    (C) a plant mitochondrion
    (D) nucleolus
    (E) a Golgi body

37. Which is most dependent on a signal transduction pathway?

    (A) the sodium-potassium pump in an axon
    (B) uptake of nitrates in the roots of a plant by active transport
    (C) hormones triggering a cell response
    (D) absorption of light by chloroplast
    (E) the Krebs cycle

38. An amoeba living in a freshwater lake is _____ relative to its environment, and therefore, water will flow _____ the amoeba.

    (A) isotonic; both into and out of
    (B) hypertonic; into
    (C) hypertonic; out of
    (D) hypotonic; into
    (E) hypotonic; out of

39. Water readily diffuses through a plasma membrane because the water molecule is

    (A) large and highly polar
    (B) large, highly polar, and electrically charged
    (C) small, polar, and electrically neutral
    (D) small, highly polar, and electrically charged
    (E) soluble in lipids, small, and highly polar

40. Which of the following stimulates the growth of the long bones in humans?

    (A) thyroid
    (B) spleen
    (C) anterior pituitary
    (D) posterior pituitary
    (E) thymus

41. Which is true of T cells but not of B cells?

    (A) They are produced in the bone marrow.
    (B) They release histamine.
    (C) They are responsible for humoral immunity.
    (D) They attack infected body cells.
    (E) They use pseudopods to phagocytose microbes.

42. A particular poison binds to certain enzymes at positions other than the active site, thus deactivating the enzyme. This is an example of

    (A) noncompetitive inhibition
    (B) competitive inhibition
    (C) denaturing
    (D) allosteric inhibition
    (E) irreversible inhibition

43. Which is homologous to a human's arm?

    (A) an insect's wing
    (B) a lobster's leg
    (C) the tail fin of a shark
    (D) the lateral fin of a whale
    (E) a reptile's front leg

44. Which organ in humans is most closely involved with ADH?

    (A) small intestine
    (B) pancreas
    (C) liver
    (D) kidney
    (E) large intestine

45. At which level of protein structure are interactions between R groups most important?

    (A) primary
    (B) secondary
    (C) tertiary
    (D) quaternary
    (E) the R groups are not related to the overall structure of a protein

46. Which gland is responsible for releasing a hormone that causes uterine contractions during labor?

    (A) posterior pituitary
    (B) thyroid
    (C) anterior pituitary
    (D) thymus
    (E) adrenal gland

47. Here is a graph of the absorption spectrum for a particular photosynthetic pigment. What color does the pigment appear?

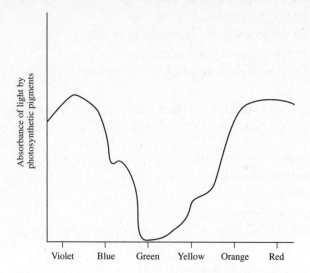

(A) red
(B) orange
(C) yellow
(D) green
(E) violet

48. In the frog embryo, gastrulation

(A) produces a blastocoel
(B) proceeds as ectoderm cells roll over the dorsal lip of the blastopore
(C) occurs along a primitive streak in the animal hemisphere
(D) occurs only in the vegetal hemisphere
(E) continues as the individual cells grow larger

49. Which of the following is an example of positive feedback?

(A) An increase in blood sugar increases the release of insulin, which lowers blood sugar.
(B) An infant suckling at a mother's breast stimulates an increase in the levels of oxytocin, which causes more milk to flow.
(C) When the level of thyroxin in the blood goes below a set point, the pituitary stimulates the thyroid to produce more thyroxin.
(D) A decrease in calcium levels in the blood increases the levels of hormone that causes the release of calcium from the bones.
(E) A decrease in blood sugar levels causes an increase in glucagon levels, which stimulates release of glucose from the liver.

50. Which is not correctly matched?

    (A) closed circulatory system; grasshopper
    (B) external respiratory surface; earthworm
    (C) tracheal tubes; arthropods
    (D) breathing by negative pressure; humans
    (E) aortic arches; earthworms

51. This molecule is

$$
\begin{array}{c}
R \\
| \\
O \\
| \\
O - P - O^- \\
| \\
O \\
| \\
CH_2 - CH - CH_2 \\
| \qquad | \\
O \qquad O \\
| \qquad | \\
C - O \quad C - O \\
| \qquad | \\
CH_2 \quad CH_2 \\
| \qquad | \\
CH_2 \quad CH_2 \\
| \qquad | \\
CH_2 \quad CH_2 \\
| \qquad | \\
CH_2 \quad CH_2 \\
| \qquad | \\
CH_2 \quad CH_2 \\
| \qquad | \\
CH_2 \quad CH_2 \\
| \qquad \backslash \\
CH_2 \quad CH \\
| \qquad \backslash \backslash \\
CH_2 \quad CH \\
| \qquad \backslash \\
CH_2 \quad CH_2 \\
| \qquad \backslash \\
CH_2 \quad CH_2 \\
| \qquad \backslash \\
CH_2 \quad CH_2 \\
| \qquad \backslash \\
CH_2 \quad CH_2 \\
| \qquad \backslash \\
CH_2 \quad CH_2 \\
| \qquad \backslash \\
CH_2 \quad CH_2 \\
| \qquad | \\
CH_3 \quad CH_3
\end{array}
$$

    (A) a major component of a plasma membrane
    (B) a component of DNA
    (C) a subunit of starch
    (D) a protein
    (E) ATP

52. Which is NOT matched correctly?

    (A) earthworm-urea
    (B) birds-urea
    (C) humans-urea
    (D) grasshopper-uric acid
    (E) hydra-ammonia

53. The penetrance for a particular autosomal recessive trait *a* is 90%. In a mating between two animals that are both hybrid for this trait, what percent of the offspring would be expected to show the recessive trait *a*?

    (A) 10%
    (B) 22.5%
    (C) 25%
    (D) 50%
    (E) 90%

54. Which gland is the bridge between the nervous system and the endocrine system?

    (A) anterior pituitary
    (B) posterior pituitary
    (C) thyroid
    (D) hypothalamus
    (E) adrenal gland

Questions 55–57

    (A) Krebs cycle
    (B) glycolysis
    (C) chemiosmosis
    (D) all of the above
    (E) none of the above

55. Releases carbon dioxide as a by-product.

    (A) A
    (B) B
    (C) C
    (D) D
    (E) E

56. ATP is required.

    (A) A
    (B) B
    (C) C
    (D) D
    (E) E

57. CO_2 is required.

    (A) A
    (B) B
    (C) C
    (D) D
    (E) E

Questions 58–59

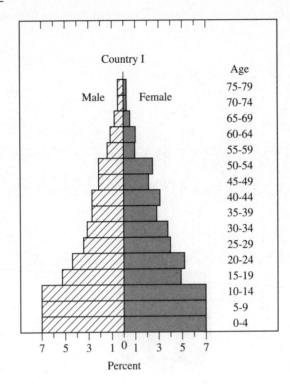

Here is an age structure pyramid for the human population in one country. The sexes are represented separately, and the percentages in the different age classes are shown by the relative widths of the horizontal bars.

58. Approximately what percentage of the individuals are younger than 20 years of age?

    (A) 7%
    (B) 21%
    (C) 35%
    (D) 52%
    (E) 72%

59. Which of the following best approximates the ratio of males to females among individuals between the ages of 65 and 69?

    (A) 1:1
    (B) 1:2
    (C) 2:1
    (D) 1:4
    (E) 4:1

Questions 60–62

The following questions refer to this sketch of an important structure in any eukaryotic cell.

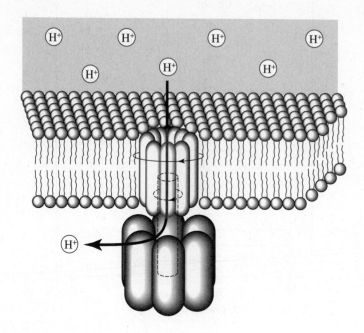

60. Where would this structure most likely be found?

    (A) Golgi apparatus
    (B) nucleus
    (C) nucleolus
    (D) chloroplast
    (E) ribosome

61. The purpose of this structure is most directly related to which of the following?

    (A) production of energy
    (B) replication of DNA
    (C) intake of materials into a cell
    (D) cell division
    (E) packaging substances for the cell

62. Which of the following would have the most profound effect on the normal functioning of this mechanism?

    (A) blocking the formation of microtubules
    (B) blocking of the sarcoplasmic reticulum
    (C) increasing the release of neurotransmitter
    (D) blocking of the signal transduction pathway
    (E) allowing protons to diffuse through the membrane

63. Some varieties of bacteria are now resistant to antibiotics. These varieties of bacteria most likely developed the resistance as a result of

    (A) divergent evolution
    (B) convergent evolution
    (C) adaptive radiation
    (D) the founder effect
    (E) natural selection

64. The site of lipid synthesis in the cell is the

    (A) lysosomes
    (B) ribosomes
    (C) cytoplasm
    (D) rough endoplasmic reticulum
    (E) smooth endoplasmic reticulum

Questions 65–66

Here is a chart describing two newly discovered organisms. To which phylum do they most likely belong?

Organisms	Protostome	Deuterostome	Acoelomate	Coelomate	Skeleton	Symmetry of Embryo	Number of Germ Layers
65.	X		X		None	Bilateral	3
66.		X		X	Endoskeleton	Bilateral	3

    (A) Cnidarians
    (B) Porifera
    (C) Echinodermata
    (D) Platyhelminthes
    (E) Mollusk

Questions 67–70

You carry out an experiment involving bacteria and antibiotic-resistant plasmids. While using a culture of *Escherichia coli* that has been growing for 24 hours at optimal temperature, you make all four tubes of bacteria competent. Then you add a plasmid carrying a gene for resistance to the antibiotic ampicillin to two of the tubes only. You do not add plasmid to the remaining two tubes. After allowing the bacteria time to recover, you inoculate four plates, each with the contents of a different tube. Two plates contain only Luria broth agar, and two plates contain Luria broth with ampicillin. You seal the plates, incubate them overnight, and examine them in the morning. Here is a drawing of the four plates. LB stands for Luria broth and AMP stands for ampicillin.

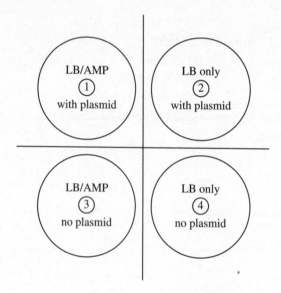

67. Which plate or plates would contain only bacteria that are resistant to antibiotic?

    (A) 1
    (B) 2
    (C) 3
    (D) 4
    (E) all the plates

68. Which plate or plates would be covered with bacteria?

    (A) 1 only
    (B) 2 only
    (C) 1 and 2
    (D) 2 and 4
    (E) 1, 2, and 4

69. Which plate would have only a few colonies growing on it?

    (A) 1
    (B) 2
    (C) 3
    (D) 4
    (E) cannot be determined

70. Which would be the best technique to make the bacteria competent?

    (A) Spin them in a centrifuge.
    (B) Place them into an incubator for 1 hour at 140°F (60°C).
    (C) Place them into the freezer for 1 hour.
    (D) Add calcium chloride and quickly transfer them from ice to a warm water bath and back to ice again.
    (E) Shake them briskly for 30 minutes.

Questions 71–72

A male fruit fly with red eyes and long wings was mated with a female fruit fly with purple eyes and vestigial wings. All the offspring in the F_1 generation had red eyes and long wings. These F_1 flies were testcrossed with purple-eyed, vestigial-winged flies. Here are the results of that testcross.

F_2 Generation

172 Red eyes, long wings
194 Purple eyes, vestigial wings
 15 Purple eyes, long wings
 _19_ Red eyes, vestigial wings
400 Total

71. Given the results of the testcross, it is most likely that

    (A) the genes for eye color do not follow traditional Mendelian inheritance
    (B) the genes are sex-linked
    (C) codominance is the pattern of inheritance
    (D) the traits for eye color and wing length are linked
    (E) there are many genes controlling these two traits

72. What is a possible explanation for the disparity in the phenotype ratios?

    (A) In this case, the laws of probability do not apply.
    (B) Crossover has occurred.
    (C) Penetrance for the purple eye trait is very low.
    (D) The trait for purple eyes and vestigial wings is referred to as superdominant.
    (E) Experimental error has occurred.

<u>Questions 73–75</u>
The following three questions refer to this sketch of a root of a plant.

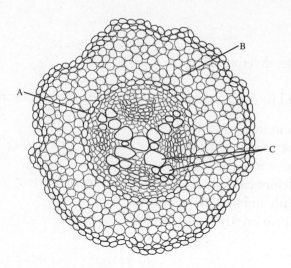

73. What is the major function of area A?

    (A) to transport sugar around the plant
    (B) to control what enters the plant
    (C) to transport water upward from the roots to the leaves
    (D) support and storage
    (E) sugar production

74. What is the major function of area C?

    (A) to transport sugar around the plant
    (B) to control what enters the plant
    (C) to transport water upward from the roots to the leaves
    (D) support and storage
    (E) sugar production

75. Identify the cell type at label B.

    (A) sieve
    (B) transport
    (C) collenchyma
    (D) sclerenchyma
    (E) parenchyma

Questions 76–77

Answer the following two questions based on this pedigree for the biochemical disorder known as alkaptonuria. Affected individuals are unable to break down a substance called alkapton, which colors the urine black and stains body tissues. Otherwise, the condition is of no consequence.

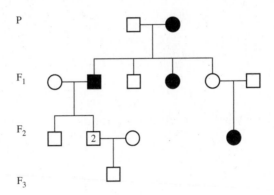

76. What is the most probable pattern of inheritance?

   (A) sex-linked dominant
   (B) sex-linked recessive
   (C) autosomal dominant
   (D) autosomal recessive
   (E) holandric

77. What is the genotype for person #2?

   (A) *A/a*
   (B) *A/A*
   (C) *a/a*
   (D) X–Y
   (E) XY

Questions 78–80

Use these choices to answer questions 78–80.

   (A) Gonadotropin-releasing hormone
   (B) Luteinizing hormone
   (C) Follicle-stimulating hormone
   (D) Progesterone
   (E) Oxytocin

78. Hormone produced by the corpus luteum that is responsible for thickening the uterine wall

79. Hormone from the hypothalamus that triggers the release of follicle-stimulating hormone

80. Hormone that triggers ovulation of the secondary oocyte

Questions 81–83
Refer to the figure below for questions 81–83.

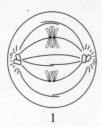 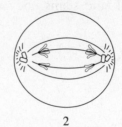 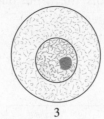  

1   2   3   4   5

81. Identify the phase when recombination occurs.

(A) 1
(B) 2
(C) 3
(D) 4
(E) 5

82. Identify the cell in interphase.

(A) 1
(B) 2
(C) 3
(D) 4
(E) 5

83. Identify the cell in metaphase II.

(A) 1
(B) 2
(C) 3
(D) 4
(E) 5

Questions 84–86
Refer to the figure below for questions 84–86.

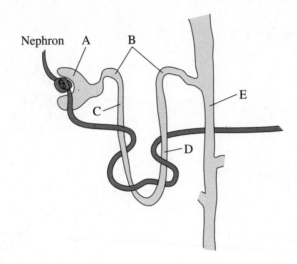

84. Identify the region of the nephron under control of ADH.

   (A) A
   (B) B
   (C) C
   (D) D
   (E) E

85. Identify the portion of the nephron where only simple diffusion occurs.

   (A) A
   (B) B
   (C) C
   (D) D
   (E) E

86. Identify the region of the nephron that is not permeable to water.

   (A) A
   (B) B
   (C) C
   (D) D
   (E) E

Questions 87–88

Here is a sketch of prokaryotic DNA undergoing replication and transcription simultaneously.

87. If 1 is thymine, what is a?

   (A) adenine
   (B) thymine
   (C) cytosine
   (D) guanine
   (E) uracil

88. If 4 is adenine, what is D?

   (A) adenine
   (B) thymine
   (C) cytosine
   (D) guanine
   (E) uracil

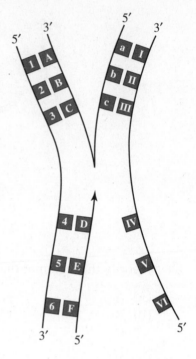

Questions 89–93

Choose from the following choices for the next five questions.

   (A) Transposon
   (B) Transformation
   (C) Transcription
   (D) Transduction
   (E) Translation

89. DNA is transferred from one cell to another by a virus

90. DNA from one strain of bacteria is taken up by another strain of bacteria

91. A segment of DNA that can jump from one place to another

92. RNA makes a polypeptide

93. DNA makes RNA

Questions 94–96
The following three questions refer to this sketch.

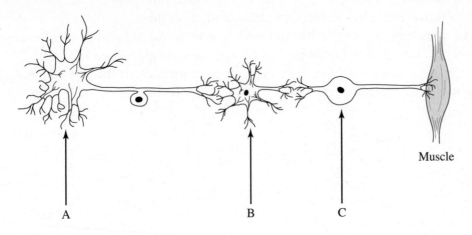

94. Which statement is true about these three cells?

    (A) A is a motor neuron.
    (B) B is an interneuron.
    (C) C is a sensory neuron.
    (D) A is an interneuron.
    (E) B is a sensory neuron.

95. Where is cell B located?

    (A) spinal cord
    (B) skeletal muscle
    (C) chest cavity
    (D) smooth muscle
    (E) skin

96. The three neurons constitute

    (A) a reflex arc
    (B) a neuromuscular junction
    (C) the central nervous system
    (D) the Hatch-Slack pathway
    (E) a signal transduction pathway

Questions 97–100

Intact chloroplasts are isolated from dark green leaves by low-speed centrifugation and are suspended into six tubes, containing cold buffer. A blue dye, DPIP, which turns clear when reduced, is also added to all tubes.

Then each tube is exposed to different wavelength(s) of light. A measurement of the amount of decolorization is made and the data plotted on the graph. Although the wavelengths of light vary, the light intensity is the same. Here is a graph of the collected data.

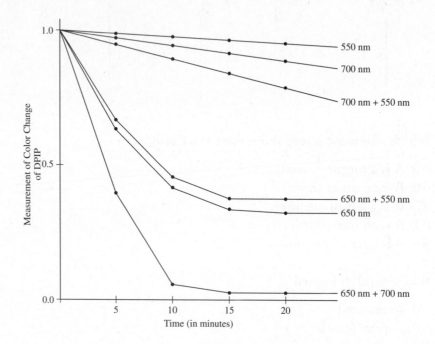

97. Which statement best describes the results of the experiment?

(A) The lower the wavelength of light, the greater the rate of photosynthesis.
(B) The highest wavelength of light provides the greatest rate of photosynthesis.
(C) The highest rate of photosynthesis results from the exposure to two different wavelengths of light.
(D) The data are too confusing to determine a clear relationship between wavelengths of light and the rate of photosynthesis.
(E) The data are clear. However, the sample is too small and more data are required to make an accurate determination.

98. What instrument would be best to quantify the color change of the DPIP?

    (A) a light microscope
    (B) the naked eye is adequate
    (C) a spectrophotometer
    (D) an electron microscope
    (E) a magnifying glass

99. The greatest reduction of the DPIP by two different wavelengths of light suggests that

    (A) plants absorb various wavelengths of green light
    (B) UV light is responsible for the greatest amount of photosynthesis
    (C) different portions of the plant absorb different wavelengths of light
    (D) the plant contains two separate photosynthetic pigments that absorb light in different wavelengths
    (E) the light shining on the plant contains the entire spectrum of light

100. Which is the dependent variable?

    (A) time
    (B) both time and the different wavelengths of light
    (C) the change in color of the DPIP
    (D) the different wavelengths of light only
    (E) temperature

STOP

If there is still time remaining, you may review your answers.

READING PERIOD

10 minutes

Read all four essays, organize your thoughts, make any notes you need to assist you on your question sheet. Following this 10-minute period, you will be directed to answer the four essays formally.

You must answer all four essays. Because each question will be weighted equally, you are advised to divide your time equally among them. Do not spend too much time on any one essay. You must keep track of the time yourself; no one will do that for you.

You are to write your answers in the answer booklet only. Use black or blue ink. Be sure to write clearly and legibly. If you make a mistake, you may save time by crossing it out with a single line rather than trying to erase it. Number each answer as the question is numbered in the examination. Begin each answer on a new page. *You may answer the essays in any order.*

Each answer should be organized, well-balanced, and as comprehensive as time permits. You must write in prose form; outline form is not acceptable. Do not spend time restating the questions; simply answer the question. If a specific number of examples are asked for, no credit will be given for extra examples. Diagrams may be used to help explain your answer, but in no case will a diagram alone suffice.

WRITING PERIOD

1 hour and 30 minutes
Suggested writing time per question—20 minutes

Question 1

A major theme in biology is that structure and function are related. This is true at all levels of life. For each of the following, give one example and discuss how the form relates to function in each case.

   a At the molecular
   b. At the cellular level
   c. At the organ level

Question 2

Living things strive to maintain homeostasis. Describe what a feedback mechanism is. Choose one positive and one negative feedback mechanism in humans, and explain how they help maintain homeostasis.

Question 3

You carry out an experiment to study how different environmental conditions affect the rate of transpiration. You set up four respirometers, in two water baths, one at 0°C and one at 25°C. The organism you use is sweet pea seeds. Some have been moistened overnight and are sprouting, and some are dry and dormant. You measure the respiration rate by measuring the volume of oxygen consumed when the respirometers are submerged in water. The sprouting peas, having absorbed water, are much larger than the dormant peas. Therefore, you add glass beads, which, you previously established, are inert to the vial with dormant peas in order to make the volumes in both vials the same.

Vial 1—Sprouting seeds at 0°C
Vial 2—Dormant seeds plus beads at 0°C
Vial 3—Sprouting seeds at 25°C
Vial 4—Dormant seeds plus beads at 25°C

   a. Label the *x*-axis and *y*-axis appropriately. Predict what the graph of this data would look like. Draw and label the lines.
   b. If you measured the rates of respiration of a 25 g reptile and a 25 g mouse at 10°C, how would they compare? Explain your reasoning.
   c. If the respiration of the mouse were monitored at 30°C, how would it differ from the rate of respiration of the mouse at 10°C? Explain your results.

Question 4

a. Genetic variation is the raw material for evolutionary change. Explain four cellular or molecular mechanisms that introduce variation into the gene pool of a population.

b. Explain the evolutionary mechanisms that can change the composition of a gene pool.

STOP

If there is still time remaining, you may review your answers.

Answer Key

MODEL TEST 2

Multiple-Choice Questions

1. A	26. B	51. A	76. D
2. A	27. E	52. B	77. A
3. C	28. C	53. B	78. D
4. D	29. D	54. D	79. A
5. B	30. C	55. A	80. B
6. E	31. C	56. B	81. D
7. E	32. B	57. E	82. C
8. D	33. C	58. D	83. E
9. B	34. E	59. C	84. E
10. B	35. B	60. D	85. A
11. E	36. E	61. A	86. D
12. B	37. C	62. E	87. B
13. A	38. B	63. E	88. E
14. B	39. C	64. E	89. D
15. D	40. C	65. D	90. B
16. B	41. D	66. C	91. A
17. B	42. D	67. A	92. E
18. E	43. D	68. D	93. C
19. D	44. D	69. A	94. B
20. B	45. C	70. D	95. A
21. A	46. A	71. D	96. A
22. C	47. D	72. B	97. C
23. C	48. B	73. B	98. C
24. D	49. B	74. C	99. D
25. D	50. A	75. E	100. C

ANSWER EXPLANATIONS TO MODEL TEST 2

Multiple-Choice Questions

1. **(A)** Selection acts on the phenotype of a population. Selection cannot act on something that does not react with the environment. The genotypes, alleles, and entire genome may be hidden and unexpressed.

2. **(A)** The oceans make up the marine biome. It is the largest biome and provides the greatest proportion of the earth's oxygen (due to the photosynthesis of algae and archaea) and nutrition.

3. **(C)** The cotyledon ($3n$) results from the fusion of one sperm nucleus (n) and two polar bodies ($2n$). A seed consists of an embryo ($2n$), which resulted from fertilization of the ovum (n) and a sperm nucleus (n). Pollen contains haploid nuclei. The fruit is a ripened ovary and consists of cells with the diploid chromosome number.

4. **(D)** Human enzymes are most active at body temperature, about 37°C. However, there is no single optimal pH for enzyme function; gastric enzymes are most active at pH 3, while hydrolytic enzymes of the small intestine are most active in the alkaline pH, around 8. Enzymes are proteins that speed up reactions by lowering the energy needed to start the reaction, and they catalyze reactions in both directions. Human enzymes begin to denature at 40°C, which is why a high fever is dangerous.

5. **(B)** The photoreceptor responsible for keeping track of the length of day and night is the pigment phytochrome. There are two forms of phytochrome, Pr (red light absorbing) and Pfr (infrared light absorbing). Phytochrome is synthesized in the Pr form. When the plant is exposed to light, Pr converts to Pfr. In the dark, Pfr reverts back to Pr. The conversion from one to the other enables the plant to keep track of time.

6. **(E)** All digestion is completed in the small intestine. One function of the large intestine is water reabsorption. Bile is produced by the liver and stored in the gallbladder. It is an emulsifier that breaks apart fats. The lacteal is located inside the villus in the small intestine. It is part of the lymphatic system and absorbs fatty acids and glycerol from the small intestine. Secretin is a hormone released by the duodenal wall that stimulates the pancreas to release bicarbonate to neutralize acid in the duodenum.

7. **(E)** The grasslands in the United States are located in the central part of the states. Availability of water is the major limiting factor for plant growth. There is adequate sunlight anywhere in North America.

8. **(D)** Runoff from sewage and fertilizer from the land increases nutrients in a lake and causes excessive growth of algae. Shallow areas become choked with weeds, and a cycle of death, decomposition, and oxygen depletion begins. In the end, so much dead organic matter accumulates on the lake bottom that the lake gets shallower and shallower until it dries up completely.

9. **(B)** A pollen grain, the mature male gametophyte, consists of three haploid nuclei. One sprouts a pollen tube and is called the tube nucleus. The other two, the sperm nuclei, fertilize the ovum and the two polar bodies during double fertilization. One sperm nucleus fuses with the ovum and becomes the embryo ($2n$). The other sperm nucleus fuses with two polar bodies and becomes the cotyledon, food for the growing embryo ($3n$).

10. **(B)** Neither parent has the condition, but the fact that a child has the autosomal recessive condition means that each parent must be a carrier, *C/c*. Therefore, their chance of producing a child with cystic fibrosis (*c/c*) is ¼. Here is the cross.

	C	*c*
C	*CC*	*Cc*
c	*Cc*	*cc*

11. **(E)** Free oxygen was not present in the early earth's atmosphere. Oxygen is a highly reactive and corrosive molecule. If it had been present in the early earth's atmosphere, it would have reacted with the molecules that were present and would have degraded them. With free oxygen, the early earth would have evolved very differently.

12. **(B)** All plants carry out photosynthesis, while some Protista, like euglena, are also photosynthetic. All animals and fungi are heterotrophs.

13. **(A)** When cones mature, they open and allow the wind to blow the seeds away. Gymnosperm means "naked seed." When mature, the seeds are exposed and naked on the cones.

14. **(B)** Hox genes control the overall body plan of an animal by controlling the developmental fate of groups of cells. Hox stands for homeotic genes or homeobox. Hox genes have been found in insects, nematodes, mollusks, fish, frogs, birds, and mammals. These genes evolved early in the history of life and have remained virtually unchanged.

15. **(D)** Normal blood clotting requires the following: thromboplastin, thrombin, fibrinogen, fibrin, calcium ions, thrombocytes, and platelets. In addition, other clotting factors are required. Macrophages fight infection.

16. **(B)** Darwin was not able to explain the origin of variation in the populations he saw. Mendel, who published his theories of the law of segregation and law of independent assortment, could have explained some of the variation that Darwin saw. However, although Mendel made his work public, no one understood it until 50 years later.

17. **(B)** Glycoproteins sticking up from the plasma membrane form part of the glycocalyx, which is important for cell identification. Cholesterol makes the plasma membrane itself more stable and less fluid. Protein channels, such as the ATP synthase channels, enable large polar proteins to pass through the membrane. Single ions channels, such as those found at the terminal branch of a neuron, allow calcium ions to flood into the terminal branch and cause the vesicles to release neurotransmitter into the synapse. Phospholipids make up 60% of most plasma membranes. In diagrams, they look like balloons with two tails.

18. **(E)** Since $p + q = 1$, if the frequency of the allele for blue eyes is 0.3, then the frequency of the allele for brown eyes is 0.7. The percentage of the population with pure brown eyes is p^2, which is $0.7 \times 0.7 = 0.49$ or 49%.

19. **(D)** Density-dependent factors are those caused by population density. When there is overpopulation, there is scarcity of food and starvation. Density-independent factors are those that do not depend on population, such as natural disasters like earthquakes.

20. **(B)** Genetic drift is any change in allelic frequency due to chance. Gene flow is the movement of alleles into or out of a population. Divergent evolution is change in allelic frequencies due to isolation of a population and exposure to new selective pressures. Balanced polymorphism is the presence of two or more distinct forms of a trait in a single population. One morph is better adapted to one area, while the other morph has the advantage in a different area. The differences are not due to chance but due to different selective pressures in different environments.

21. **(A)** Golgi apparatus consists of layers of membranes. Basal bodies, spindle fibers, and centrioles consist of 9 triplets of microtubules organized around an open central area. Cilia consist of 9 pairs of microtubules organized around 2 single microtubules.

22. **(C)** Polyploidy is having extra sets of chromosomes, $3n$, $4n$, or $5n$. This is common in plants. Plants that exhibit polyploidy cannot breed with diploid plants. This isolation can result in sympatric speciation. Behavioral isolation applies only to animals because only they can move from one place to another. Genetic drift, flow of genes due to chance, and mutation occur in both animals and plants.

23. **(C)** Human skeletal muscles convert pyruvate into lactic acid or lactate when muscles are deprived of oxygen during strenuous exercise. Lactic acid buildup in the muscles causes fatigue and burning. Cells can continue to generate ATP anaerobically only if there is an adequate supply of NAD^+ to accept electrons. During the process of converting pyruvate to lactate, NADH is oxidized to NAD^+ so that ATP production can continue.

24. **(D)** A population is the smallest biological unit that can evolve. A single individual or cell can never evolve. A population is one type of organism living in one place. A species can be spread over a great distance and can consist of thousands of isolated populations worldwide that do not interbreed and are not part of the same population.

25. **(D)** Embryonic induction is the ability of one group of embryonic cells to influence the development of another group. Spemann proved that the dorsal lip of the blastopore normally initiates a chain of inductions that results in the formation of a neural tube. He named the dorsal lip the primary organizer.

26. **(B)** Organisms at higher trophic levels have a greater concentration of accumulated toxins stored in their bodies than those at lower levels. When humans eat high on the food chain (when we eat meat), we are ingesting all the chemicals that all the animals that we eat ingested during their lifetimes. The bald eagle almost became extinct because DDT, a pesticide, got into the food chain and prevented the eagles from laying normal eggs.

27. **(E)** Polygenic inheritance refers to traits that are controlled by more than one or several genes. Traits that show a range of phenotypes like height and skin and hair color are controlled by several genes. Pink four o'clock flowers exhibit blending inheritance. ABO blood type is an example of multiple alleles. Marfan's syndrome is an example of pleiotropy, where one gene controls the expression of several genes. Epistasis is the pattern of inheritance when one trait blocks the expression of another trait.

28. **(C)** The release of histamine dilates capillaries, thus increasing blood supply to a region. This brings more phagocytes to the area to fight infection. The increased blood supply is responsible for the redness, swelling, and perhaps itchiness accompanying the increased blood supply. Fever is a response to speed up your body processes and to help destroy the pathogens that cannot tolerate high temperatures. Fever occurs later in the immune response.

29. **(D)** According to Hardy-Weinberg, the population must be large in order for the laws of probability to apply. All the other choices define a population in Hardy-Weinberg equilibrium.

30. **(C)** Facilitated diffusion is a type of diffusion, so it requires no energy. It makes use of membrane channels, such as channels to assist glucose to diffuse into a cell.

31. **(C)** *Acer sucre* (sugar maple) and *Acer rubrum* (red maple) are in the same genus. *Pseudotriton rubrum* is a red salamander and not a close relative to *Acer rubrum*. The species name is often merely an adjective.

32. **(B)** Since macrophages are giant white blood cells that phagocytose and digest huge numbers of microbes; they must contain many lysosomes, which are sacs of digestive enzymes. Plant cells do not contain lysosomes. Liver cells have large amounts of smooth endoplasmic reticulum because they synthesize steroids like cholesterol and detoxify poisons.

33. **(C)** The first prokaryote cells on earth developed 1 billion years after the earth cooled down. After another 2 billion years, the first eukaryotic cells appeared as a result of endosymbiosis.

34. **(E)** Desmosomes are junctions that keep cells from separating under physical stress. They are not involved with cell division. For example, they secure cells of the human cervix together during labor and childbirth when the cervix must stretch to accommodate the head of the fetus.

35. **(B)** Pioneer organisms are the first ones to colonize a region. Lichens are a symbiont made of an algae and a fungus. They can tolerate very hostile conditions and are the pioneer organisms in many biomes.

36. **(E)** Chloroplasts and mitochondria were once free-living prokaryotic cells (theory of endosymbiosis), and they contain DNA. Golgi bodies do not.

37. **(C)** A signal transduction pathway is a mechanism that converts an extracellular chemical signal to a specific cellular response. It involves a receptor on the cell surface and a secondary messenger, such as cAMP, within the cytoplasm to trigger a response from the nucleus. A majority of hormones trigger a cell response using signal transduction pathways.

38. **(B)** Freshwater is 100% water, while the amoeba has many solutes dissolved in it. Hypertonic means having more solutes dissolved in it than are dissolved in the surrounding area. The amoeba is hypertonic to the lake water. Since there is a higher concentration of water outside the amoeba, water will flow down the gradient and into the cell. To keep from bursting, freshwater Protista have contractile vacuoles to pump out excess water.

39. **(C)** A molecule of water is small, electrically uncharged, highly polar, and not soluble in lipids. Since the phosphate end of a phospholipid is also very polar, water can diffuse through the phospholipid layer. In addition membranes have aquaporins, special channels that assist in the transport of water across a membrane.

40. **(C)** The anterior pituitary releases human growth hormone, which stimulates growth in the long bones. The thyroid releases thyroxin, which controls metabolic rate. The spleen acts as a reservoir for extra blood and is a place where blood gets filtered. The posterior pituitary releases antidiuretic hormone and oxytocin. The thymus in humans is responsible for maturing T lymphocytes in childhood and is critical for normal development of the immune system.

41. **(D)** T cells attack infected body cells, while B cells produce antibodies and provide humoral immunity. Both T and B lymphocytes mature in the bone marrow. Histamine is secreted, not by either B or T lymphocytes, but by another type of white blood cell.

42. **(D)** The question does not say that the poison competes with a substrate for one active site. Therefore, this is not competitive inhibition. The poison bonds to the enzyme at a site other than the active site for the substrate and deactivates the enzyme. This is allosteric inhibition, choice D. When the poison bonds to the allosteric site, it changes the shape of the enzyme and deactivates it. The question also does not describe noncompetitive inhibition. In noncompetitive inhibition, there are two active sites on one enzyme and two substrates that do not resemble each other. The binding of one substrate to the enzyme may block the binding of the second substrate.

43. **(D)** The lateral fin of the whale consists of the same bones as a human's arm, the radius, ulna, and humerus. These structures are considered to be homologous because the underlying structure is similar and, therefore, humans and whales share a common ancestor. The lobster's leg, the reptile's front leg, and the insect's wing are analogous to the human's arm. They have a common function but no common structure, and they do not share a common ancestor. The tail fin of the shark is not related in any way to any of the others.

44. **(D)** ADH initiates a change in the kidneys in response to a change in concentration of the blood. The nephron responds to stimulation by ADH by increasing the reabsorption of water and decreasing the output of urine. The response is rapid.

45. **(C)** Tertiary structure of a protein refers to its intricate three-dimensional shape, known as conformation. Interactions between R groups of the amino acids that make up the protein contribute to this three-dimensional shape and to its particular function. The primary structure of a protein refers to the amino acid sequence. Secondary structure refers to intramolecular hydrogen bonding. Quaternary structure refers to a polypeptide consisting of more than one chain. The R groups are all-important in the overall conformation of the protein.

46. **(A)** The posterior pituitary secretes oxytocin, which stimulates uterine contractions as well as the production of milk from mammary glands. The posterior pituitary also secretes antidiuretic hormone, which targets the kidney.

47. **(D)** Everything is being absorbed except green, which is being reflected. The color reflected is the color observed. Therefore, the pigment is green. Green plants reflect green light and therefore cannot utilize green light as an energy source for photosynthesis.

48. **(B)** The end result of gastrulation is the formation of three primary cell layers: the endoderm, ectoderm, and mesoderm. The mesoderm will form as a consequence of the ectoderm migrating inward over the dorsal lip of the blastopore, as described in the question.

49. **(B)** Positive feedback is characterized by an enhancement of something that is already occurring and continues to an endpoint. Negative feedback maintains homeostasis (it does not reach an endpoint) by increasing or decreasing levels. Choices A, C, D, and E are all examples of negative feedback.

50. **(A)** Arthropods, like the insect the grasshopper, have open circulatory systems with hemocoels for exchange of nutrients and respiratory gases between the cells and the environment. Arthropods have spiracles, openings in the abdomen, that lead to tracheal tubes that lead into the hemocoels. Earthworms exchange respiratory gases through moist skin; so we say they have an external respiratory surface. The earthworm has a closed circulatory system. Breathing in humans is by negative pressure. Air is drawn into the lungs as the pressure of the chest cavity decreases because the volume is expanded.

51. **(A)** This molecule is a phospholipid; it consists of one glycerol molecule, two fatty acid molecules, and a phosphate. Along with proteins, it makes up the plasma membrane.

52. **(B)** Birds must conserve water and excrete nitrogenous waste as uric acid, a solid crystal. The other choices are correctly matched.

53. **(B)** The cross between two hybrids results in the homozygous recessive trait appearing 25 percent of the time. However, the penetrance for this trait is only 90 percent. That means that when the trait is present, it shows in the phenotype only 90 percent of the time. So, .90 × .25 = 22.5%

54. **(D)** The hypothalamus is part of the nervous system when it stimulates a gland electrically, releasing neurotransmitters. As an endocrine gland, it produces the hormones oxytocin and antidiuretic hormone, which it sends to the posterior pituitary for storage and from where it is released to the body.

55. **(A)** Of the choices, only the Krebs cycle releases CO_2. The end products of glycolysis are pyruvic acid and ATP. Chemiosmosis is the process by which ATP is produced during the electron transport chain.

56. **(B)** Two molecules of ATP are the energy of activation for glycolysis. The other two processes produce ATP.

57. **(E)** Carbon dioxide is not used during cell respiration at all; it is given off. It is, however, required for the light-independent reactions of photosynthesis.

58. **(D)** Add up the values of the right *and* left side of the graph *below* age 20.

59. **(C)** There are twice as many males (1%) as females (.5%) between the ages of 65 and 69.

60. **(D)** This sketch represents the ATP synthetase molecule that produces ATP during the electron transport chains of aerobic respiration and the light-dependent reactions of photosynthesis. It is found in the cristae membrane of mitochondria and the thylakoid membrane of the chloroplast.

61. **(A)** Most ATP in a cell is produced from chemiosmosis at this ATP synthetase channel.

62. **(E)** ATP is produced as protons flow down a steep gradient from one side of the membrane through the ATP synthetase molecule to the other side. This proton motive force exists because of the steep proton gradient. If protons could diffuse through the membrane at any point, there could be no gradient and no ATP produced. Therefore, allowing protons to diffuse through the membrane would destroy the gradient and destroy the ability of the cell to make ATP.

63. **(E)** The appearance of antibiotics selected against susceptible bacteria by killing them. Only resistant individuals survived to reproduce, passing their resistant genes on to the next generation. This is the process of natural selection.

64. **(E)** The smooth endoplasmic reticulum also synthesizes lipids, including steroids like cholesterol. In addition, the smooth E.R. detoxifies the cell. The liver is known to be the detoxifying organ, and liver cells are packed with smooth E.R.

65. **(D)** Platyhelminthes are triploblastic, protostome acoelomates with bilateral symmetry and no skeleton. The mollusk is a triploblastic coelomate with bilateral symmetry but it has a shell. Hydra and sponges are diploblastic. The hydra has radial symmetry, and the sponge has no symmetry.

66. **(C)** Echinodermata are triploblastic, deuterostome coelomates with bilateral symmetry. They have an endoskeleton that grows with the animal; they do not shed a shell. The thing that we see in seaside shops that we call a starfish (sea star) is merely the endoskeleton; that remains after the animal dies and decomposes.

67–69. Here is a general explanation for the results of the experiment with plasmids. Only bacteria that have taken up a plasmid with resistance to ampicillin can grow on a plate containing ampicillin. Bacteria containing no plasmid will die when they are poured onto a plate containing ampicillin.

    Plates 2 and 4 contain no ampicillin, so nothing is going to hamper the growth of any bacteria. They will be covered with bacteria. Plates 1 and 3, by contrast, contain antibiotic. The only bacteria that can grow on those plates are bacteria that carry a plasmid for resistance to that antibiotic.

67. **(A)** Since Plate 1 contains ampicillin, any bacteria growing on that plate must be ampicillin resistant.

68. **(D)** Only plates 2 and 4 would be covered with bacteria because neither plate contains ampicillin. There is no impediment to their growth.

69. **(A)** Plate A contains ampicillin and was plated with bacteria that had the opportunity to uptake a plasmid. However, not every bacterium actually absorbs the plasmid and survives. The transformation rate is generally not very high. You should expect about 3–12 colonies growing on plate 1. These will all be resistant to ampicillin.

70. **(D)** The procedure to make a cell competent is heat shock. Apply ice, then heat, then ice again, all in the presence of calcium ions, which help disrupt the bacterial membrane to allow plasmids to enter.

71. **(D)** There are four phenotypes in the results from the testcross, but the ratios are not exactly what you have seen before. These ratios are close to 1:1 with some odd exceptions (purple eyes, long wings and red eyes, vestigial wings). A ratio of 1:1 in a testcross, results from crossing the monohybrid with the pure recessive, such as *A/a* × *a/a*. A testcross of a dihybrid when the traits are on separate chromosomes, *T/t Y/y* (law of independent assortment), would yield four different phenotypes in a ratio of 1:1:1:1. In this case, crossing *R/r L/l* × *r/r l/l* produces a 1:1 ratio in the offspring and means the traits for eye color and wing length must be linked.

72. **(B)** The two small numbers of animals (15 and 19) are recombinants and result from crossover. To get a recombinant rate, take the number of recombinants, divide by the total number of organisms, and multiply that answer by 100.

73. **(B)** Arrow A points to the endodermis surrounding the stele that contains the vascular cylinder. Each endodermal cell is surrounded by an impermeable Casparian strip. The endoderm controls what enters the vascular cylinder and, thus, what enters the entire plant.

74. **(C)** The letter C points to the large, thick-walled, empty-looking structures. These are xylem tubes that carry water and nutrients from the soil up to the leaves.

75. **(E)** The arrow points at the ground tissue whose function is support and storage. These are parenchymal cells. They have a thin primary cell wall only and provide support because they swell and become turgid.

76. **(D)** Approach these pedigree problems by trying to eliminate any inheritance patterns you can. In this case, the F_2 daughter is the key to the puzzle. First, you can eliminate sex-linked. If the trait were sex-linked, you could not account for the F_2 daughter with the condition. In order for her to have a sex-linked recessive trait, she would have had to have received two mutant traits from her parents. Additionally, her father does not have the condition. Can you eliminate autosomal dominant? Yes, look again at the same F_2 daughter. She has the condition because she got a mutant gene from each parent, even though they do not express it. That only leaves autosomal recessive as the pattern of inheritance.

77. **(A)** Person #2 must be hybrid because his father has two mutant genes and he must have inherited one. If person #2 had inherited another mutant gene from his mother, he would have the condition, but he does not.

78. **(D)** The corpus luteum forms from the follicle left behind after ovulation. It secretes progesterone and estrogen, which both thicken the endometrial wall of the uterus.

79. **(A)** Gonadotropin-releasing hormone is released by the hypothalamus and does just what its name describes. It stimulates the anterior pituitary to release its gonad-stimulating hormones: follicle-stimulating hormone (FSH) and luteinizing hormone (LH). These, in turn, stimulate the ovaries.

80. **(B)** Luteinizing hormone is a gonadotropin; it stimulates the activities of the male and female gonads. In males, it promotes production of sperm. In females, it stimulates the maturing of the secondary oocyte.

81. **(D)** Recombination occurs as a result of crossing-over, which happens during prophase of meiosis I when homologous chromosomes pair up. This process of pairing up is called synapsis; it occurs during prophase I.

82. **(C)** The nucleolus is visible during interphase and the nuclear membrane is intact. The chromosomes appear as chromatin network, as dots, not as linear, condensed chromosomes.

83. **(E)** Metaphase II is like mitosis, chromosomes line up *single file* on the metaphase plate. Picture 1 shows meiosis I, with chromosomes lined up double file. Picture 2 shows anaphase 1, with replicated chromosomes being pulled apart by the spindle fibers.

84. **(E)** The collecting tube is under the control of antidiuretic hormone from the posterior pituitary. ADH controls the permeability of the collecting tube. If ADH is released, the collecting tube becomes more permeable to water and more water passes into the surrounding tissue and back into the bloodstream. As a result, less urine is produced. If no ADH is released, the reverse situation occurs; large amounts of dilute urine are produced.

85. **(A)** Simple diffusion occurs during filtration, as molecules that are small enough diffuse from the glomerulus to Bowman's capsule. This is the least selective process in the formation of urine in the nephron.

86. **(D)** The lower portion of the ascending loop of Henle is impermeable to water. The control of the flow of water in and out of the nephron is critical to the normal functioning of the nephron.

87. **(B)** Replication is occurring at the top of the molecule, while transcription is occurring at the bottom of the molecule. Therefore, the strands *ABC* and *abc* are new DNA strands. If 1 is thymine, then I must be adenine and a must be thymine also. Therefore A must be adenine.

88. **(E)** If 4 is adenine, then D, which is a newly forming strand of RNA, must be uracil.

Numbers 89–93 are all definitions.

89. **(D)** Transduction describes the transfer of DNA from one cell to another by a virus. Phage viruses acquire bits of bacterial DNA as they infect one cell after another. This process leads to genetic recombination. There are two types of transduction, generalized and restricted, or specialized. Generalized transduction moves random pieces of bacterial DNA from one cell to another. Restricted transduction involves the transfer of specific pieces of DNA.

90. **(B)** Bacterial transformation was discovered by Frederick Griffith when he performed experiments with several different strains of the bacterium *Diplococcus pneumoniae*. Some strains were virulent and caused pneumonia in humans and mice, and some strains were harmless. Griffith discovered that something in the heat-killed virulent bacteria could be taken up by the live harmless bacteria and transform them.

91. **(A)** Transposons are movable genetic elements and are often called jumping genes. They were discovered by Barbara McClintock, who was studying the genetics of corn. Some transposons jump, in a cut-and-paste fashion, from one part of the genome to another. Others make copies of themselves that then move to another region of the genome, leaving the original behind. There are two types: insertion sequences and complex transposons.

92. **(E)** Translation is the process by which the codons of mRNA sequence are changed into an amino acid sequence. This occurs at the ribosome and consists of three parts: initiation, elongation, and termination.

93. **(C)** Transcription is the process by which DNA makes RNA. Transcription occurs in three stages: initiation, elongation, and termination.

94. **(B)** The interneuron is also called the association neuron. It integrates messages from sensory and motor neurons and sends impulses to the brain.

95. **(A)** The interneuron is a compact cell and sits within the spinal cord.

96. **(A)** The reflex arc is automatic and protective. When you step on a sharp object, you jump away and only later realize what happened because the impulse passed from the sensory neuron to the interneuron to the motor neuron and directly to your skeletal muscle. The interneuron simultaneously sent the message to your brain. Processing the message takes more time than does sending the impulse to your limb that pulls your foot or hand away from the cause of the pain.

97. **(C)** The lower the line on the graph, the less blue is the DPIP. The less DPIP indicates that more DPIP has been reduced, which means that the rate of photosynthesis is the fastest.

98. **(C)** A spectrophotometer measures either how much light passes through or is absorbed by a solution, depending on how you set up the instrument.

99. **(D)** Plants use chlorophyll *a* plus antenna pigments to broaden the wavelengths of light that a plant can use for photosynthesis. That is why combined wavelengths of light were more effective in this experiment.

100. **(C)** The dependent variable in this case is the decolorization of DPIP in response to light.

What Topics Do You Need to Work On?

Table 22.1 shows an analysis by topic for each question on Model Test 2.

Table 22.1

Topic Analysis			
Biochemistry and Enzymes (Ch. 3)	**Cells and Cell Division (Ch. 4, 7)**	**Cell Respiration and Photosynthesis (Ch. 5, 6)**	**Heredity and Molecular Genetics (Ch. 8, 9)**
4, 42, 45, 51	17, 21, 30, 34, 36, 38, 39, 64, 81, 82, 83	23, 37, 47, 55, 56, 57, 60, 61, 62, 97, 98, 99, 100	10, 27, 53, 67, 68, 69, 70, 71, 72, 76, 77, 87, 88, 89, 90, 91, 92, 93
Classification and Evolution (Ch. 10, 11)	**Animals (Ch. 13, 14, 15, 17)**	**Plants (Ch. 12)**	**Ecology (Ch. 16)**
1, 11, 12, 16, 18, 20, 22, 24, 29, 31, 33, 43, 63, 65, 66	6, 14, 15, 25, 28, 32, 40, 41, 44, 46, 48, 49, 50, 52, 54, 78, 79, 80, 84, 85, 86, 94, 95, 96	3, 5, 9, 13, 73, 74, 75	2, 7, 8, 19, 26, 35, 58, 59

How to Score Your Essay

After you have written the best essay you can write in about 20 minutes, you are ready to grade it. First, though, take a short break to clear your head. When you are ready, reread your essay. Put a 1 to the left or right of any line where you explain any point that is listed in the following **scoring standard**. If you simply list examples, like "induced fit" or "sodium-potassium pump," you get NO points. *You must explain each answer.* Also, if you try to explain something without using the scientific term, you get NO credit. Here is an example. If you say, "There are protein channels in the plasma membrane" without saying what they do or how they function, you get no credit.

After reading and analyzing each part of the question, add up all the points you placed at the end of the line. If a question has three parts, the maximum number of points you can earn in any one part is 3–4. If, in your essay, you happen to include every point listed, you will get a maximum of 10 points. Be honest. You must use and explain scientific terminology and explain all concepts clearly so that anyone would understand what you are trying to say.

Scoring Standard for Free-Response Questions

Question 1
Total—10 points

At the molecular level

1 pt.   Enzyme/induced fit/denaturing—A specific enzyme works on a substrate based on shape.

1 pt.   Allosteric interactions/modifiers—These are molecules that fit into an allosteric site and cause a change in the shape of the enzyme's active site where the substrate binds, thus activating or inactivating the enzyme. An example of such an allosteric inhibitor is ATP, which inhibits phosphofructo-kinase in step 3 of glycolysis.

1 pt.   Hemoglobin/Cooperativity in picking up oxygen molecules— Once adult hemoglobin successfully binds to one oxygen molecule, the hemoglobin molecule undergoes a change in conformation and then readily binds to three other oxygen molecules. The ease with which hemoglobin bonds to oxygen is a function of its shape.

1 pt.   Sodium-potassium pump changes its shape (conformation) and carries Na^+ or K^+ across a membrane—The pump can carry three sodium ions in one direction and two potassium ions in the other direction.

1 pt.   ATP synthetase channel—Like all channels, this channel is specific and based on its shape. It allows only protons to flow through. This is the basis of ATP production during cellular respiration.

1 pt.   Protein channels in plasma membrane are specific—There are calcium gated ion channels in the terminal branches of neurons. There are channels that will allow only potassium but not sodium, which is a smaller ion, to cross a membrane. These can be found in nephrons.

1 pt.   Aquaporins are channels in plasma membranes that assist in the transport of water. These allow water to diffuse through a membrane in great quantities.

1 pt.   The glycocalyx of the plasma membrane consists of carbohydrates attached to proteins within the membrane. The shape of the glycocalyx determines how it functions. It is important to the normal functioning of cell-to-cell recognition.

4 pts.   Maximum

At the cellular level

1 pt.   A motor neuron is long and spindly because it must send a message a distance. A fat cell consists mostly of a large vacuole that stores lipids.

1 pt.   An interneuron is compact because it resides in the confines of the spinal cord.

1 pt.   Parenchymal cells have a thin, flexible cell wall and no secondary cell wall. These cells support the plant because they become turgid.

1 pt.   Sclereid cells have very thick cell walls fortified with lignin and very little cytoplasm. Their function is physical support of the plant. They make up hemp and rope, tough seed coats, and pits.

1 pt.   Normal shape of red blood cell vs. the misshapen sickling cell—Normal red blood cells are oval shaped and carry oxygen. When they sickle, as a result of a mutation in the hemoglobin, they cannot carry oxygen.

4 pts.   Maximum

At the organ level

1 pt.   Long, convoluted small intestine for complex digestion and absorption

1 pt.   Spongelike structure of the lungs that must exchange respiratory gases

1 pt.   Long tubule and collecting duct of the nephron in the kidney—its length is critical for its function. The longer the loop of Henle is, the greater the reabsorption of water

1 pt.   Vascular system—Thick muscular walls in the arteries withstand enormous pressure. The veins have thin walls because the blood pressure in the veins is very low.

1 pt.   The walls of the heart vary in thickness to accommodate function. The left ventricle has the thickest wall because it must pump blood the farthest.

4 pts.   Maximum

Question 2
Total—10 points

> *Note: Even though there are no letters separating one part of this question from another, there are clearly three parts to the question. First, explain what a feedback mechanism is, then give two examples in humans, one positive one and one negative one.*

1 pt. Define feedback mechanism—A feedback mechanism is a self-regulating mechanism that increases or decreases the level of a particular substance.

2 pts. Distinguish positive feedback from negative feedback. Negative feedback maintains homeostasis; while, positive feedback increases something that is already going on until a goal is achieved.

3 pts. Maximum

Examples of negative feedback: Discuss one of the following in some detail. You can discuss almost any hormone for negative feedback.

1 pt. Anterior pituitary → thyroid → thyroxin when thyroxin levels are low

1 pt. Increase in glucagon when blood sugar is too low

1 pt. Increase in insulin when blood sugar is too high

1 pt. When the body is dehydrated, pituitary pumps out ADH to increase water reabsorption in kidney

1 pt. Referring to cell respiration—when ATP levels are high, the allosteric enzyme PFK does not catalyze step 3 of glycolysis and ATP production ceases.

5 pts. Maximum

Examples of positive feedback: These are less common. You should memorize one.

1 pt. Levels of oxytocin, which cause uterine contractions and induces labor, increase when contractions begin. The pressure of the baby's head on the cervix increases oxytocin levels, which in turn, increase contractions. The goal is delivery of a baby and the afterbirth. Soon after, the contractions cease.

1 pt. Helper T cells release interleukin-1 to stimulate greater helper T cell activity.

5 pts. Maximum

Answer Explanations

Question 3
Total—10 points

a. **Graph of 4 vials in a respirometer.**

1 pt.   Proper labeling of the *x*-axis and *y*-axis. The *x*-axis is labeled "Time (minutes)." The *y*-axis is labeled "Oxygen Consumed (mL)."

2 pts.   Drawing the lines in the correct fashion and in the correct order. See lab #5.

3 pts.   Maximum

b. **Elaborate on this theme**
The respiration rate of the reptile would be less than that of the mouse because the reptile is an ectotherm and has a lower rate of respiration because it does not have to keep its body warm. The lower ambient temperature would slow down the rate of respiration of the reptile. The mouse, a mammal, is an endotherm and has a high rate of respiration, much higher than the reptile. In addition, small mammals have high metabolic rates and, therefore, high rates of respiration.

4 pts.   Maximum

c. **Elaborate on this theme**
The main point is that at a higher temperature, the rate of respiration in the mouse would not change because endotherms maintain a homeostatic body temperature regardless of the external temperature. However, you might add another point. If you assume and state that a temperature of 86°F (30°C) is very hot, you could say this would stress the animal severely and might cause an increased need for oxygen and, therefore, an increased rate of respiration.

4 pts.   Maximum

Question 4
Total—10 points

> *Note: Review this section in the chapter "Evolution." All of these are listed and explained in that section.*

---

a. **Four mechanisms that introduce variation into the gene pool.**

1 pt.	Mutation—change in the DNA
1 pt.	Mutagens—things that cause mutation, such as radiation
1 pt.	Point mutations/deletions, additions, frameshift
1 pt.	Chromosome mutations/nondisjunction/translocations/ inversions aneuploidy, any error caused by nondisjunction/ polyploidy
1 pt.	Crossover
1 pt.	Transposon/jumping genes
1 pt.	Transduction/general and restricted
1 pt.	Genetic engineering
1 pt.	Sexual reproduction
1 pt.	Recombination of alleles
1 pt.	Random fertilization
1 pt.	Independent assortment
1 pt.	Outbreeding

5 pts. Maximum

b. **Evolutionary mechanisms that can change the composition of a gene pool**

1 pt.	Natural selection—explain the process—give examples
1 pt.	Gene flow
1 pt.	Genetic drift/founder effect/bottleneck effect—you must explain each of them
1 pt.	Mutation
1 pt.	Sexual selection
1 pt.	Artificial selection

5 pts. Maximum

APPENDICES
AND
GLOSSARY

Appendix A

Bibliography

These textbooks are valuable resources for any basic college biology course.

Campbell, Neil A., and Jane Reece. *Biology*, 6th ed., San Francisco, California: Pearson, Benjamin Cummings, 2005

Freeman, Scott. *Biological Science*, 2nd ed., Upper Saddle River, New Jersey: Pearson Prentice Hall, 2005

Klug, William S., and Michael R. Cummins. *Concepts of Genetics,* 6th ed., Upper Saddle River, New Jersey: Prentice Hall, 2000

Krogh, David. *Biology, A Guide to the Natural World*, Upper Saddle River, New Jersey: Prentice Hall, Inc., 2000

Mader, Sylvia S. *Biology*, 9th ed., Dubuque, Iowa: McGraw Hill, 2007

McFadden, C., and W. Keeton. *Biology—Exploration of Life*. New York: Norton, 1995

Micklos, David A. and Freyer, George A. *DNA Science, A First Course*, 2nd ed., Cold Spring Harbor, New York: Cold Spring Harbor Press, 2003

Morgan, J., and M. Carter. *Investigating Biology*, 2nd ed., Redwood, California: Benjamin Cummings, 1996

Nelson, David L., and Cox, Michael M. *Lehninger Principles of Biochemistry*, 4th ed., New York: W. H. Freeman, 2005

Purves, William K., Gordon H. Orians, and H. Craig Heller. *Life: The Science of Biology*, 6th ed., New York: W. H. Freeman, 2000

Raven, Peter H. and George B. Johnson, *Biology*, 5th ed., Dubuque, Iowa: William C. Brown/McGraw Hill Publishers, 1999

Solomon, Eldra, Linda R. Berg, and Diane W. Martin. *Biology*, 5th ed., Orlando, Florida: Harcourt, 1999

Starr, Cecie, and Ralph Taggart. *Biology: The Diversity of Life*, 12th ed., Belmont, California: Wadsworth Publishing, 2009

Tobin, Allan J. and Jennie Dushek. *Asking About Life*, 2nd ed. Orlando, Florida: Harcourt College Publishers, 2001

Wallace, Robert A., Gerald P. Sanders, and Robert J. Ferl. *Biology: The Science of Life*, 4th ed., New York: Addison-Wesley Publishing Co., 1996

Appendix B

Measurements Used in Biology

Measurements

Quantity	Name of Unit	Symbol	Conversion
length	meter	m	$1\,m = 1{,}000\,mm = .001\,km = 100\,cm$
	kilometer	km	$1\,km = 1{,}000\,m$
	centimeter	cm	$1\,cm = 1/100\,m$
	millimeter	mm	$1\,mm = 1{,}000\,\mu m$
	micrometer	μm	$1\,\mu m = 1{,}000\,nm$
	nanometer	nm	$1\,nm = 1/1{,}000\,\mu m$
area	square meter	m^2	area encompassed by a square, each side of which is 1 meter in length
	hectare	ha	$1\,ha = 10{,}000\,m^2 = 2.47\,acres$
	square centimeter	cm^2	$1\,cm^2 = 1/10{,}000\,m^2$
volume	liter	L	$1\ liter = 1/1{,}000\,m^3 = 1.057\,quarts$
	milliliter	mL	$1\,mL = 0.0001\,L$
	microliter	μL	$1\,\mu L = 0.0001\,mL$
mass	kilogram	kg	$1\,kg = 1{,}000\,g$
	gram	g	$1\,g = 1{,}000\,mg$
	milligram	mg	$1\,mg = 1{,}000\,mg$
	microgram	μg	$1\,\mu g = 1/1{,}000\,mg$
temperature	Kelvin	K	$0\,K = absolute\ zero$
	degrees Celsius	°C	$°C + 273 = Kelvin$

Measurements

Quantity	Name of Unit	Symbol	Conversion
heat, work	calorie	cal	1 calorie = the amount needed to raise 1 gram of pure water 1° Celsius (from 14.5°C to 15.5°C = 4.184 Joules)
	kilocalorie	kcal	1 kcal = 1,000 cal
	joule	J	1 cal = 4.184 J
electric potential	volt	V	a unit of potential difference or electromotive force
	millivolt	mV	$1\,mV = 1/1{,}000\,V = 10^{-3}\,V$
time	second	s.	1 s. = 1/60 min.
	minute	min.	1 min. = 1/60 hr.
	hour	hr.	1 hr. = 1/24 d.
	day	d.	1 d. = 24 hr.

Glossary

ABA (abscisic acid) Plant hormone that inhibits growth, closes stomates during times of water stress and counteracts breaking of dormancy.

Abiotic Nonliving and includes temperature, water, sunlight, wind, rocks, and soil.

Abscission The process of leaves falling off a tree or bush.

Acetylcholine One of many neurotransmitters.

Acid rain Caused by pollutants in the air from combustion of fossil fuels. The pH is less than 5.6.

Actin Thin protein filaments that interact with myosin filaments in the contraction of skeletal muscle.

Action potential A rapid change in the membrane of a nerve or muscle cell when a stimulus causes an impulse to pass.

Active immunity The type of immunity when an individual makes his or her own antibodies after being ill and recovering or after being given an immunization or vaccine.

Adaptive radiation The emergence of numerous species from one common ancestor introduced into an environment.

Adenine A nucleotide that binds to thymine and uracil. It is a purine.

Adipose Fat tissue.

Allopatric speciation The formation of new species caused by separation by geography, such as mountain ranges, canyons, rivers, lakes, glaciers, altitude, or longitude.

Allosteric A type of enzyme that changes its conformation and its function in response to a modifier.

Amoebocytes Found in sponges, these cells are mobile and perform numerous functions, including reproduction, transport of food particles to nonfeeding cells, and secretion of material that forms the spicules.

Amphipathic A molecule with both a positive and negative pole.

Anaerobic respiration The anaerobic breakdown of glucose into pyruvic acid with the release of a small amount of ATP.

Analogous structures Structures, such as a bat's wing and a fly's wing, that have the same function, but the similarity is superficial and reflects an adaptation to similar environments, not a common ancestry.

Aneuploidy Any abnormal number of a particular chromosome.

Angiosperms Flowering plants.

Anode The positive pole in an electrolytic cell.

Antenna pigment Accessory photosynthetic pigment that expands the wavelengths of light that can be used to power photosynthesis.

Anterior pituitary Gland in the brain that releases many hormones, including growth hormone, luteinizing hormone, thyroid-stimulating hormone, adrenocorticotropic hormone, and follicle-stimulating hormone.

Anther Part of a flowering plant that produces male gametophytes.

Antheridia   Structures in plants that produce male gametes.

Antibodies   Produced by B lymphocytes and destroy antigens.

Anticodon   The three-base sequence of nucleotides at one end of a tRNA molecule.

Antidiuretic hormone   Released by the posterior pituitary, its target is the collecting tube of the nephron.

Apical dominance   The preferential growth of a plant upward (toward the sun), rather than laterally.

Apoplast   The network of cell walls and intercellular spaces within a plant body that permits extensive extracellular movement of water within a plant.

Apoptosis   Programmed cell death.

Aposematic coloration   The bright, often red or orange coloration of poisonous animals as a warning that predators should avoid them.

Archegonia   Structures in plants that produce female gametes.

Artificial selection   The intentional selection of specific individuals with desired traits for breeding.

Associative learning   One type of learning in which one stimulus becomes linked, through experience, to another.

ATP-synthase channels   Located in the cristae of mitochondria and thylakoids of chloroplasts, these are membrane channels that allow protons to diffuse down a gradient in the production of ATP.

Autonomic nervous system   The branch of the vertebrate peripheral nervous system that controls involuntary muscles.

Autotrophs   Organisms that synthesize their own nutrients.

Auxin   A plant hormone that stimulates stem elongation and growth, enhances apical dominance, and is responsible for tropisms.

Back cross   See test cross.

Bacteriophage   A virus that attacks bacteria.

Balanced polymorphism   The presence of two or more phenotypically distinct forms of a trait in a single population, such as two varieties of peppered moths, black ones and white ones.

Barr body   An inactivated X chromosome seen as a condensed body lying just inside the nuclear envelope.

Batesian mimicry   The copycat coloration where one harmless animal mimics the coloration of one that is poisonous. An example is the viceroy butterfly, which is harmless but looks similar to the monarch butterfly.

Binomial nomenclature   A scientific naming system where every organism has a unique name consisting of two parts: a genus name and a species name.

Biological magnification   A trophic process in which substances in the food chain become more concentrated with each link of the food chain.

Biomes   Very large regions of the earth, named for the climatic conditions and for the predominant vegetation. Examples are marine, tropical rain forest, and desert.

Biosphere   The global ecosystem.

Biotic potential   The maximum rate at which a population could increase under ideal conditions.

B lymphocyte   A lymphocyte that produces antibodies.

Bottleneck effect   An example of genetic drift that results from the reduction of a population, typically by natural disaster. The surviving population is no longer genetically representative of the original one.

Botulinum   The genus name for the bacterium that produces botulism, a very serious form of food poisoning.

Bryophytes   Nonvascular plants like mosses.

Bulk flow   The general term for the overall movement of a fluid in one direction in an organism, such as sap flowing in a tree or blood flowing in a human.

Bundle sheath cell A type of photosynthetic plant cell that is tightly packed around the veins in a leaf.

C-3 plant The common type of plant, different from C-4 and CAM plants.

C-4 plant A plant with anatomical and biochemical modifications for a dry environment that differ from C-3 and CAM plants. Examples are sugarcane and corn.

Calvin cycle A cyclical metabolic pathway in the dark reactions of photosynthesis that fixes or incorporates carbon into carbon dioxide and produces phosphoglyceraldehyde (PGAL), a three-carbon sugar.

CAM (crassulacean acid metabolism) A form of photosynthesis that is an adaptation to dry conditions; stomates remain closed during the day and open only at night.

Capsid The protein shell that encloses viral DNA or RNA.

Carbon fixation Carbon becomes fixed or incorporated into a molecule of PGAL. This happens during the Calvin cycle.

Carbonic acid anhydrase An enzyme found in red blood cells that catalyzes the conversion of carbon dioxide and water into carbonic acid as part of the system that maintains blood pH at 7.4.

Carotenoid Accessory photosynthetic pigment that is yellow or orange.

Carrying capacity The limit to the number of individuals that can occupy one area at a particular time.

Catalase An enzyme produced in all cells to decompose hydrogen peroxide, a by-product of cell respiration.

Cathode The negative pole in an electrolytic cell.

CDKs (cyclin-dependent kinases) A kinase whose activity depends on the level of cyclins and that controls the timing of cell division.

Cell plate A double membrane down the midline of a dividing plant cell between which the new cell wall will form.

Centriole One of two structures in animal cells involved with cell division.

Centromere A specialized region in a chromosome that holds the two chromatids together.

Chemiosmosis The process by which ATP is produced from the flow of protons through an ATP-synthetase channel in the thylakoid membrane of the chloroplast during the light reactions of photosynthesis and in the cristae membrane of the mitochondria during cell respiration.

Chemokines A chemical secreted by blood vessel endothelium and monocytes during an immune response to attract phagocytes to an area.

Chiasma/chiasmata The site at which a crossover and recombination occurs.

Chitin A structural polysaccharide found in the cell walls.

Chlorophyll *a* One type of chlorophyll that participates directly in the light-dependent reactions of photosynthesis.

Chlorophyll *b* One type of chlorophyll that acts as an antenna pigment, expanding the wavelengths of light that can used to power photosynthesis.

Chloroplast The site of photosynthesis in plant cells.

Choanocytes Collar cells that line the body cavity and have flagella that circulate water in sponges.

Chromatid Either of the two strands of a replicated chromosome joined at the centromere.

Chromatin network The complex of DNA and protein that makes up a eukaryotic chromosome. When the cell is not dividing, chromatin exists as long, thin strands and is not visible with the light microscope.

Cilia Hairlike extensions from the cytoplasm used for cell locomotion.

Citric acid cycle Another name for the Krebs cycle.

Cladogenesis Branching evolution occurs when a new species branches out from a parent species.

Classical conditioning One type of associative learning that is widely accepted because of the ingenious work of Ivan Pavlov associating a novel stimulus with an innately recognized one.

Cleavage furrow A shallow groove in the cell surface in an animal cell where cytokinesis is taking place.

Climax community The final, stable community in an ecosystem.

Cline A variation in some trait of individuals coordinated with some gradual change in temperature or other factor over a geographic range.

Clonal selection A fundamental mechanism in the development of immunity. Antigenic molecules select or bind to specific B or T lymphocytes, activating them. The B cells then differentiate into plasma cells and memory cells.

Cnidocytes Stinging cells in all cnidarians.

Codominance The type of inheritance when there is no trait that dominates over another; both traits show.

Codons The three-base sequence of nucleotides in mRNA.

Coelom The body cavity that arises from within the mesoderm and is completely surrounded by mesoderm tissue.

Coevolution Evolution that is caused by two species that interact and influence each other. All predator-prey relationships are examples.

Cohesion tension Force of attraction between molecules of water due to hydrogen bonding.

Collaboration Two genes interact to produce a novel phenotype.

Collenchyma cells Plant cells with unevenly thickened primary cell walls that are alive at maturity and that function to support the growing stem.

Commensalism A symbiotic relationship where one organism benefits and one is unaware of the other organism (+/o).

Community All the organisms living in one area.

Companion cell Connected to each sieve tube member in the phloem and nurtures the sieve tube elements.

Complement An important part of the immune system, a group of about twenty proteins that assists in lysing cells.

Complementary genes The expression of two or more genes where each depends upon the alleles of the other in order for a trait to show.

Conformation The particular three-dimensional shape of a protein molecule.

Conjugation A primitive form of sexual reproduction that is characteristic of bacteria and some algae.

Convergent evolution Evolution that occurs when unrelated species occupy the same environment and are subjected to similar selective pressures and show similar adaptations.

Countercurrent mechanism A mechanism or strategy to maximize the rate of diffusion. This is a major strategy to transport substances across membranes passively, such as in the nephron.

Cristae The internal membranes of mitochondria that are the site of the electron transport chain.

Crop Part of the digestive tract of many animals where food is temporarily stored until it can continue to the gizzard.

Crossing-over The reciprocal exchange of genetic material between nonsister chromatids during synapsis of meiosis I.

Cutin The main component of the waxy cuticle covering leaves to minimize water loss.

Cyclic photophosphorylation Part of the light-dependent reactions in photosynthesis where electrons travel on a short-circuit pathway to replenish ATP levels only.

Cyclin A regulatory protein whose levels fluctuate cyclically in a cell, in part, related to the timing of cell division.

Cystic fibrosis The most common lethal genetic disease in the United States; characterized by a buildup of extracellular fluid in the lungs and digestive tract.

Cytochrome An iron-containing pigment present in the electron transport chain of all aerobes.

Cytokines Chemicals that stimulate helper T cells, B cells, and killer T cells.

Cytokinesis Division of the cytoplasm.

Cytokinins Plant hormone that stimulates cell division and delays senescence (aging).

Cytosine A nucleotide that binds with guanine. A pyrimidine.

Cytoskeleton A complex network of protein filaments that gives a cell its shape and helps it move.

Cytotoxic T cells A type of lymphocyte that kills infected body cells and cancer cells.

Decomposers Organisms, like bacteria and fungi, that recycle nutrients back to the soil.

Deletion A chromosomal mutation where a fragment is lost during cell division.

Dendrites The sensory processes of a neuron.

Denitrifying bacteria Convert nitrates (NO_3) into free atmospheric nitrogen.

Density-dependent factors Factors, such as starvation, that increase directly as the population density increases.

Density-dependent inhibition A characteristic of normal cells grown in culture that causes cell division to cease when the culture becomes too crowded.

Density-independent factors Factors, such as earthquakes, whose occurrence is unrelated to the population density.

Depolarization An electrical state where the inside of an excitable cell is made less negative compared with the outside. If an axon is depolarized, an impulse is passing.

Detrivores Consumers that derive their nutrition from nonliving, organic matter.

Deuterostomes Animals in which the blastopore becomes the anus during early embryonic development.

Dicotyledon A subdivision of flowering plants whose members possess an embryonic seed leaf made of two halves or cotyledons.

Dihybrid cross A cross between individuals that are hybrid for two different traits, such as height and seed color.

Diploblastic An organism whose body is made of only two cell layers, the ectoderm and the endoderm. The two are connected by a noncellular layer called the mesoglea. Animal phyla that are diploblastic are the Porifera (sponges) and the Cnidaria (jellyfish and hydra).

Directional selection Selection where one phenotype replaces another in the gene pool.

Disruptive selection Selection that increases the extreme types in a population at the expense of intermediate forms.

Divergent evolution Evolution that occurs when a population becomes isolated (for any reason) from the rest of the species, becomes exposed to new selective pressures, and evolves into a new species.

DNA ligase An enzyme that permanently attaches pieces of DNA together.

Dopamine A neurotransmitter.

Down syndrome A genetic condition caused by trisomy 21.

Duodenum The first 12 inches (30 cm) of the human small intestine.

Ecdysone A hormone that helps control metamorphosis in insects.

Ecological succession The sequential rebuilding of an entire ecosystem after a disaster.

Ecosystem All the organisms in a given area as well as the abiotic (nonliving) factors with which they interact.

Ectoderm The germ layer that gives rise to the skin and nervous system.

Effectors Muscles or glands.

Electron transport chain A sequence of membrane proteins that carry electrons through a series of redox reactions to produce ATP.

Endergonic Any process that absorbs energy.

Endoderm The embryonic germ layer that gives rise to the viscera, the digestive tract, and other internal organs.

Endodermis The tightly packed layer of cells that surrounds the vascular cylinder in the root of a plant.

Endoplasmic reticulum A system of transport channels inside a eukaryotic cell.

Endosperm The food source for the growing embryo in monocots.

Endosymbiosis This theory states that mitochondria and chloroplasts were once free-living prokaryotes that took up residence inside larger prokaryotic cells in a permanent, symbiotic relationship.

Endotherms Animals that can raise their body temperature, although they cannot maintain a stable body temperature.

Envelope Cloaks the capsid of a virus and aids the virus in infecting the host. The envelope is derived from membranes of host cells.

Enzyme A protein that serves as a catalyst.

Epicotyl Part of the developing embryo that will become the upper part of the stem and the leaves of a plant.

Epinephrine A neurotransmitter.

Epiphytes Photosynthetic plants that grow on other trees rather than supporting themselves.

Epistasis Two separate genes control one trait, but one gene masks the expression of the other gene.

Esterase An enzyme that breaks down excess neurotransmitter.

Ethylene A gaseous plant hormone that promotes fruit ripening and opposes auxins in its actions.

Eukaryotes Cells with internal membranes.

Eutrophication Translates as "true feeding." A process begun by the entrance of large amounts of nutrients into a lake, ultimately ending with the death of the lake.

Exocytosis The process by which cells expel substances.

Exons Stands for expressed sequences of DNA. These are genes.

Exothermic Any process that gives off energy.

Expressivity The range of expression of mutant genes.

Extranuclear genes Genes outside the nucleus, in the mitochondria and chloroplasts.

Facultative anaerobes Organisms that can live without oxygen in the environment.

FAD (flavin adenine dinucleotide) A coenzyme that carries protons or electrons from glycolysis and the Krebs cycle to the electron transport chain.

Fermentation A synonym for anaerobic respiration. The anaerobic breakdown of glucose into pyruvic acid.

Fixed action pattern An innate, highly stereotypic behavior, which when begun, is continued to completion, no matter how useless.

Flagella The tail-like structure that propels some single-celled organisms. Flagella consist of microtubules.

Follicle-stimulating hormone (FSH) A hormone released from the anterior pituitary that stimulates the ovarian follicle.

Food chain The pathway along which food is transferred from one trophic level to the next.

Food pyramid A model of the food chain that demonstrates the interaction of the organisms and the loss of energy.

Food web The interconnected feeding relationships of organisms in an ecosystem.

Founder effect An example of genetic drift, when a small population breaks away from a larger one to colonize a new area; it is most likely not genetically representative of the original larger population.

Frameshift One type of mutation caused by a deletion or addition where the entire reading sequence of DNA is shifted. AAA TTT CCC GGG could become AAT TTC CCG GG.

Frequency-dependent selection A form of selection that acts to decrease the frequency of the more-common phenotypes and increase the frequency of the less-common ones.

Fruit A ripened ovary of a flowering plant.

Fungi The kingdom that consists of heterotrophs that carry out extracellular digestion and have cell walls made of chitin; includes mushrooms and yeast.

GABA (gamma-aminobutyric acid) A neurotransmitter.

Gametangia A protective jacket of cells that prevents some plants' gametes and zygotes from drying out.

Gametophyte The monoploid generation of a plant.

Gastrodermis Cells that line the gastrovascular cavity in cnidarians.

Gastrovascular cavity A digestive cavity with only one opening, characteristic of cnidarians.

Gated-ion channel A channel in a plasma membrane for one specific ion, such as sodium or calcium. In the terminal branch of a neuron, it is responsible for the release of neurotransmitter into the synapse.

Gene flow The movement of alleles into or out of a population.

Genetic drift Change in the gene pool due to chance.

Genetic engineering The technology of manipulating genes for practical purposes.

Genomic imprinting Certain traits whose expression varies, depending on the parent from which they are inherited. Diseases that result from imprinting are Prader-Willi and Angelman syndromes.

Genotype The types of genes an organism has.

Gibberellins Plant hormone that promotes stem elongation.

Gizzard Part of the digestive tract of many animals. It is the site of mechanical digestion.

Glial cells Cells that nourish neurons.

Gluteraldehyde A chemical fixative often used in the preparation of tissue for electron microscopy.

Glycocalyx The external surface of a plasma membrane that is important for cell-to-cell communication.

Glycolysis A nine-step, anaerobic process that breaks down one glucose molecule into two pyruvic acid molecules and four ATP.

Golgi apparatus An organelle in eukaryotes that lies near the nucleus and that packages and secretes substances for the cell.

Gonadotropic-releasing hormone (GgRH) A hormone released by the hypothalamus that stimulates other glands to release their hormones.

Gradualism The theory that organisms descend from a common ancestor gradually, over a long period of time, in a linear or branching fashion.

Grana Membranes in the chloroplast where the light reactions occur.

Greenhouse effect The warming of the planet because of the accumulation of atmospheric carbon dioxide.

Ground tissue The most common tissue type in a plant, functions mainly in support and consists of parenchyma, collenchyma, and sclerenchyma cells.

GTP (guanosine triphosphate) A molecule closely related to ATP that provides the energy for translation.

Guanine A nucleotide that binds with cytosine. A purine.

Guttation Due to root pressure, droplets of water appear in the morning on the leaf tips of some herbaceous leaves.

Gymnosperms Conifers or cone-bearing plants.

Habitat isolation Separation of two or more organisms of the same species living in the same area but in separate habitats, such as in the water and on land.

Habituation One of the simplest forms of learning in which an animal comes to ignore a persistent stimulus.

Halophiles (halobacteria) Aerobic bacteria that thrive in environments with very high salt concentrations.

Hatch-Slack pathway An alternate biochemical pathway found in C-4 plants; its purpose is to remove CO_2 from the airspace near the stomate.

Head-foot The part of the body of mollusks that contains both sensory and motor organs.

Helicase An enzyme that untwists the double helix at the replication fork.

Helper T cells One type of T lymphocyte that activates B cells and other T lymphocytes.

Hemocoels Blood-filled cavities within the body of arthropods and mollusks with open circulatory systems.

Hemophilia An inherited genetic disease caused by the absence of one or more proteins necessary for normal blood clotting.

Hermaphrodites Organisms possessing both male and female sex organs.

Heterosis See hybrid vigor.

Heterosporous A plant that produces two kinds of spores, male and female.

Heterotroph hypothesis This theory states that the first cells on earth were heterotrophic prokaryotes.

Heterotrophs Organisms that must ingest nutrients rather than synthesize them.

Histamine A chemical released by the body during an inflammatory response that causes blood vessels to dilate.

Homeotherms Organisms that maintain a consistent body temperature.

Homologous structures Structures in different species that are similar because they have a common origin.

Homosporous A plant that produces a single bisexual spore.

Huntington's disease A degenerative, inherited, dominant disease of the nervous system that results in certain and early death.

Hybrid vigor A phenomenon in which the hybrid state is selected because it has greater survival and reproductive success. Also known as heterosis.

Hydrophilic Having an affinity for water.

Hyperpolarized An electrical state where the inside of the excitable cell is made more negative compared with the outside of the cell and the electric potential of the membrane increases (gets more negative).

Hypertonic Having a greater concentration of solute than another solution.

Hypocotyl Part of the developing embryo that will become the lower part of the stem and roots.

Hypothalamus Gland located in the brain above the pituitary that is the bridge between the endocrine and nervous systems.

Hypotonic Having a lesser concentration of solute than another solution.

Immunoglobins See antibodies.

Immunological memory The capacity of the immune system to generate a secondary immune response against a specific antigen for a lifetime.

Imprinting A type of learning that is responsible for the bonding between mother and off-spring. Common in birds, it occurs during a sensitive or critical period in early life.

Incomplete dominance The type of inheritance that is characterized by blending traits. For instance, one gene for red plus one gene for white results in a pink four o'clock flower.

Indoleacetic acid IAA. A naturally occurring auxin.

Inflammatory response A nonspecific defensive reaction of the body to invasion by a foreign substance that is accompanied by the release of histamine, fever, and red, itchy areas.

Interferons A class of chemicals that block viral infections.

Interneuron Also known as association neuron, resides within the spinal cord and receives sensory stimuli and transfers the information directly to a motor neuron or to the brain for processing.

Interphase The longest stage of the life cycle of a cell; it consists of G_1, S, and G_2.

Introns Intervening sequences, the noncoding regions of DNA, that are sometimes referred to as junk.

Inversion A chromosome mutation where a chromosomal fragment reattaches to its original chromosome but in the reverse orientation.

In vitro In the laboratory.

In vivo In the living thing.

Isotonic Two solutions containing equal concentrations of solutes.

Karyotype A procedure that analyzes the size, number, and shape of chromosomes.

Kinase An enzyme that transfers phosphate ions from one molecule to another.

Kinetochore A disc-shaped protein on the centromere that attaches the chromatid to the mitotic spindle during cell division.

Klinefelter's syndrome A genetic condition in males in which there is an extra X chromosome; the genotype is XXY.

Kranz anatomy Refers to the structure of C-4 leaves and differs from C-3 leaves. In C-4 leaves, the bundle sheath cells lie under the mesophyll cells, tightly wrapping the vein deep within the leaf, where CO_2 is sequestered.

Krebs cycle Also known as the citric acid cycle, it completes the breakdown of pyruvic acid into carbon dioxide, with the release of a small amount of ATP.

Lactic acid fermentation The process by which pyruvate from glycolysis is reduced to form lactic acid or lactate. This is the process that the dairy industry uses to produce yogurt and cheese.

Lateral meristem Growth region of a plant that provides secondary growth, increase in girth.

Law of dominance One of Mendel's laws. It states that when two organisms, each pure for two opposing traits, are crossed, the offspring will be hybrid but will exhibit only the dominant trait.

Law of independent assortment States that each allelic pair separates during gamete formation. Applies when genes for two traits are not on the same chromosome.

Law of segregation During the formation of gametes, allelic pairs for two traits separate.

Learning A sophisticated process in which the responses of the organism are modified as a result of experience.

Linked genes Genes that are on the same chromosome.

Luteinizing hormone Triggers the ovulation of the secondary oocyte from the ovary.

Lysosomes Sacs of hydrolytic enzymes and the principal site of intracellular digestion.

Lytic cycle A type of viral infection that results in the lysing of the host cell and the release of new phages that will infect other cells.

Macroevolution The development of an entirely new species.

Macrophage While acting as an antigen-presenting cell, it engulfs bacteria by phagocytosis and presents a fragment of the bacteria on the cell surface by an MHC II molecule.

Malpighian tubules The organ of excretion in insects.

Mantle The part of the body of mollusks that contains specialized tissue that surrounds the visceral mass and secretes the shell.

Map unit The distance on a chromosome within which recombination occurs 1 percent of the time.

Marsupials Animals whose young are born very early in embryonic development and where the joey completes its development nursing in the mother's pouch. Includes kangaroos.

Matrix The inner region of a mitochondrion, where the Krebs cycle occurs.

Medusa The free-swimming, upside-down, bowl-shaped stage in the life cycle of the cnidarians. An example is jellyfish.

Megaspores In flowering plants, these produce the ova.

Meiosis Occurs in sexually reproducing organisms and results in cells with half the chromosome number of the parent cell.

Membrane potential A measurable difference in electrical charge between the cytoplasm (negative ions) and extracellular fluid (positive ions).

Memory cells A long-lived form of a lymphocyte that bear receptors to a specific antigen and that remains circulating in the blood in small numbers for a lifetime.

Meristem Actively dividing cells that give rise to other cells such as xylem and phloem.

Mesoderm The germ layer that gives rise to the blood, bones, and muscles.

Methanogens Prokaryotes that synthesize methane from carbon dioxide and hydrogen gas.

MHC (major histocompatibility complex) A collection of cell surface markers that identify the cells as self. No two people, except identical twins, have the same MHC markers. Also known as HLA (human leukocyte antigens).

Microevolution Refers to the changes in one gene pool of a population.

Microfilaments Solid rods of the protein actin that make up part of the cytoskeleton.

Micropyle The opening to the ovule in a flowering plant.

Microspores In flowering plants, these produce sperm.

Microtubules A hollow rod of the protein tubulin in the cytoplasm of all eukaryote cells that make up cilia, flagella, spindle fibers, and other cytoskeletal structures of cells.

Middle lamella A distinct layer of adhesive polysaccharides that cements adjacent plant cells together.

Minority advantage See frequency-dependent selection.

Mitchell hypothesis An attempt to explain how energy is produced during the electron transport chain by oxidative phosphorylation.

Mitochondria The site of cell respiration and ATP synthesis in all eukaryotic cells.

Mitosis Produces two genetically identical daughter cells and conserves the chromosome number ($2n$).

Monera No longer used as the name of the kingdom that contains all the prokaryotes, including bacteria.

Monoclonal antibodies Antibodies produced by a single B cell that produces a single antigen in huge quantities. They are important in research and in treating and diagnosing certain diseases.

Monocotyledon A subdivision of flowering plants whose members possess one embryonic seed leaf or cotyledon.

Monocytes A type of white blood cell that transforms into macrophages, extends pseudopods, and engulfs huge numbers of microbes over a long period of time.

Monohybrid cross This is the cross between two organisms that are each hybrid for one trait.

Monotremes Egg-laying mammals where the embryo derives nutrition from the yolk, like the duck-billed platypus.

Motor neuron A neuron that stimulates effectors (muscles or glands).

Mucosa The innermost layer of the human digestive tract. In some parts of the digestive system, it contains mucus-secreting cells and glands that secrete digestive enzymes.

Müllerian mimicry Two or more poisonous species resemble each other and gain an advantage from their combined numbers. Predators learn more quickly to avoid any prey with that appearance.

Multiple alleles More than two allelic forms of a gene.

Mutagenic agents Substances that cause mutations.

Mutation Changes in DNA.

Mutualism Asymbiotic relationship where both organisms benefit (+/+).

Mycorrhizae The symbiotic structures consisting of the plant's roots intermingled with the hyphae (filaments) of a fungus that greatly increase the quantity of nutrients that a plant can absorb.

Myosin Thick protein filaments that interact with actin filaments in the contraction of skeletal muscle.

NAD **(nicotinamide adenine dinucleotide)** A coenzyme that carries protons or electrons from glycolysis and the Krebs cycle to the electron transport chain.

NADP **(nicotinamide nucleotide phosphate)** Carries hydrogen from the light reactions to the Calvin cycle in the dark reactions of photosynthesis.

Natural killer (NK) cells Part of the nonspecific immune response that destroys virus-infected body cells (as well as cancerous cells).

Natural selection A theory that explains how populations evolve and how new species develop.

Neuromuscular junction The place where a neuron synapses on a muscle.

Neurotransmitter The chemical held in presynaptic vesicles of the terminal branch of the axon that are released into a synapse and that excite the postsynaptic membrane.

Neutrophils A type of white blood cell that engulfs microbes by phagocytosis.

Niche Organisms that live in the same area and use the same resources.

Nitric oxide Acts as a local signaling molecule.

Nitrifying bacteria Convert the ammonium ion into nitrites and then into nitrates.

Nitrogen-fixing bacteria Convert free nitrogen into the ammonium ion.

Nondisjunction Homologous chromosomes fail to separate as they should during meiosis.

Norepinephrine A neurotransmitter.

Notochord A rod that extends the length of the body and serves as a flexible axis in all chordates.

Nucleoid Nuclear region in prokaryotes.

Nucleolus Located in the nucleus and is the site of protein synthesis.

Nucleotides The building blocks of nucleic acids. They consist of a five-carbon sugar, a phosphate, and a nitrogenous base: adenine, thymine (in DNA), cytosine, guanine, or uracil (in RNA).

Obligate anaerobes Prokaryotes that cannot live in the presence of oxygen.

Okazaki fragments In DNA replication, the segments in which the 3' to 5' lagging strand of DNA is synthesized.

Omnivores Organisms, like humans, that eat both plants and animals.

Operant conditioning A type of associative learning where an animal learns to associate one of its own behaviors with a reward or punishment and then repeats or avoids that behavior. Also called trial-and-error learning.

Operator In an operon, the binding site for the repressor.

Operon Functional genes and their switches that are found in bacteria.

Osmotic potential The tendency of water to move across a permeable membrane into a solution.

Outbreeding Mating of organisms that are not closely related; it is a major mechanism of maintaining variation within a species.

Oxidative phosphorylation The production of ATP using energy derived from the electron transport chain.

Oxytocin Hormone released by the posterior pituitary that stimulates labor and the production of milk from mammary glands.

Parallel evolution   Evolution that occurs when two related species have made similar evolutionary adaptations after their divergence from a common ancestor.

Parasitism   A symbiotic relationship (+/−) where one organism, the parasite, benefits while the other organism, the host, is harmed.

Parasympathetic   One of two branches of the autonomic nervous system that has a relaxing effect.

Parenchyma cells   Traditional plant cells with primary cell walls that are thin and flexible and that lack secondary cell walls.

Passive immunity   Immunity is transferred to an individual from someone else.

Pathogens   Organisms that cause disease.

Pedigree   A family tree that indicates the phenotype of one trait being studied for every member of a family and will help determine how a particular trait is inherited.

Peroxisomes   Organelles in both plants and animals that break down peroxide, a toxic by-product of cell respiration.

Phage   Short form of bacteriophage, the virus that attacks bacteria.

Phagocytes   A type of white blood cell that ingests invading microbes.

Phagocytosis   The process by which a cell engulfs large particles using pseudopods.

Phenotype   The appearance of an organism.

Phenylketonuria   An inborn inability to break down the amino acid phenylalanine. It requires elimination of phenylalanine from the diet, otherwise serious mental retardation will result.

Phloem   Transport vessels in plants that carry sugars from the photosynthetic leaves to the rest of the plant by active transport.

Phosphodiester linkages   The bonds that join the nucleotides in DNA.

Phosphofructokinase   (PFK)   An allosteric enzyme important in glycolysis.

Phosphoglyceraldehyde   (PGAL)   A three-carbon sugar, the first stable carbohydrate to be produced by photosynthesis.

Photolysis   The process of splitting water, providing electrons to replace those lost from chlorophyll *a* in P680. This is powered by the light energy absorbed during the light-dependent reactions.

Photoperiod   The environmental stimulus a plant uses to detect the time of year and the relative lengths of day and night.

Photophosphorylation   The process of generating ATP by means of a proton motive force during the light reactions of photosynthesis.

Photorespiration   A process that occurs when rubisco binds with O_2 instead of CO_2. It is a dead-end process because no ATP is produced and no sugar is formed.

Photosynthesis   The process by which light energy is converted to chemical bond energy.

Photosystem I (P700)   Energy, with an average wavelength of 700nm, is absorbed in this photosystem and transferred to electrons that move to a higher energy level.

Photosystem II (P680)   Energy, with an average wavelength of 680nm, is absorbed in this photosystem and transferred to electrons that move to a higher energy level.

Photosystems   Light-harvesting complexes in the thylakoid membranes of chloroplasts. They consist of a reaction center containing chlorophyll *a* and a region containing several hundred antenna pigment molecules that funnel energy into chlorophyll *a*.

Phycobilin   A photosynthetic antenna pigment common in red and blue-green algae.

Phytochrome   The photoreceptor responsible for keeping track of the length of day and night. There are two forms of phytochrome, Pr (red light absorbing) and Pfr (infrared light absorbing).

Pili   Cytoplasmic bridges that connect one cell to another and that allow DNA to move from one cell to another in a form of primitive sexual reproduction called conjugation.

Pinocytosis   A type of endocytosis in which a cell ingests large, dissolved molecules.

Pioneer organisms   The first organisms, such as lichens and mosses, to inhabit a barren area.

Pistil   Part of a flowering plant that produces female gametes.

Placental mammals   Animals whose young are born and where the embryo develops internally in a uterus connected to the mother by a placenta where nutrients diffuse from mother to embryo. Also called eutherians.

Plasma cells   A short-lived form of a lymphocyte that secretes antibodies.

Plasmid   Foreign, small, circular, self-replicating DNA molecule that inhabits a bacterium and imparts characteristics to the bacterium such as resistance to antibiotics.

Plasmodesmata   An open channel in the cell walls of plant cells allowing for connections between the cytoplasm of adjacent cells.

Plasmolysis   Cell shrinking.

Plastids   Organelles in plant cells, including chloroplasts, chromoplasts, and leucoplasts.

Plastoquinone   A proton and electron carrier in the electron transport chain during the light reactions of photosynthesis.

Pleiotropy   One single gene affects an organism in several or many ways.

Poikilotherms   Cold-blooded animals.

Point mutation   A change in one nucleotide in DNA.

Polarized membrane   An axon membrane at rest where the inside of the cell is negative compared with the outside of the cell.

Pollen   One pollen grain contains three monoploid nuclei, one tube nucleus, and two sperm nuclei.

Pollination   The transfer of pollen from the stamen to the pistil.

Polygenic   Genes that vary along a continuum, like skin color or height.

Polymerase chain reaction (PCR)   A cell-free, automated technique by which a piece of DNA can be rapidly copied or amplified.

Polyp   A vase-shaped body or the sessile phase in the life cycle of cnidarians.

Polyploid   A chromosome mutation in which the organism possesses extra sets of chromosomes; the cell becomes $3n$, $4n$, $5n$, and so on.

Population   A group of individuals of one species living in one area.

Predation   One animal eating another animal, or it can also refer to animals eating plants.

Primary consumer   The animal that eats the producer.

Primary immune response   The initial immune response to an antigen.

Primase   An enzyme that joins RNA nucleotides to make the primer.

Prions   Infectious proteins that cause several brain diseases: scrapie in sheep, mad cow disease in cattle, and Creutzfeldt-Jakob disease in humans.

Producer   Those photosynthetic organisms at the bottom of any food chain.

Prokaryotes   Cells with no internal membranes. Bacteria are one example.

Promoter   The binding site of RNA polymerase in an operon.

Prophage   A phage genome that has been inserted into a specific site in a bacterial chromosome.

Prostaglandin   A hormone that promotes blood supply to an area.

Protista   The kingdom that consists of single-celled and primitive multicelled organisms, such as paramecium and amoeba.

Proton pump   A mechanism in cells that uses ATP to pump protons across a membrane to generate a membrane or electric potential.

Protostome   An animal in which the blastopore becomes the mouth during early embryonic development. Literally, first opening.

Pseudocoelomate   A body cavity with mesoderm on only one side, characteristic of nematodes.

Pseudopods   Cellular extensions of amoeboid cells used in moving and feeding.

Punctuated equilibrium   A theory that proposes that new species appear suddenly after long periods of stasis.

Purines   A class of nucleotides that includes adenine and guanine.

Pyrimidines   A class of nucleotides that includes cytosine, thymine, and uracil.

Pyrogens   A chemical released by certain leukocytes that increases body temperature to speed up the immune system and make it more difficult for microbes to function.

Pyruvate   A variant of pyruvic acid.

Pyruvic acid   A three-carbon molecule that is the product of glycolysis and is the raw material for the Krebs cycle.

Radicle   In the embryonic root, the first organ to emerge from the germinating seed.

Radula   A movable, tooth-bearing structure that acts like a tongue in mollusks.

Receptor-mediated endocytosis   The uptake of specific molecules by a cell's receptors.

Recessive trait   The trait that remains hidden in the hybrid state.

Recognition sequence   A specific sequence of nucleotides at which a restriction enzyme cleaves a DNA molecule.

Recombinant chromosomes   Chromosomes that combine genes from both parents due to crossing-over.

Recombination   The result of a cross-over.

Reflex arc   The simplest nerve response; it is inborn, automatic, and protective.

Refractory period   The period of time during which a neuron cannot respond to another stimulus because the membrane is returning to its polarized state.

Replication bubbles   There are thousands of replication bubbles along the DNA molecule that speed up the process of replication.

Replication fork   A Y-shaped region where the new strands of DNA are elongating.

Repressor   Binds to the operator of an operon and prevents RNA polymerase from binding to the promoter, thus blocking transcription.

Resolution   A measure of the clarity of an image; the ability to see two objects as separate.

Resource partitioning   The exploitation of environmental resources by organisms living in the same area so that each group of organisms can occupy a different niche.

Restriction enzymes   Enzymes, naturally occurring in bacteria, that cut DNA at certain specific recognition sites.

Restriction fragment length polymorphisms (RFLPs)   Noncoding regions of human DNA that vary from person to person. They can be used to identify a single individual. Pronounced "riflips."

Restriction fragments   Fragments of DNA that result from the cuts made by restriction enzymes.

Reverse transcriptase   An enzyme found in retroviruses that facilitates the production of DNA from RNA.

Rhizobium   A symbiotic bacterium that lives in the nodules on roots of specific legumes and that incorporates nitrogen gas from the air into a form of nitrogen the plant requires.

Ribosomes   The site of protein synthesis in the cytoplasm.

RNA polymerase   The enzyme that binds to the promoter in DNA and that begins transcription.

RNA primer   An already existing chain of RNA attached to DNA to which DNA polymerase adds nucleotides during DNA synthesis.

Rubisco (ribulose biphosphate carboxylase)   The enzyme that catalyzes the first step in the Calvin cycle: the addition of RuBP (ribulose biphosphate) to CO_2.

Sarcolemma   The modified plasma membrane surrounding a skeletal muscle cell and that can propagate an action potential.

Sarcomere   The basic functional unit of skeletal muscle.

Sarcoplasmic reticulum Modified endoplasmic reticulum in skeletal muscle cells.

Satellite DNA Short sequences of DNA that are tandemly repeated as many as 10 million times in the DNA. Much of it is located at the telomeres.

Schwann cells Glial cells that are located in the peripheral nervous system and that form the myelin sheath around the axon of a neuron.

Sclerenchyma cells Plants cells with very thick primary and secondary cell walls fortified with lignin.

Secondary consumer The animal that eats the primary consumer.

Seed After fertilization, the ovule becomes the seed.

Semiconservative replication The way DNA replicates, each double helix separates and forms two new strands of DNA. Each new molecule of DNA consists of one old strand and one new strand.

Senescence Aging.

Sertoli cells The cells found in the mammalian testes that nourish the developing sperm cells, which contain no cytoplasm.

Sessile Nonmotile.

Sex-influenced trait The inheritance of a trait influenced by the sex of the individual carrying the trait.

Sexual selection Selection based on variation in secondary sexual characteristics related to competing for and attracting mates.

Sieve tube members Along with companion cells, these make up the phloem.

Single-stranded binding proteins Proteins that act as scaffolding, holding two DNA strands apart during replication.

snRNPs (small nuclear ribonucleoproteins) Help to process RNA after it is formed and before it moves to the ribosome.

Sodium-potassium pump A protein pump within a plasma membrane of an axon that restores the membrane to its original polarized condition by pumping sodium and potassium ions across the membrane.

Solute The substance dissolved.

Solvent The substance doing the dissolving.

Somatic cell A body cell.

Somatic nervous system The branch of the vertebrate peripheral nervous system that controls skeletal (voluntary) muscles.

Sori Structures on the underside of the fern leaves that are clusters of sporangia containing monoploid spores.

Specific heat The amount of heat a substance must absorb to increase 1 gram of the substance by 1°C.

Spicules Found in sponges, these consist of inorganic materials and support the animal.

Spindle fibers Made of microtubules that connect centrioles to kinetochores of chromosomes and that separate sister or homologous chromosomes during cell division.

Spiracles Openings in the exoskeleton of arthropods, such as the grasshopper, that connect to internal cavities called hemocoels where respiratory gases are exchanged.

Splicesomes Enzymes that (along with SnRPs) help process RNA after it is formed and before it moves to the ribosome.

Spongocoel Found in sponges, it is the central cavity into which water is drawn to filter nutrients.

Sporopollenin A tough polymer that protects plants in a harsh terrestrial environment.

Stabilizing selection Selection that eliminates the extremes and favors the more common intermediate forms.

Stele The vascular cylinder of the root, consisting of vascular tissue.

Stroma The site of the light-independent (dark) reactions in chloroplasts.

Submucosa A layer of the human digestive system that contains nerves, blood vessels, and lymph vessels.

Survivorship curves Show the size and composition of a population.

Sympathetic nervous system One of two branches of the autonomic nervous system that is generally excitatory.

Sympatric speciation The formation of new species without geographic isolation; such as polyploidy or behavioral isolation.

Symplast A continuous system of cytoplasm of cells interconnected by plasmodesmata.

Synapsis The process of pairing replicated homologous chromosomes during prophase I of meiosis.

Systematics Scientific study of the classification of organisms and their relationships to one another.

Taq polymerase A heat-stable form of DNA polymerase extracted from bacteria that live in hot environments, such as hot springs, that is used during PCR technique.

Taxa A particular group at a category level; such as kingdom or genus.

Taxonomy The study of classification of organisms.

Tay-Sachs disease An inherited genetic disease that is caused by lack of an enzyme necessary to break down lipids necessary for normal brain function and results in seizures, blindness, and early death. Common in Ashkenazi Jews.

Telomerase An enzyme that catalyzes the lengthening of the telomeres at the ends of eukaryotic chromosomes.

Telomeres The protective ends of eukaryotic chromosomes.

Tertiary consumer The third trophic level of consumer in a food chain.

Testcross A cross done to determine whether an individual plant or animal showing the dominant trait is homozygous dominant (*B/B*) or heterozygous (*B/b*). The individual in question (*B/_*) is crossed with a homozygous recessive individual.

Tetanus The smooth, sustained contraction of a skeletal muscle.

Thermophiles Prokaryotes that thrive in very high temperatures.

Theta replication The way in which prokaryotes replicate their DNA.

Thylakoids Membranes in chloroplasts that make up the grana, the site of the light reactions.

Thymine A nucleotide that binds with adenine. It is a pyrimidine and is not present in RNA.

T lymphocytes One type of lymphocyte that fights pathogens by cell-mediated response.

Tracheids Long, thin cells that overlap and are tapered at the ends and that, along with vessel elements, make up xylem in a plant.

Tracheophytes Plants that have transport vessels, xylem and phloem.

Transcription The process by which DNA makes RNA.

Transduction Transfer of bacterial DNA by phages from one bacterium to another.

Transformation The transfer of genes from one bacterium into another.

Translation The process by which the codons of an mRNA sequence are changed into an amino acid sequence.

Translocation A chromosome mutation where a fragment of a chromosome becomes attached to a nonhomologous chromosome; the transport of sugar in a plant from source to sink.

Transpiration Loss of water from stomates in leaves.

Transpirational pull–cohesion tension theory This theory describes the passive transport of water up a tree. For each molecule of water that evaporates from a leaf by transpiration, another molecule of water is drawn in at the root to replace it.

Transposons Transposable genetic elements, sometimes called jumping genes.

Triploblastic Having three cell layers: ectoderm, mesoderm, and endoderm.

Triploid A chromosome mutation where an organism has three sets of chromosomes (3*n*) instead of two (2*n*).

Trisomy A chromosome condition in which a cell has an extra copy of one chromosome. The cell has three of that chromosome, instead of two.

Trophic level   Any level of a food chain based on nutritional source.

Tropic hormones   Hormones released by one endocrine gland that stimulate other endocrine glands to release their hormones.

Tropism   The growth of a plant toward or away from a stimulus, for example, phototropism.

T system   A set of tubules that traverse the skeletal muscle, conduct the action potential deep into the cell, and stimulate the sarcoplasmic reticulum to release calcium ions.

Turgid   Firm. Plant cells swollen because they have absorbed water.

Turner's syndrome   A genetic condition in females caused by a deletion of one of the two X chromosomes.

Typhlosole   A large fold in the upper surface of the intestine of the earthworm that increases surface area to increase absorption.

Ultramicrotome   An instrument used to cut very thin sections of tissue for use in the transmission electron microscope.

Uracil   A nucleotide in only RNA that binds with adenine. It is a pyrimidine.

Vacuoles   A membrane-enclosed sac for storage in all cells, particularly in plant cells.

Vasodilation   The enlargement of blood vessels to increase blood supply.

Vegetative propagation   Plants can clone themselves or reproduce asexually from any vegetative part of the plant; the root, stem, or leaf.

Vesicles   Small vacuoles.

Vessel elements   Wide, short tubes that, along with tracheids, make up the xylem.

Vestigial structures   Structures of no importance, such as the appendix, that were once important to ancestors.

Visceral mass   The part of the body of mollusks that contains the organs of digestion, excretion, and reproduction.

Water potential   The tendency of water to move across a semipermeable membrane, ψ. The water potential of pure water is zero. Any solution has a water potential less than zero.

Wave of depolarization   The wavelike reversal of the polarity of the membrane when an impulse passes.

Wobble   Refers to the translation of mRNA to protein. The relaxation of base-pairing rules where the pairing rules for the third base of a codon are not as strict as they are for the first two bases. UUU and UUA both code for the amino acid phenylalanine.

Xanthophyll   A photosynthetic antenna pigment common in algae that is a structural variant of a carotenoid.

Xylem   Transport vessels in plants that carry water and minerals from the soil to the leaves.

Z lines   These define the edges of the sarcomere in a muscle cell.

Zone of cell division   The region of a plant's root with actively dividing cells that grow down into the soil.

Zone of differentiation   The region of root tip where cells undergo specialization.

Zone of elongation   The region of root tip where cells elongate and that are responsible for pushing the root cap downward, deeper into the soil.

Index